Medicinal Plants of Bangladesh and West Bengal

Natural Products Chemistry of Global Plants

This unique book series focuses on the natural products chemistry of botanical medicines from different countries such as Turkey, Sri Lanka (in print), Borneo, Brazil, Cambodia, Vietnam, Yunnan (China), Africa, Thailand, Turkey and the Silk Road Countries. The series will focus on the pharmacognosy, covering recognized areas rich in folklore as well as botanical medicinal uses as a platform to present the natural products and organic chemistry. Where possible, the authors will link these molecules to pharmacological modes of action. The series intends to trace a route through history from ancient civilizations to the modern day showing the importance to man of natural products in medicines, in foods and a variety of other ways.

Medicinal Plants of Bangladesh and West Bengal: Botany, Natural Products, and Ethnopharmacology
Christophe Wiart

Medicinal Plants of Bangladesh and West Bengal

Botany, Natural Products, and Ethnopharmacology

Christophe Wiart

CRC Press
Taylor & Francis Group
Boca Raton London New York

CRC Press is an imprint of the
Taylor & Francis Group, an **informa** business

CRC Press
Taylor & Francis Group
6000 Broken Sound Parkway NW, Suite 300
Boca Raton, FL 33487-2742

Printed on acid-free paper

International Standard Book Number-13: 978-1-138-73516-3 (Hardback)

Library of Congress Cataloging-in-Publication Data

Names: Wiart, Christophe, author.
Title: Medicinal plants of Bangladesh and West Bengal : botany, natural products, and ethnopharmacology / Christophe Wiart.
Other titles: Natural products chemistry of global plants.
Description: Boca Raton, Florida : CRC Press, [2019] | Series: Natural products chemistry of global plants | Includes bibliographical references and index.
Identifiers: LCCN 2019004440| ISBN 9781138735163 (hardback : alk. paper) | ISBN 9781315186443 (e-book)
Subjects: LCSH: Medicinal plants—Bangladesh. | Medicinal plants—India—West Bengal. | Materia medica, Vegetable—Bangladesh. | Materia medica, Vegetable—India—West Bengal.
Classification: LCC RS179 .W532 2019 | DDC 615.3/21095492—dc23
LC record available at https://lccn.loc.gov/2019004440

Visit the Taylor & Francis Web site at
http://www.taylorandfrancis.com

and the CRC Press Web site at
http://www.crcpress.com

"The human body is the temple of God.
One who kindles the light of awareness within
gets true light.
The sacred flame of your inner shrine
is constantly bright.
The experience of unity
is the fulfillment of human endeavors.
The mysteries of life are revealed."

Rig Veda (1500 and 1200 BC)

"Verily, knowledge is a lock and its key is the question."

Jaʿfar al-Sadiq (702–765)

"Let me tell you the secret that has led me to my goal. My strength lies solely in my tenacity."

Louis Pasteur (1822–1895)

"Sit down before fact as a little child, be prepared to give up every preconceived notion, follow humbly wherever and to whatever abysses nature leads, or you shall learn nothing. I have only begun to learn content and peace of mind since I have resolved at all risks to do this."

Thomas Henry Huxley (1825–1895)

"He who studies without passion will never become anything than a pedant"

Stefan Zweig (1881–1942)

Contents

1 Family Acanthaceae A.L. de Jussieu 1789

1.1 *Acanthus ilicifolius* L.

Synonyms: *Acanthus ebracteatus* var. *xiamenensis* (R.T. Zhang) C.Y. Wu & C.C. Hu; *Acanthus ilicifolius* var. *xiamenensis* (R.T. Zhang) Y.F. Deng, N.H. Xia & H.B. Chen; *Acanthus xiamenensis* R.T. Zhang; *Dilivaria ilicifolia* (L.) Juss.

Local name: Harkuch kanta

Common names: Sea holly; holly-leaved acanthus

Botanical description: It is an erect shrub growing to about 1.5 m tall. The stem is terete and stout. The leaves are decussate, simple, and exstipulate. The petiole is 0.5 cm long. The blade is coriaceous, oblong, 5–15 cm × 2–5 cm, glabrous, and holly-like. The inflorescence is a terminal spike, which is about 15 cm long. The calyx is 4-lobed and about 1 cm long. The corolla is bluish, tubular, 2–4 cm long, and producing a conspicuous recurved limb. The androecium includes 4 stamens adnate to the corolla tube and 1.5 cm long. The fruit is a dehiscent capsule that is 2.5 cm long.

Medicinal use: snake bite

Pharmacology: Extracts of the plant demonstrated anti-HBV activity[1]. The plant contains an anti-inflammatory alkaloid,[2] an antibacterial benzoate, and phenylethanoid derivatives.[3]

1.2 *Andrographis paniculata* (Burm. f.) Wall. ex Nees

Synonym: *Justicia paniculata* Burm.f.

Local name: Kalmegh

Common name: Creat

Botanical description: It is an upright herb that grows up to 40 cm in height. The stems are finely quadrangular. The leaves are simple, decussate, and exstipulate. The petiole is minute. The blade is lanceolate, bitter about 1.5–5 cm × 0.5–2 cm, and shortly acuminate. The inflorescence is a terminal, leafy panicle. The calyx has 5 tiny lobes. The corolla is tubular, white, and up to about 1 cm long cm. The upper lip of the corolla is 3-lobed and the lower lip is entire. The lower lip presents purple spots. The androecium includes 2 stamens coming out from the corolla tube. The fruit is a fusiform upright dehiscent capsule, which is up to 1.5 cm long, and containing numerous tiny seeds.

Medicinal uses: syphilis, intestinal worms, fever, cholera, dysentery

Pharmacology: The plant has been well studied pharmacologically.[4] It contains andrographolide and 14-deoxyandrographolide which have anti-diabetic,[5] anti-tumor,[6] antibacterial,[7] anti-HIV[8] and anti-fungal[9] properties.

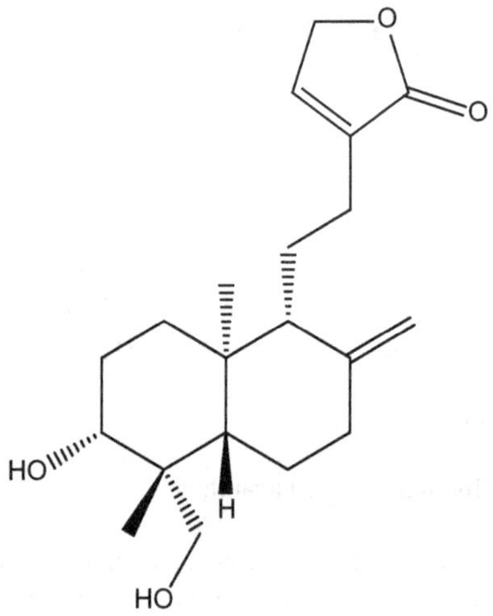

Structure of 14-Deoxyandrographolide

1.3 *Barleria longifolia* L.

Synonyms: *Asteracantha longifolia* (L.) Nees; *Hygrophila auriculata* Heine; *Ruellia longifolia* Roxb.

Local name: Kanta kulika

Common name: Hygrophila

Botanical description: It is a stout, erect herb that grows up to 1 m in height. The stem is quadrangular. The leaves are simple, exstipulate, decussate, and form spiny pseudo-whorls. The blade is sessile, 5–10 cm × 1.5–2.5 cm, hairy, and elongated. The flowers are arranged in axillary fascicles. The calyx is 4-lobed. The corolla is tubular, bluish, and about 2 cm long. The upper lip is 2-lobed whereas the lower lip is 3-lobed. The androecium includes 4 stamens. The ovary is oblong. The fruit is a dehiscent capsule that is oblong and contains 4–8 tiny seeds.

Medicinal use: diuretic

Pharmacology: Extracts of the plant exhibited hepatoprotective,[10] anti-diabetic,[11] and anti-inflammatory[12] effects.

1.4 *Barleria prionitis* L.

Synonyms: *Barleria coriacea* Oberm.; *Barleria methuenii* Turrill; *Barleria prionitis* subsp. *angustissima* (Hochr.) Benoist; *Barleria prionitis* subsp. *madagascariensis* Benoist; *Barleria prionitis* var. *angustissima* Hochr.

Local name: Kanta jati

Common name: Common yellow nail dye plant

Botanical description: It is a spiny shrub that grows to about 1.5 m in height. The stem is terete. The leaves are simple, exstipulate, and decussate. The petiole is 1–2.5 cm long. The blade is membranous, narrowly elliptic, 4–10 cm × 1–5 cm, margin entire, and acute at apex. The flowers are arranged in clusters which are axillary. The calyx comprises 4 lobes. The corolla is plain yellow and 2–5 cm long. The he upper lip is 4-lobed, the lower lip is entire. The androecium protrudes from the corolla and includes 4 stamens. The gynoecium is ovoid and the stigma bifid. The fruit is a dehiscent capsule, which is in fusiform, about 1.5 cm long and contains 2 seeds.

Medicinal uses: boils, fever, diuretic, cough

Pharmacology: The extract of the plant exhibited anti-inflammatory[13] activity. The plant contains iridoids with gastroprotective[14] and antiviral[15] activities.

1.5 *Ecbolium linneanum* Kurz.

Synonyms: *Ecbolium viride* (Forsk.) Alston; *Justica ecbolium* L.

Local name: Udujati

Common name: Blue fox tail

Botanical description: It is an herb that grows up to 50 cm tall. The stem is terete. The leaves are opposite, simple, and exstipulate. The petiole is 2 cm long. The blade is elliptic, base and apex acute, and 5–15 cm × 2–5 cm. The flowers are arranged in terminal and bracteate spikes up to 15 cm long. The calyx comprises 5 lobes. The corolla is tubular, up to 3.5 cm long and bluish-jade colored, and with 3 conspicuous lobes. The fruit is an elongated capsule up to 1.5 cm long and containing small tuberculate seeds.

Medicinal use: diuretic

Pharmacology: Extracts of the plant exhibited anti-inflammatory activity.[16] The plant produces insecticidal lignans.[17]

1.6 *Justicia adhatoda* L.

Synonyms: *Adhatoda vasica* Nees; *Adhatoda zeylanica* Medik.

Local names: Arusa; bakas

Common name: Malabar nut

Botanical description: It is a handsome shrub that grows to 3 m tall. The stems are quadrangular. The leaves are simple, opposite, and exstipulate. The petiole is 0.5–2.5 cm long. The blade is broadly lanceolate, wavy at the margin, dark green above, 5–20 cm × 2–10 cm, and with 9–12 pairs of secondary nerves. The flowers are arranged in terminal or axillary and bracteate spikes which are 10 cm long. The calyx presents 5 lobes which are linear and about 1 cm long. The corolla is pure white, yellow at the throat, showy, up to 3 cm long. The upper lip is oblong, 1.5 cm long, 2-lobed and the lower lip is 3-lobed and colored with purplish lines. The androecium comprises 2 stamens that are 2 cm long. The gynoecium is 2.5 cm long. The fruit is a capsule which is obovoid and about 2.5 cm long.

Justicia adhatoda L

Medicinal uses: cough, asthma, inflammation

Pharmacology: The plant produces alkaloids with hepatoprotective,[18] antibacterial,[19,20] and anti-inflammatory[21] activities.

1.7 *Justicia gendarussa* L.

Synonyms: *Gendarussa vulgaris* Bojer; *Gendarussa vulgaris* Nees

Local name: Jogut-mudun

Common name: Gendarussa

Botanical description: It is an erect herb that grows up to 1.2 m tall. The stem is terete, swollen at the nodes and somewhat dark purplish and glossy. The leaves are simple, opposite, and exstipulate. The petiole is up to 1 cm long. The blade is narrowly lanceolate, 5–10 cm × 1–1.5 cm, with 5–8 pairs of discrete secondary nerves. The flowers are arranged in terminal and axillary spikes which are 10 cm long. The calyx is 0.5 cm long and 5-lobed. The corolla is tubular, whitish, about 1.5 cm long. The lower lip is marked with purple lines, and somewhat 3-lobed. The gynoecium is 1 cm long. The fruit is a capsule that is elongated and about 1 cm long.

Medicinal uses: rheumatism, fever, cough

Pharmacology: Extracts of the plant demonstrated anti-inflammatory[22,23] and insecticidal[24] properties.

1.8 *Lepidagathis incurva* Buch.-Ham. ex D.Don

Synonym: *Lepidagathis hyalina* Nees

Common name: Curved Lepidagathis

Local name: Vangvattur

Botanical description: It is an herb that grows up to 50 cm in height. The leaves are simple, opposite, and exstipulate. The petiole is 0.5–3.5 cm long. The blade is elliptic, 2.5–10 cm × 1–5 cm, pubescent and with 4–9 pairs of secondary nerves. The flowers are arranged in terminal spikes that are pubescent and bracteate. The calyx is 5-lobed and about 1 cm long. The corolla is tubular, white, and 1 cm long. The lower lip is 3-lobed and the upper lip is bifid. The androecium includes 4 stamens. The fruit is an oblong capsule that is 0.5 cm long.

Medicinal uses: cough, bleeding

Pharmacology: Extracts of the plant exhibited cytotoxic effects.[25]

1.9 *Ruellia tuberosa* L.

Local name: Potpoti

Common names: Manyroot; bluebell

Botanical description: It is a tuberous herb that grows to 30 cm tall. The stems are swollen at the nodes and slightly quadrangular. The petiole is about 1 cm long. The blade is obovate, about 3–8 cm × 1.5–5 cm, at apex cuneate and tapering at base, the margin is wavy, and acute. The flowers are grouped in axillary cymes. The calyx comprises 5 lobes and up to 2 cm long. The corolla is bluish, funnel-like, 5-lobed, membranous, and 2–5 cm long. The androecium consists of 4 stamens. The gynoecium is about 2 cm long. The fruit is a dehiscent, linear, and dark brown capsule containing numerous tiny seeds.

Medicinal uses: bronchitis, fever, leprosy, cystitis

Pharmacology: Extracts of the plant demonstrated anti-inflammatory[26] and antiviral[27] effects.

2 Family Acoraceae Martynov 1820

2.1 *Acorus calamus* L.

Synonyms: *Acorus americanus* (Raf.) Raf.; *Acorus angustatus* Raf.; *Acorus angustifolius* Schott; *Acorus asiaticus* Nakai; *Acorus calamus* var. *americanus* Raf.; *Acorus calamus* var. *angustatus* Besser; *Acorus calamus* var. *angustifolius* (Schott) Engl.; *Acorus calamus* var. *spurius* (Schott) Engl.; *Acorus calamus* var. *verus* L.; *Acorus calamus* var. *vulgaris* L.; *Acorus cochinchinensis* (Lour.) Schott; *Acorus griffithii* Schott; *Acorus spurius* Schott; *Acorus triqueter* Turcz. ex Schott; *Acorus verus* (L.) Houtt.; *Calamus aromaticus* Garsault; *Orontium cochinchinense* Lour.

Local name: Bach

Common name: Sweet flag

Botanical description: It is a rhizomatous herb that grows in marshy places. The rhizome is up to 20 cm long, aromatic, and presents long roots. The leaves are erect, sword-shaped, light green in color, and about 70–100 cm × 1–2.5 cm. The flowers are arranged in a spike which is elongated and up to 7 cm long. The flowers are yellowish green, fragrant when bruised, minute and present 6 tepals. The fruits are tiny berries, which are oblong.

Medicinal uses: fever, dyspepsia, cough, diarrhea, sore throat

Pharmacology: The plant produces an essential oil containing β-asarone,[28] which is cytotoxic,[29] neuroprotective,[30] and anti-fungal.[31,32]

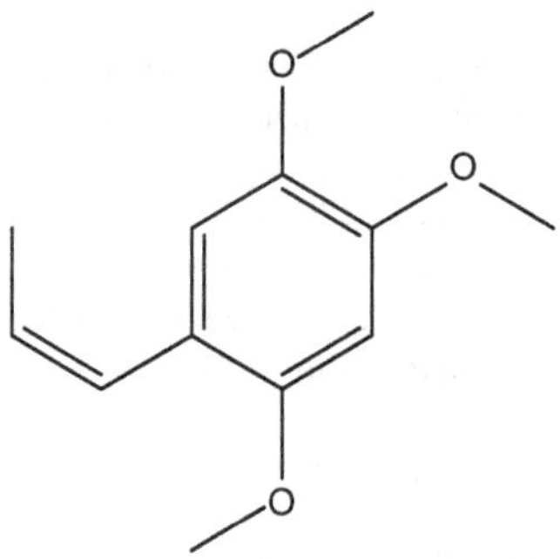

Structure of β-asarone

3 Family Agavaceae Dumortier 1829

3.1 *Agave americana* L.

Synonyms: *Agave complicata* Trel. ex Ochot.; *Agave felina* Trel.; *Agave gracilispina* Engelm. ex Trel.; *Agave melliflua* Trel.; *Agave rasconensis* Trel.; *Agave subzonata* Trel.; *Agave zonata* Trel. ex Bailey

Local name: Murga muji

Common names: American aloe; century plant

Botanical description: It is a massive fleshy plant that consists of a rosette of 30 or more lanceolate leaves. The blades are spiny at the margin and apex, 1–2 m × 10–20 cm, sappy, fibrous and light green. The inflorescence is a panicle that is about 10 m tall. The flowers are numerous, 7–10.5 cm long, with a bright yellow perianth. The androecium consists of conspicuous stamens protruding out of the corolla. The fruit is a capsule.

Medicinal uses: gonorrhea, diuretic

Pharmacology: The plant contains saponins and sapogenins[33,34] with anti-inflammatory activities.[35] Extracts of the plant exhibited anti-fungal activity.[36]

4 Family Alangiaceae A.P. de Candolle 1828

4.1 *Alangium lamarckii* Thwaites

Synonym: *Alangium salvifolium* (L. f.) Wangerin

Local name: Akar kanta

Common name: Lamarck Alangium

Botanical description: It is a tree that grows up to 15 m tall. The leaves are simple, alternate, and exstipulate. The petiole is about 1 cm long. The blade is broadly lanceolate, 5–15 cm × 2.5–7 cm, acute at the base and apex, glossy, light green, and presents 3–5 pairs of secondary nerves. The flowers are fragrant and are arranged in cauliflorous clusters of 4–8 flowers. The calyx is tubular, minute, and presents 5–10 lobes. The corolla consists of 4–6 petals that are white, recurved and up to 2.5 cm long. The androecium is showy and comprises 10–30 white stamens. A lobed disc is present. The gynoecium is up to 2 cm long and presents a capitate stigma. The fruit is a globose drupe that is red and about 2 cm in diameter.

Medicinal uses: intestinal worms, leprosy, fever, snake bites

Pharmacology: Extracts of the plant demonstrated anti-inflammatory and analgesic activity.[37,38]

5 Family Alismataceae Ventenat 1799

5.1 *Alisma plantago-asiatica* L.

Synonyms: *Alisma subcordatum* Raf.; *Alisma triviale* Pursh

Local name: Pani kola

Common name: Water plantain

Botanical description: It is an aquatic herb that grows from tubers. The leaves are simple and arranged in a rosette. The petiole is of variable length and up to 30 cm long. The blade is broadly lanceolate, 2–10 cm × 1.5–7 cm, and with secondary nerves parallel to the midrib. The flowers are arranged in a lax panicle that can reach up to 50 cm long. The calyx consists of 3 sepals that are broadly ovate and minute. The corolla comprises 3 petals, which are white and membranous. The androecium includes 6 stamens. The gynoecium comprises numerous carpels. The fruits are tiny achenes.

Medicinal uses: wounds and ulcers, scar, insect repellent, laxative

Pharmacology: The plant produces cytotoxic triterpenes[39] and anti-inflammatory sesquiterpenes and phenolics.[40]

5.2 *Sagittaria trifolia* L.

Synonym: *Sagittaria sagittifolia* L.

Local names: Sagudana; kuka

Common name: Arrowhead

Botanical description: It is an elegant aquatic herb that grows from a globose corm. The leaves are simple and arranged in a rosette. The petiole is elongated and 60–70 cm long. The blade is sagittate and up to 8–15 cm × 5–10 cm. The inflorescences are racemose and present whorls of flowers. The calyx consists of 3 sepals that are ovate and minute. The calyx comprises 3 petals that are white tinged with pink at the base, membranous, broadly spathulate, and about 1.5 cm long. The androecium includes numerous tiny stamens. The fruits are tiny-winged achenes.

Medicinal uses: wounds, skin diseases

Pharmacology: The plant produces anti-inflammatory and antibacterial diterpenes.[41,42]

6 Family Amaranthaceae A.L. de Jussieu 1789

6.1 *Achyranthes aspera* L.

Synonyms: *Achyranthes argentea* Lam.; *Achyranthes aspera* var. *indica* L.; *Achyranthes aspera* var. *obtusifolia* Suess.; *Achyranthes indica* (L.) Mill.; *Achyranthes obtusifolia* Lam.; *Achyranthes robusta* C.H. Wright; *Achyranthes sicula* Roth; *Cadelaria indica* (L.) Raf.; *Cadelaria sicula* Raf.; *Centrostachys aspera* (L.) Standl.; *Centrostachys indica* (L.) Standl.; *Cyathula geniculata* Lour.; *Stachyarpagophora aspersa* (L.) M. Gómez

Local name: Apang

Common name: Prickly chaff flower

Botanical description: It is an invasive herb that grows up to 1.2 m tall. The stems are articulated and quadrangular. The leaves are simple, opposite, and exstipulate. The petiole is 0.5–1.5 cm long. The blade is dull green, elliptic, 1.5–10 cm × 1–6 cm, and the apex is obtuse. The inflorescences are terminal and slender spikes that can reach 30 cm long. The perianth consists of 5 tepals that are lanceolate and minute. The androecium includes five stamens which are of about 2.5–3.5 mm long. The gynoecium is oblong. The fruits are ovoid utricles containing numerous tiny seeds.

Medicinal uses: diuretic, gonorrhea, fever, boils

Pharmacology: The plant contains anti-inflammatory saponins,[43] polyphenolic anti-neoplastic, and immunomodulatory compounds,[44] and antiviral triterpenes.[45]

6.2 *Aerva lanata* (L.) A.L. Juss. ex Schultes

Synonyms: *Achyranthes lanata* L.; *Achyranthes lanata* Roxb.; *Aerva elegans* Moq.; *Illecebrum lanatum* (L.) L.

Local name: Chaya

Common name: Mountain knot grass

Botanical description: It is a prostrate herb that can reach up to 2 m long. The stem is terete and hairy. The leaves are hairy, simple, alternate, and exstipulate. The petiole can reach 2 cm long. The blade is orbicular, cuneate at base, rounded at the apex, and 1–5 cm × 0.5–3.5 cm. The inflorescence is an axillary, up to 1.5 cm long, and hairy spike. The flowers are minute. The perianth comprises 5 tepals, the androecium includes 5 stamens, and a gynoecium presents 2 stigmas. The fruits are tiny utricles containing numerous seeds.

Medicinal uses: headache, diuretic, strangury

Pharmacology: Extracts of the plant showed immunomodulatory[46] and antibacterial activities.[47]

6.3 *Aerva sanguinolenta* (L.) Bl.

Synonyms: *Achyranthes sanguinolenta* L.; *Achyranthes scandens* Roxb.; *Aerva scandens* (Roxb.) Wall.; *Aerva timorensis* Moq.; *Aerva velutina* Moq.

Local names: Nuriya; chaya

Common name: Kapok bush

Botanical description: It is an herb that grows up to 1 m long. The stem is terete and somewhat hairy. The leaves are simple, exstipulate, alternate, or subopposite. The petiole can reach up to 2 cm in length. The blade is oblong or lanceolate and is about 1.5–10 cm × 0.5–3.5 cm. The flowers are arranged in axillary or terminal spikes that can reach up to 5 cm long. The perianth consists of 5 tiny membranaceous tepals that are white or pink. The androecium comprises 5 minute stamens. The gynoecium is ovoid with 2 stigmas. The fruits are utricles that are ovoid and contain tiny seeds.

Medicinal use: wounds

Pharmacology: The plant contains anti-inflammatory cerebrosides[48] as well as bakuchiol, which is antibacterial.[49]

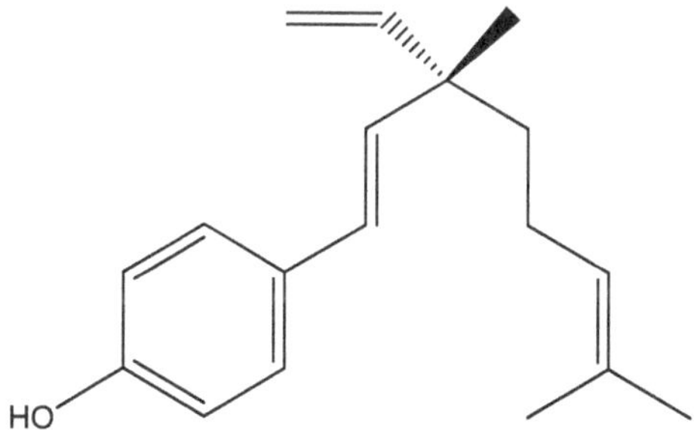

Structure of bakuchiol

6.4 *Alternanthera paronychioides* A. St. –Hil.

Synonyms: *Achyranthes ficoidea* (L.) Lam.; *Alternanthera amoena* Back. & Sloot.; *Alternanthera bettzickiana* (Regel) G. Nicholson; *Alternanthera ficoidea* (L.) R. Br.; *Alternanthera ficoidea* (L.) Sm.; *Alternanthera polygonoides* (L.) R. Br. ex Sweet; *Alternanthera polygonoides* var. *glabrescens* Griseb.; *Alternanthera versicolor* (Lcm.) Regel; *Bucholzia ficoidea* (L.) Mart.; *Bucholzia polygonoides* var. *diffusa* Mart.; *Bucholzia polygonoides* var. *erecta* Mart.; *Bucholzia polygonoides* var. *radicans* Mart.; *Gomphrena ficoidea* L.; *Illecebrum ficoideum* Jacq.; *Illecebrum ficoideum* L.; *Paronychia ficoidea* (L.) Desf.; *Steiremis ficoidea* (L.) Raf.; *Telanthera ficoidea* (L.) Moq.; *Telanthera polygonoides* var. *brachiata* Moq.; *Telanthera polygonoides* var. *diffusa* Moq.; *Teleianthera manillensis* Walp.

Local name: Jal sachiba

Common name: Smooth joyweed; Smooth chaff flower

Botanical description: It a prostrate herb that grows up to 80 cm long. The stem is terete and fleshy. The leaves are simple, exstipulate, and in clusters. The blade is sessile, elliptic, 0.5–2.5 cm × 0.3–1.1 cm, acute at the apex, and hairy. The flowers are arranged in axillary heads, which are globose and about 1 cm in diameter. The perianth consists of 5 white, lanceolate, tepals that are minute. The androecium includes 5 stamens. The stigma is capitate. The fruits are tiny utricles containing numerous seeds.

Medicinal use: diabetes

Pharmacology: Extracts of the plant protected β-cells of the pancreas against glucotoxicity.[50]

6.5 *Alternanthera pungens* Kunth

Synonyms: *Achyranthes leiantha* (Seub.) Standl.; *Achyranthes lorentzii* (Uline) Standl.; *Achyranthes mucronata* Lam.; *Achyranthes radicans* B. Heyne ex Roth; *Achyranthes repens* L.; *Alternanthera achyrantha* (L.) R. Br.; *Alternanthera achyrantha* (L.) R. Br. ex Sweet; *Alternanthera achyrantha* var. *leiantha* Seub.; *Alternanthera echinata* Sm.; *Alternanthera lorentzii* Uline; *Alternanthera pungens* fo. *Pauciflora* Suess.; *Alternanthera pungens* var. *leiantha* Suess.; *Alternanthera repens* (L.) J.F. Gmel.; *Alternanthera repens* (L.) Kuntze; *Alternanthera repens* (L.) Link; *Celosia echinata* Willd. ex Roem. & Schult.; *Desmochaeta sordida* Bunbury; *Guilleminea procumbens* Rojas Acosta; *Illecebrum achyranthum* L.; *Illecebrum pungens* (Kunth) Spreng.; *Pityranthus crassifolius* Mart.; *Pupalia sordida* Moq.; *Telanthera pungens* (Kunth) Moq.

Local name: Ahoara

Common name: Creeping chaffweed

Botanical description: It is a prostrate herb. The stem is terete and hairy. The leaves are simple, exstipulate, and arranged in clusters. The blade is obovate, 1.5–4.5 cm × 0.3–2.5 cm, rounded to acute at the apex and hairy. The flowers are axillary. The perianth consists of 5 tepals which are minute. The androecium includes 5 stamens and the gynoecium is minute. The fruit is an orbicular utricle containing discoid seeds.

Medicinal uses: diuretic, gonorrhea, dysentery

Pharmacology: Unknown. It contains flavones.[51]

6.6 *Alternanthera sessilis* (L.) DC.

Synonyms: *Achyranthes ficoidea* var. *sessilis* (L.) Pers.; *Achyranthes sessilis* (L.) Desf. ex Steud.; *Achyranthes triandra* Roxb.; *Alternanthera achyranth* Forssk.; *Alternanthera achyranthoides* Forssk.; *Alternanthera denticulata* R. Br.; *Alternanthera ficoidea* (L.) P. Beauv.; *Alternanthera nodiflora* R. Br.; *Alternanthera polygonoides* (L.) R. Br.; *Alternanthera tenella* Moq.; *Alternanthera triandra* Lam.; *Gomphrena polygonoides* L.; *Gomphrena sessilis* L.; *Illecebrum indicum* Houtt.; *Illecebrum sessile* (L.) L.; *Paronychia sessilis* (L.) Desf.

Local name: Sarhanchi

Common name: Sessile joyweed

Botanical description: It is an herb that grows to 30 cm in height. The stem is terete, somewhat purplish and hairy. The leaves are simple, exstipulate, and opposite. The petiole is about 0.5 cm long. The blade is elliptic, tapering at the base, acute at the apex, 1–8 cm × 0.2–1.5 cm, and subglabrous. The inflorescence is an axillary, sessile, and ovoid spike, which is about 0.5 cm in diameter. The perianth includes 5 tepals that are white, ovate, and minute. The androecium consists of 5 stamens. The gynoecium is ovoid. The fruits are tiny bottle-shaped utricules containing minute seeds.

Medicinal uses: boils, abscesses

Pharmacology: Extracts of the plant demonstrated anti-inflammatory,[52] hepatoprotective,[53] and antibacterial effects.[54]

6.7 *Amaranthus blitum* L.

Synonyms: *Albersia blitum* (L.) Kunth; *Amaranthus ascendens* Loisel.; *Amaranthus blitum* var. *polygonoides* Moq.; *Amaranthus lividus* L.; *Amaranthus lividus* subsp. *Polygonoides* (Moq.) Probst; *Amaranthus lividus* var. *ascendens* (Loisel.) Hayw. & Druce; *Amaranthus lividus* var. *ascendens* Thell.; *Amaranthus lividus* var. *polygonoides* (Moq.) Thell.; *Euxolus ascendens* (Loisel.) H. Hara; *Euxolus blitum* Gren.; *Euxolus viridis* var. *ascendens* (Loisel.) Moq.

Local name: Gobura nutya

Common names: Purple amaranth; livid amaranth

Botanical description: It is an erect herb that grows up to 30 cm tall. The stem is terete and glabrous. The leaves are simple, spiral, edible and exstipulate. The petiole is 1–3.5 cm long. The blade is broadly lanceolate to rhombic, 1.5–4.5 cm × 1–3 cm, cuneate at base, margin entire, and notched at apex. The inflorescences are axillary or terminal spikes, which are about 3–8.5 cm long. The corolla comprises 5 tepals. The androecium includes 5 stamens. The gynoecium develops 2 stigmas. The fruits are tiny utricles that are ovoid and contain circular seeds.

Medicinal use: ringworm

Pharmacology: Unknown.

6.8 *Amaranthus cruentus* L.

Synonyms: *Amaranthus dussii* Sprenger.; *Amaranthus flavus* L.; *Amaranthus hybridus* subsp. *cruentus* (L.) Thell.; *Amaranthus hybridus* subsp. *paniculatus* (L.) Heijný; *Amaranthus hybridus* var. *cruentus* (L.) Mansf.; *Amaranthus hybridus* var. *paniculatus* (L.) Thell.; *Amaranthus hybridus* var. *paniculatus* (L.) Uline & W.L. Bray; *Amaranthus hybridus* var. *sanguineus* (L.) Farw.; *Amaranthus paniculatus* L.; *Amaranthus paniculatus* var. *cruentus* (L.) Moq.; *Amaranthus paniculatus* var. *purpurascens* Moq.; *Amaranthus paniculatus* var. *sanguineus*

(L.) Moq.; *Amaranthus parisiensis* Schkuhr; *Amaranthus sanguineus* L.; *Amaranthus speciosus* Sims; *Galliaria patula* Bubani.

Local name: Cholai

Common names: Blood amaranth; purple amaranth; caterpillar amaranth

Botanical description: It is an erect herb that grows to 1.5 m tall. The stem is terete, fleshy, and striated. The leaves are simple, spiral, and exstipulate. The petiole can reach about 7.5 cm long. The blade is broadly lanceolate, cuneate at base, margin wavy, acute or acuminate at apex, and 3–15 cm × 1.5–10 cm. The inflorescence is a terminal, elongated, red or purplish and showy spike. The perianth is minute and comprises 5 tepals. The androecium comprises 5 stamens. The fruits are ovoid utricles containing tiny seeds.

Medicinal uses: diuretic, blood purifier

Pharmacology: Extracts of the plant exhibited hematopoietic effects in rats.[55]

6.9 *Amaranthus gangeticus* L.

Synonym: *Amaranthus tricolor* L.

Local name: Lal shak

Common name: Garden amaranth

Botanical description: It is an erect herb that grows to 1.5 m in height. The stem is terete, fleshy, and striated. The leaves are simple, spiral, and exstipulate. The petiole is 1.5–5 cm long. The blade is green and purple, 5–10 cm × 2–6.5 cm, broadly lanceolate to rhombic, wavy at the margin, tapering at base, and acute at apex. The inflorescences are axillary or terminal elongated spikes that can reach 15 cm in length. The androecium comprises three stamens. The gynoecium comprises three stigmas. The fruits are ovate utricles containing tiny seeds.

Medicinal uses: dry skin, leucorrhea, diarrhea

Pharmacology: The plant exhibited anti-cancer activity[56] and protected mice against radiations.[57]

6.10 *Amaranthus spinosus* L.

Synonyms: *Amaranthus caracasanus* Kunth; *Amaranthus diacanthus* Raf.; *Amaranthus spinosus* fo. *Inermis* Lauterb. & K. Schum.; *Amaranthus spinosus* var. *basiscissus* Thell.; *Amaranthus spinosus* var. *circumscissus* Thell.; *Amaranthus spinosus* var. *indehiscens* Thell.; *Amaranthus spinosus* var. *purpurascens* Moq.; *Amaranthus spinosus* var. *pygmaeus* Hassk.; *Amaranthus spinosus* var. *rubricaulis* Hassk.; *Amaranthus spinosus* var. *viridicaulis* Hassk.; *Galliaria spinosa* L.) Nieuwl.

Local name: Kanta maris

Common name: Prickly amaranth

Botanical description: It is an erect herb that grows up to 1 m tall. The stem is terete, fleshy, and subglabrous. The leaves are simple, spiral, and exstipulate. The petiole 1–8 cm, with a pair of spines at the base. The blade is lanceolate, tapering at the base along the petiole, acute at the apex, margin is wavy, and presents 3 to 10 pairs of secondary nerves. The inflorescence is an axillary or terminal spike which is about 8–15 cm long. The perianth includes 5 tepals that are linear, minute, and green. The androecium present 5 stamens. The gynoecium is ovoid and develops three stigmas. The fruits are tiny utricles containing subglobose seeds.

Medicinal uses: blenorrhea, menorrhagia, colic, gonorrhea, boils, diuretic

Pharmacology: The plant produces antibacterial fatty acids.[58]

7 Family Anacardiaceae R. Brown 1818

7.1 *Anacardium occidentale* L.

Synonyms: *Acajuba occidentalis* (L.) Gaertn.; *Anacardium microcarpum* Ducke; *Anacardium occidentale* var. *luteum* Bello; *Anacardium occidentale* var. *rubrum* Bello; *Cassuvium pomiferum* Lam.

Local names: Kajul; hijli badam

Common name: Cashew

Botanical description: It is a resinous tree, which can reach 10 m tall. The bole is not straight. The leaves are simple, spiral, and exstipulate. The petiole is 1–1.5 cm long. The blade is spathulate, coriaceous, 8–11 cm × 6–8.5 cm, rounded, truncate to retuse at apex, and with 6–13 pairs of conspicuous secondary nerves. The inflorescence is a terminal panicle, which is 10–20 cm long. The calyx comprises 5 minute lobes. The corolla consists of 5 petals, which are up to 1 cm long. The androecium includes 7–10, stamens. The gynoecium is minute. The fruit is a kidney-shaped drupe on a fleshy stalk which is about 3–8 cm × 4–5 cm and purplish. The seeds are edible after being roasted.

Medicinal uses: leprosy, warts, syphilis

Pharmacology: The plant contains agathisflavone, which is antibacterial.[59] It also produces anacardic acid, which has manifold pharmacological effects[60] including antibacterial effects.[61–63]

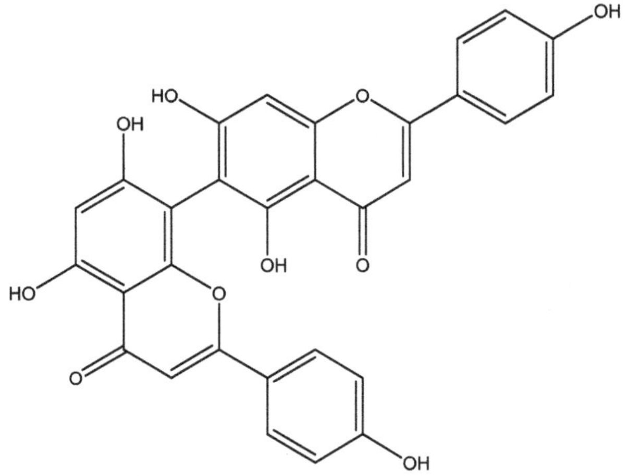

Structure of agathisflavone

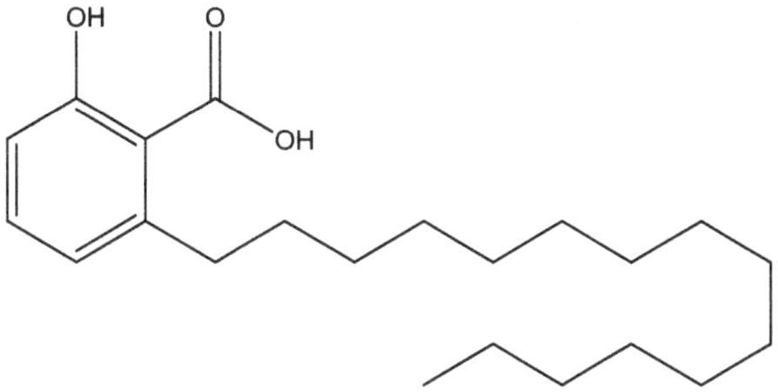

Structure of anacardic acid

7.2 *Lannea coromandelica* (Houtt.) Merr.

Synonyms: *Calesiam grande* (Dennst.) Kuntze; *Lannea grandis* (Dennst.) Engl.; *Odina odier* Roxb.

Local name: Jiwul

Botanical description: It is a resinous tree that grows up to 10 m in height. The leaves are compound, spiral, and exstipulate. The petiole is hairy. The blade is imparipinnate and includes 3–4 pairs of oblong folioles, which are membranaceous, 5.5–9 cm × 2.5–4 cm, asymmetrical, with 5–10 pairs of secondary nerves, and acuminate at the apex. The inflorescence is racemose, at the apex of stems and up to 30 cm long. The calyx is minute and presents 4 lobes. The corolla has 4 petals that are yellow to purple. The androecium includes 8 stamens. The receptacle presents a disc. The gynoecium is ovoid, and 4 locular. The fruit is a hard drupe and is purple-red.

Medicinal uses: ulcers, wounds

Pharmacology: Extracts of the plant demonstrated anti-inflammatory,[64,65] anti-nociceptive,[66] and anti-diarrheal[67] effects.

7.3 *Mangifera indica* L.

Synonyms: *Mangifera austroyunnanensis* H.H. Hu; *Mangifera indica* var. *armeniaca* Bello; *Mangifera indica* var. *intermedia* Bello; *Mangifera indica* var. *leiosperma* Bello; *Mangifera indica* var. *macrocarpa* Bello; *Mangifera indica* var. *viridis* Bello; *Rhus laurina* Nutt.

Local name: Am

Common name: Mango

Botanical description: It is a resinous tree that grows to 20 m in height. The leaves are simple, spiral, and exstipulate. The petiole is 1.5–6 cm channeled and swollen at base. The blade is coriaceous, dark purplish red, and glossy when young, later green and coriaceous, lanceolate, 10–30 cm × 3.5–6.5 cm, the margin wavy, and presents 20–25 pairs of secondary nerves. The inflorescence is a showy panicle that is light greenish-yellow and up to 30 cm long. The flowers are minute. The calyx consists of 5 sepals. The corolla includes 5 petals that are light greenish-yellow. The androecium comprises 5 stamens. A 5-lobed nectary disc is present. The gynoecium is minute. The fruit is a delicious drupe, which is reniform, greenish-yellow to red, and 5–10 cm × 3–4.5 cm.

Medicinal uses: diarrhea, dysentery, leucorrhea, menorrhagia

Pharmacology: The plant has been the subject of numerous pharmacological studies.[68] It produces mangiferin that has anti-diabetic effects.[69] Extracts of the plant demonstrated immunomodulatory,[70] anti-inflammatory,[71] and anti-diarrheal[72] activities.

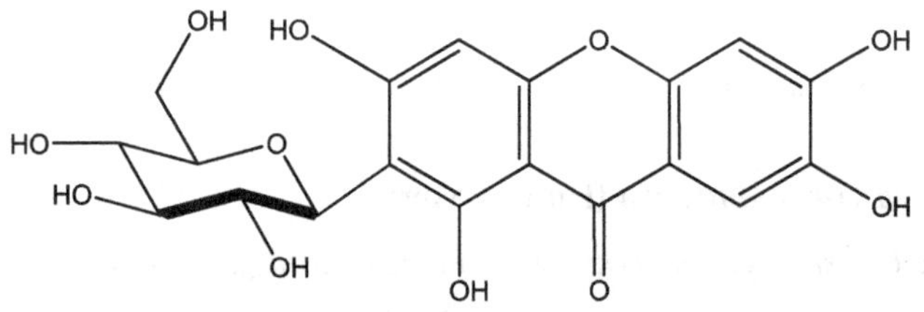

Structure of mangiferin

7.4 *Semecarpus anacardium* L.f.

Local name: Bela

Common name: Marking nut

Botanical description: It is a tree that grows up to 10 m tall. Upon incision of the bole and stem, a vesicant sap is secreted. The leaves are simple, spiral, and exstipulate. The petiole is 2–5 cm long. The blade is broadly elliptic, obovate, coriaceous, slightly glossy, wedge-shaped at the base and round at the apex, 3.5–7 cm × 10–20 cm, and presents about 10–15 pairs of conspicuous secondary nerves. The inflorescence is a terminal or axillary panicle that is pubescent. The flowers are minute. The calyx is minutely 5-lobed. The androecium includes 5 stamens. A nectary disc is present. The gynoecium consists of 3 styles. The fruit is a drupe, which is dark green, glossy, and up to 3 cm long on a fleshy stalk.

Medicinal uses: leprosy, rheumatism, pains, venereal diseases

Pharmacology: Extracts of the plant elicited anti-diabetic,[73] wound healing,[74] and anti-tumor[75] activities.

7.5 *Spondias cythera* Sonn.

Synonyms: *Spondias dulcis* G. Forst; *Spondias dulcis* Parkinson; *Spondias purpurea* L.

Local name: Bilati amra

Common names: Wild mango; hog plum

Botanical description: It is a resinous tree that grows to 20 m in height. The leaves are imparipinnate, spiral, and exstipulate. The blade comprises 4–8 pairs of folioles plus a terminal one. The folioles are broadly elliptic, 5–10 cm × 2.5–5 cm, coriaceous, acute at the base, wavy at the margin, acuminate at apex with inconspicuous secondary nerves. The inflorescence is a terminal and axillary and lax and panicle. The flowers are minute. The calyx is minutely 5-lobed. The corolla comprises 5 petals that are pure white. The nectary disc is bright yellow. The gynoecium is 5-lobed and produces 5 styles. The fruit is an olive-shaped drupe which is up to 4.5 cm long, and edible.

Medicinal uses: stomach ache, dysentery

Pharmacology: Extracts of the plant exhibited cytotoxic, thrombolytic,[76] and antimicrobial[77] activities.

8 Family Annonaceae A.L. de Jussieu 1789

8.1 *Annona reticulata* L.

Synonyms: *Annona excelsa* Kunth; *Annona humboldtiana* Kunth; *Annona humboldtii* Dun.; *Annona laevis* Kunth; *Annona longifolia* Sessé & Moc.; *Annona primigenia* Standl. & Steyerm.; *Annona reticulata* var. *primigenia* (Standl. & Steyerm.) Lundell; *Annona riparia* Kunth

Local name: Nona

Common names: Sweetsop; Bullock's heart

Botanical description: It is a tree that grows up to 6 m tall. The leaves are simple, alternate, and exstipulate. The petiole is stout and up to 1.5 cm long. The blade is lanceolate, coriaceous, 10–17.5 cm × 2.5–5 cm, cuneate at base, and acuminate at apex. The inflorescence is axillary. The calyx includes 3 sepals. The corolla consists of 2 whorls of 3 petals, which are coriaceous, yellowish-green, broadly lanceolate, and about 1.5–2 cm × 0.8–1.5 cm. The androecium comprises numerous stamens that are minute. The gynoecium includes numerous carpels. The fruit is somewhat heart-shaped and light brownish-pink, 10–15 cm × 5.5–12.5 cm, reticulate, edible, and containing numerous glossy black seeds.

Medicinal use: dysentery

Pharmacology: The plant has been well studied for pharmacology.[78] It contains notably a series of diterpenes of which kaur-16-en-19-oic acid, possesses anti-inflammatory and analgesic activity.[80,81]

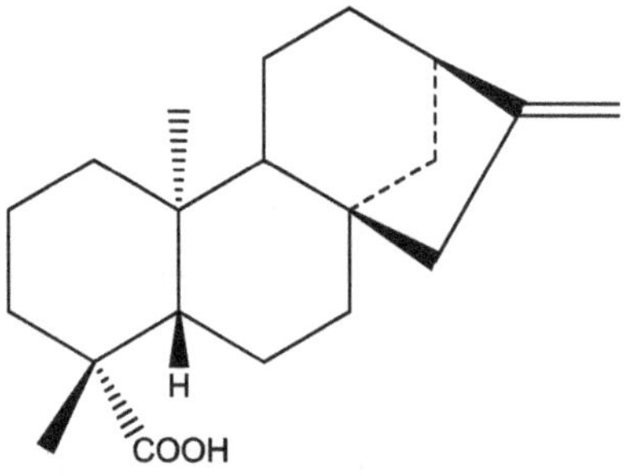

Structure of kaur-16-en-19-oic acid

8.2 *Annona squamosa* L.

Synonyms: *Annona asiatica* L.; *Annona cinerea* Dunal; *Annona forskahlii* DC.; *Annona glabra* Forssk.; *Guanabanus squamosus* (L.) M. Gómez; *Xylopia frutescens* Sieb. ex Presl; *Xylopia glabra* L.

Local name: Ata

Common names: Custard apple; sugar apple; sweetsop

Botanical description: It is a medium-sized tree. The bark is smooth and slightly aromatic. The leaves are simple, alternate, and exstipulate. The petiole is 0.4–1.5 cm long. The blade is elliptic, thinly leathery, 5–17.5 cm × 2–7.5 cm, obtuse at the base, margin entire, acute at apex, and presents 8–15 pairs of secondary nerves. The flowers are clustered and cauliflorous. The calyx comprises 3 triangular sepals. The corolla includes 3 greenish, oblong-lanceolate, 1.5–3 cm × 0.5–0.8 cm, fleshy, petals that are purplish at the base. The androecium includes numerous tiny stamens. The gynoecium includes several carpels. The fruit is heart-shaped, 5–10 cm in diameter, green to purplish, and squamose. The seeds are black, 1.4 cm long, and embedded in a white edible pulp.

Medicinal uses: lice, sores, skin infection

Pharmacology: The plant has been well studied for its pharmacology.[82] It contains acetogenins that are insecticidal[83] and anthelminthic.[84] Extracts of the plant exhibited anti-inflammatory properties.[85] Of note, the plant produces ent-16β, 17-dihydroxykauran-19-oic acid that has anti-HIV properties.[86]

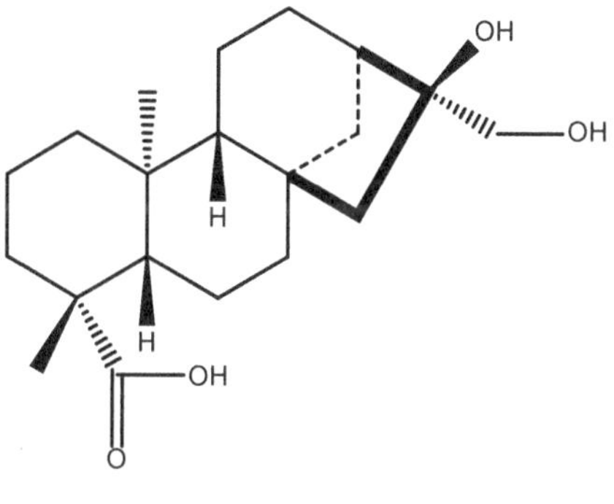

Structure of ent-16β, 17-dihydroxykauran-19-oic acid

8.3 *Polyalthia longifolia* (Sonn.) Thwaites

Synonyms: *Guatteria longifolia* Wall.; *Uvaria longifolia* Sonn.; *Unona longifolia* (Sonn.) Dunal

Local name: Debdaru

Common names: Indian mast tree; false Ashoka

Botanical description: It is a tree that grows up to 20 m tall. The bole is straight. The leaves are simple, alternate, and exstipulate. The petiole is 0.5–1 cm long. The blade is elongated-lanceolate, wavy, glossy, leathery, 11–31 cm × 2.5–8 cm, with 18–24 pairs of secondary nerves, cuneate at base, and acuminate at apex. The flowers are axillary on 0.7–1.5 cm long peduncles. The calyx includes 3 sepals that are ovate and minute. The corolla includes 6 petals that are greenish, linear-triangular, membranous, and 1.3–1.5 cm × 0.2–0.4 cm. The androecium comprises numerous minute stamens. The gynoecium includes 20–25 carpels. The fruits consist of 4–8 ripe carpels, which are purplish, ovoid, and 2–2.5 cm × 1.5 cm. Each ripe carpel encloses a single seed that is olive-shaped and deeply longitudinally grooved.

Medicinal use: fever

Pharmacology: Extracts of the plant demonstrated hypotensive effects.[87] The plant produces anti-inflammatory, cytotoxic,[88] and antimicrobial[89] diterpenes.

9 Family Apiaceae Lindley 1836

9.1 *Eryngium foetidum* L.

Synonyms: *Eryngium antihystericum* Rottb.; *Eryngium molleri* Gand.

Local name: Bilati dhonia

Common name: Spiny coriander

Botanical description: It is a dark green erect herb that grows to about 30 cm in height and with somewhat a strange architecture. The stems are green, fleshy, and striated longitudinally. The leaves are arranged into a basal rosette. The blade is oblong-lanceolate, membranous, glossy, serrate, 5–25 cm × 1.2–5 cm, and round at apex. The upper leaves are sessile and opposite. The inflorescences are axillary and terminal spikes, which are about 3 cm long and surrounded by bracts. The calyx is minute. The corolla comprises 5 petals, which are white or pale yellow. A disc is present. The androecium includes 5 stamens. The gynoecium produces 2 styles. The fruit is ovoid, minute, and covered with tubercles.

Medicinal uses: fever, hypertension, constipation, asthma

Pharmacology: Extracts of the plant demonstrated anthelmintic, anti-inflammatory, anticonvulsant, and antibacterial activities.[90,91]

9.2 *Ferula assafoetida* L.

Synonyms: *Ferula foetida* St.-Lag.; *Scorodosma foetidum* Bunge

Local name: Hing

Common name: Asafoetida

Botanical description: It is a stout, resinous, odorous herb that grows up to 3 m. The stem is stout, light green, and up to 10 cm in diameter. The leaves are arranged in a basal rosette, exstipulate, bipinnate, and up to 60 cm long. The petiole forms a sheath at the base. The inflorescences are showy yellow umbels, which are about 20 cm in diameter. The flowers are minute. The corolla includes 5 petals. The androecium presents 5 stamens. A disc is present. The fruit is a flat achene, which is striated.

Medicinal uses: indigestion, intestinal worms, laxative, anxiety

Pharmacology: The plant has been the subject of numerous pharmacological investigations.[92] Extracts of the plant exhibited antibacterial, spasmolytic,[93] and antiviral[94] effects.

10 Family Apocynaceae A.L. de Jussieu 1789

10.1 *Aganosma dichotoma* K. Schum.

Synonyms: *Aganosma heynei* (Spreng.) I.M. Turner; *Echites dichotomus* Roth

Local name: Malati

Botanical description: It is a stout laticiferous climber that grows up to 10 m long. The leaves are simple, opposite, and exstipulate. The petiole is 3–5 mm long. The blade is elliptic, 5–12 cm × 2–4 cm, coriaceous, wedge-shaped at the base, acuminate or round at the apex, and presents 4–8 pairs of secondary nerves. The inflorescences are terminal or axillary, and lax cymes that are 5–20 cm long. The calyx produces 5 lanceolate lobes. The corolla is tubular, pure white, 1.25–2 cm long, and produces 5 contorted lobes that are triangular in shape. The androecium consists of 5 stamens adnate to the corolla tube. The nectary disc is 5-lobed. The gynoecium includes 2 carpels that form a columnar stigma. The fruits are pairs of 15–40 cm × 2–5 mm follicles filled with linear-oblong and plumed seeds.

Medicinal uses: leprosy, bronchitis, indigestion

Pharmacology: Extracts displayed analgesic, and anti-inflammatory activities.[95]

10.2 *Allamanda cathartica* L.

Synonyms: *Allamanda aubletii* Pohl; *Allamanda cathartica* var. *grandiflora* (Aubl.) L.H. Bailey & Raffill; *Allamanda cathartica* var. *hendersonii* (W. Bull ex Dombrain) L.H. Bailey & Raffill; *Allamanda cathartica* var. *williamsii* (hort.) L.H. Bailey; *Allamanda grandiflora* (Aubl.) Lam.; *Allamanda hendersonii* W. Bull ex Dombrain; *Allamanda latifolia* C. Presl; *Allamanda linnei* Pohl; *Allamanda salicifolia* Hort.; *Allamanda schottii* Hook.; *Allamanda wardleyana* Lebas; *Allamanda williamsii* Hort.; *Echites verticillatus* Sessé & Moc.; *Orelia grandiflora* Aubl.

Local name: Malatilata

Common names: Yellow allamanda; golden trumpet vine

Botanical description: It is a laticiferous scandent shrub that grows up to 3 m tall. The stem is terete. The leaves are simple, verticillate, and exstipulate. The petiole is 1.5 cm long. The blade is obovate, 5.5–15 cm × 4–5 cm, and elliptic. The inflorescences are terminal cymes of showy flowers. The calyx presents 5 lobes. The perianth is tubular, 4–8 cm long, plain yellow, and produces 5 broad and orbicular lobes. The androecium includes 5 stamens attached on the corolla. A 5-lobed disc is present. The gynoecium is ovoid and develops a 2-lobed style. The fruits are spiny capsules, which are 3–7 cm × 4–5 cm, and contain winged seeds. The fruits are very seldom seen.

Medicinal use: laxative

Pharmacology: Extracts of the plant showed wound healing effects.[96]

10.3 *Alstonia scholaris* (L.) R.Br.

Synonym: *Echites scholaris* (L.)

Local name: Chattin

Common name: Devil's tree

Botanical description: It is a laticiferous tree that grows up to 25 m tall. The plant has a sinister look. The bark is greyish-dark brown and somewhat smooth. The leaves are simple, exstipulate, and whorled. The petiole is 1–2.3 cm long. The blade is elliptic, acute at the base and apex, coriaceous, dark green above, glaucous below 5–15 cm × 2–5 cm, with 25–50 pairs of parallel and straight secondary nerves. The inflorescence are terminal and showy cymes of ephemeral flowers. The corolla is white, tubular, the tube 0.5–1 cm long, and 5-lobed. The gynoecium includes 2 carpels and is pubescent. The fruits are slender, curling, light green, and pendulous follicles that can reach 25 cm long and containing minute hairy seeds.

Medicinal uses: fever, dysentery, intestinal worms, leprosy, pneumonia

Pharmacology: The plant has been well studied for its pharmacology.[97] It contains indole alkaloids with anti-inflammatory,[98,99] cytotoxic, antibacterial, antiviral, and anti-fungal activities,[100–103] as well as antibacterial triterpenes.[104] Extracts of the plant demonstrated antimalarial and anti-diabetic effects.[105,106]

10.4 *Calotropis gigantea* (L.) W.T. Aiton

Synonyms: *Asclepias gigantea* L.; *Madorius giganteus* (L.) Kuntze; *Periploca cochinchinensis* Lour.; *Streptocaulon cochinchinense* (Lour.) G. Don

Local name: Akund

Common names: Gigantic swallow-wort; giant milkweed

Botanical description: It is a laticiferous shrub that grows up to 3 m in height. The stem is terete and hairy. The leaves are simple, decussate, and exstipulate. The petiole is up to 2 cm long. The blade is broadly elliptic, 6–20 cm × 3–10 cm, coriaceous, cordate at base, acute at apex, hairy, and with 4–8 pairs of conspicuous secondary nerves. The inflorescences are terminal cymes of magnificent flowers. The calyx produces 5 lobes and is about 0.5 cm long. The corolla is 3–4.5 cm × 1.5–2.2 cm, purplish, pinkish, or pure white, 5-lobed, and fleshy. The fruits are follicles, which are ovoid, green, rough, 5–10 cm × 2.5–4 cm, recurved, and containing numerous comose seeds.

Medicinal uses: leprosy, syphilis, laxative, asthma, toothache, intestinal worms

Pharmacology: Extracts of the plant exhibited wound healing,[107] analgesic,[108] and antibacterial effects.[109]

10.5 *Carissa carandas* L.

Synonyms: *Arduina carandas* (L.) K. Schum.; *Carissa congesta* Wight.; *Damnacanthus esquirolii* H. Lév.; *Jasminonerium carandas* (L.) Kuntze

Local name: Kurumchi

Common names: Bengal currant; Christ's thorn

Botanical description: It is a laticiferous and spiny shrub that grows up to 3 m tall. The spines are woody, needle-like and up to 5 cm long. The leaves are simple, sessile, opposite, and exstipulate. The blade is coriaceous, dark green above, 3–7.5 cm × 1.5–4.5 cm, round at base, minutely apiculate at apex, broadly lanceolate, and with 8 pairs of secondary nerves. The inflorescences are terminal cymes. The calyx comprises 5 lobes and reaches 0.7 cm long. The corolla is tubular, the tube 2 cm long and pink, and produces 5 pure white contorted lobes, which are 1 cm long and lanceolate. The androecium comprises 5 stamens adnate to a corolla tube. The gynoecium is ovoid and produces an elongated stigma. The fruits are light pink, glossy, olive-shaped edible berries which are 1.5–2.5 cm × 1–2 cm.

Medicinal uses: wounds, indigestion

Pharmacology: The plant produces naringenin that demonstrated anti-inflammatory effects.[110] It demonstrated activity due to the triterpene carandinol.[111] Extracts of the plant exhibited anti-diabetic[112] and antibacterial[113] effects.

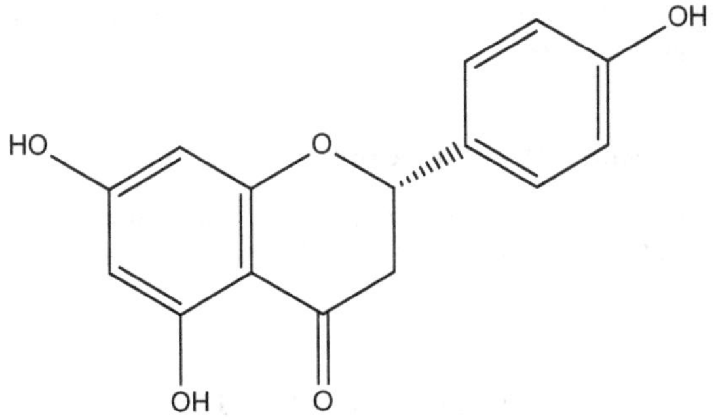

Structure of naringenin

10.6 *Ervatamia divaricata* (L.) Burkill

Synonyms: *Ervatamia coronaria* (Jacq.) Stapf; *Ervatamia cumingiana* (A. DC.) Markgr.; *Ervatamia divaricata* (L.) Burkill; *Ervatamia flabelliformis* Tsiang; *Nerium divaricatum* L.; *Taberna discolor* (Sw.) Miers; *Tabernaemontana coronaria* (Jacq.) Willd.; *Tabernaemontana cumingiana* A. DC.; *Tabernaemontana discolor* Sw.; *Tabernaemontana flabelliformis* (Tsiang) P.T. Li

Local name: Tagar

Common name: East Indian rosebay

Botanical description: It is a handsome laticiferous shrub that grows up to 3 m tall. The bark is light grey-brown and smooth. The leaves are simple, opposite, and exstipulate. The petiole is up to 1 cm long. The blade is elliptic, dark green, glossy, coriaceous, 3–12.5 cm × 1–5.5 cm, acuminate at apex, and with 5–15 pairs of secondary nerves. The inflorescences are terminal cymes. The calyx presents 5 lobes. The corolla is tubular, ephemerous, slightly fragment pure white, the tube about 2 cm long, the throat yellow, and develops 5 contorted oblong lobes that are 1.5–2.5 cm long. The androecium includes 5 stamens that are attached to the corolla tube. The gynoecium develops an elongated stigma. The fruits are a pair of ellipsoid and curved follicles that are 2–7 cm × 0.6–1.5 cm.

Medicinal uses: wounds, toothaches, inflammation, intestinal worms, fever, ophthalmia

Pharmacology: Extracts of the plant demonstrated anti-inflammatory activity[114]: The plant produces indole alkaloids, including notably tabernaemontanine, coronarine, coronaridine, and dregamine with anti-nociceptive,[115] cytotoxic,[116] and antibacterial[117] activities.

10.7 *Holarrhena pubescens* Wall. ex G. Don

Synonyms: *Chonemorpha antidysenterica* (Roth) G. Don; *Echites antidysentericus* Roth; *Echites pubescens* Buch.-Ham.; *Echites pubescens* Willd. ex Roem. & Schult.; *Holarrhena antidysenterica* (L.) Wall. ex A. DC.; *Holarrhena antidysenterica* Roth; *Holarrhena antidysenterica* Roth; *Holarrhena codaga* G. Don; *Holarrhena malaccensis* Wight; *Holarrhena villosa* Aiton ex Loudon

Local name: Kurchi

Common names: Conessi bark; Easter tree

Botanical description: It is a laticiferous tree that grows up to 8 m tall. The leaves are simple, opposite, and exstipulate. The petiole is minute and channeled. The blade is elliptic, 10–20 cm × 4–11.5 cm, membranous, pubescent, round at base, acute at apex, and with 10–15 pairs of secondary nerves. The inflorescences are terminal cymes. The calyx comprises 5 sepals which are elliptic to linear, and up to 1 cm long. The corolla is tubular, pure white, about 0.9–1.9 cm long, and develops 5 contorted lobes that are elliptic-oblong and 1–3 cm. The androecium comprises 5 stamens that are adnate to the corolla tube. The fruits are linear follicles, which are up to 30 cm long and containing numerous 0.9–1.6 cm comose seeds.

Medicinal uses: intestinal worms, dysentery, cholera

Pharmacology: The plant has been the subject of numerous pharmacological studies revealing notably antibacterial,[118] anti-amoebic,[119] and anti-inflammatory[120] properties.

10.8 *Hoya parasitica* Wall. ex Wight

Synonyms: *Hoya parasitica* Wall. ex J. Traille; *Hoya verticillata* (Vahl) G. Don

Local name: Faissa gash

Common name: Wax plant

Botanical description: It is a graceful climber that grows up to 8 m long. The stem is stout and terete. The leaves are simple, opposite, and exstipulate. The petiole is stout and up to 2.5 cm long. The blade is elliptic, coriaceous, fleshy, about 5 cm × 12.5 cm, cuneate at the base, and acute at the apex. The inflorescence is a terminal umbel-like cyme that is about 15 cm in diameter. The calyx includes 5 sepals that are elongated. The corolla is fleshy, purplish white coriaceous, 5-lobed, and 1.5 cm in diameter. The androecium includes 5 stamens which are appressed on the stigmatal head. The fruits are fusiform follicles that are about 12 cm long.

Medicinal use: fever

Pharmacology: The plant produces cytotoxic principles.[121]

10.9 *Ichnocarpus frutescens* (L.) W.T. Aiton

Synonyms: *Apocynum frutescens* L.; *Echites frutescens* (L.) Roxb.; *Echites frutescens* Wall. ex Roxb.; *Gardenia volubilis* Lour.; *Ichnocarpus moluccanus* Miq.; *Ichnocarpus ovatifolius* A. DC.; *Ichnocarpus volubilis* (Lour.) Merr.; *Micrechites sinensis* Markgr.

Local name: Shyamalata

Common names: Black creeper; red sarsaparilla

Botanical description: It is a climber that grows up to 8 m long. The leaves are simple, opposite, and exstipulate. The petiole is up to 1.5 cm long. The blade is coriaceous, dark green, 5–11 cm × 2.5–5 cm, and with 5–7 pairs of secondary nerves. The inflorescences are axillary or terminal cymes. The calyx is 5-lobed, hairy, and minute. The corolla is tubular, minute, and produces 5 strongly contorted lobes, which are hairy and about 0.5 cm long. The androecium includes 5 stamens attached to the corolla tube. The anthers are elliptic. A disc is present. The gynoecium is pubescent. The fruits are linear follicles, which are 8–15 cm × 4–5 mm, pubescent, and containing numerous comose seeds.

Medicinal uses: fever, leprosy, diuretic, bone fracture, diabetes, constipation, syphilis, cholera

Pharmacology: Phenolics from the plant demonstrated anti-diabetic[122] effects.

10.10 *Nerium indicum* Mill.

Synonym: *Nerium oleander* L.

Local name: Lal kharubi

Common names: Indian oleander; white oleander

Botanical description: It is a handsome shrub that grows 6 m tall. The stems are sappy. The leaves are simple, verticillate, and exstipulate. The petiole can reach 0.7 cm long. The blade is narrowly elliptic, coriaceous, acute at base and apex, and is about 10–15 cm × 1–2 cm. The inflorescence is a terminal cyme. The calyx is 5-lobed and about 0.7 cm long. The corolla is tubular, the tube is 1.8 cm long, hairy within, the throat narrow, producing 5 contorted lobes with scales at the base. The androecium comprises 5 stamens attached to the corolla tube. The ovary includes 2 carpels, which develop a linear style and bifid stigma. The fruits are fusiform capsules, which are 12–20 cm × 7–5 mm and contain numerous comose seeds. The fruits are seldom seen.

Medicinal uses: ringworm, leprosy, intestinal worms

Pharmacology: This plant contains cardiotoxic steroidal glycosides.[123] Extracts of this plant exhibited antibacterial[124] and anti-inflammatory[125] effects.

10.11 *Rauvolfia serpentina* (L.) Benth. ex Kurz

Synonyms: *Ophioxylon majus* Hassk.; *Ophioxylon serpentinum* L.

Local name: Chandra

Common name: Indian snakeroot

Botanical description: It is a laticiferous shrub that grows up to 2 m tall. The leaves are simple, exstipulate, and whorled. The petiole can reach up to 1.5 cm long. The blade is elliptic, 7–16 cm × 2.5–5 cm, acute at base and apex, membranous, and with 8–12 pairs of secondary nerves. The inflorescences are terminal or axillary cymes with red peduncles. The calyx is minute and 5-lobed. The corolla is tubular, 8–12 mm long, inflated in the middle, pure white, and develop 5 round lobes, which are 0.5 cm long. The androecium includes 5 stamens that are inserted in the middle of the corolla tube. The ovary is ovoid and develops a linear style and a capitate stigma. The fruit is black, olive-shaped, and glossy drupe, which is about 7 mm long.

Medicinal uses: fever, snake bite, difficult delivery

Pharmacology: This plant has been well studied for its pharmacology. It produces the hypotensive and central nervous depressant indole alkaloid reserpine, which has been used in clinical practice.[126–128]

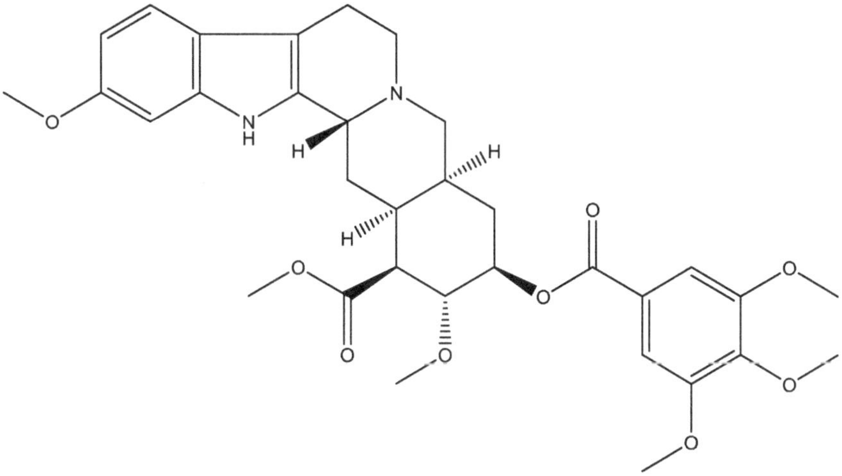

Structure of reserpine

10.12 *Tylophora indica* (Burm.f.) Merr.

Synonyms: *Apocynum reticulatum* Lour.; *Asclepias asthmatica* L. f.; *Cynanchum indicum* Burm. f.

Local name: Untomul

Common name: Emetic swallow wort; Indian ipecacuanha

Botanical description: It is a stout climber that grows up to 10 m long. The leaves are simple, opposite, and exstipulate. The petiole is about 1 cm long. The blade is broadly lanceolate, 3–10 cm × 2–6 cm, cordate at base, apiculate at apex, glossy and fleshy. The inflorescence is a terminal cyme. The calyx is hairy. The corolla is greenish purple, hairy, tubular, about 1 cm in diameter, and pink at the throat. The androecium includes 5 stamens pressed to a columnar stigma. The fruits are follicles, which are up to 10 cm × 1–2 cm and containing numerous comose seeds.

Medicinal uses: asthma, dysentery, laxative

Pharmacology: The plant contains tylophorinidine,[129] which is anti-leukemic[130] and anti-fungal.[131] Extracts of the plant demonstrated anti-inflammatory effects.[132]

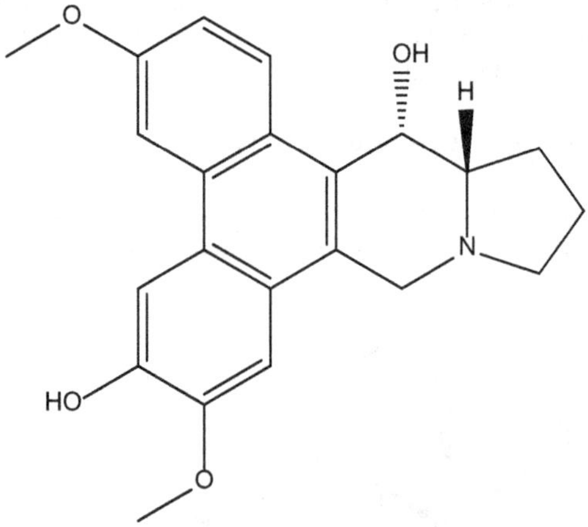

Structure of tylophorinidine

11 Family Araceae A.L. de Jussieu 1763

11.1 *Aglaonema commutatum* Schott

Synonyms: *Aglaonema marantifolium* var. *commutatum* (Schott) Engl.; *Scindapsus cuscuaria* (Aubl.) C. Presl

Local name: Shondori kuchoo

Botanical description: It is a fleshy herb that grows to 1.5 m in height. The petiole is 5–25 cm long and form sheaths. The blade is variegated, oblong-elliptic, fleshy, 10–30 cm × 2.5–15 cm, acum with 4–7 primary lateral veins. The flower peduncle is 4.5–20 cm long. The spathe is 3–7 cm × 2.5–5 cm and light green. The spadix is stipitate and 2–6.5 cm long. The ovary is ovoid and develops a short style and a broad discoid stigma. The fruit is red, olive-shaped and 1.5–2 cm × 0.5–1.5 cm in size.

Medicinal use: counterirritant

Pharmacology: Unknown.

11.2 *Aglaonema hookerianum* Schott

Synonym: *Aglaonema* clarkei Hook.f.

Local names: Sikachalal; horina shak

Botanical description: It is a fleshy herb that grows up to 50 cm tall. The petiole is 15–25 cm long and form sheaths. The blade is elliptic, 6.5–12 cm × 20–25 cm, round at base, acuminate at apex with 7–13 lateral nervations. The inflorescence peduncle is 10–20 cm long. The spathe is 3.5–6.5 cm long. The spadix is 2.5–4.5 cm long. The staminate zone is 2.0–3.5 cm long. The female portion is 0.3–0.5 cm long. The fruits are red, glossy, olive-shaped, and 2–3 cm × 0.9–1.5 cm in size.

Medicinal uses: conjunctivitis, constipation

Pharmacology: Extracts of the plant displayed antibacterial activity.[133]

11.3 *Alocasia acuminata* Schott

Local name: Bonnyo kochu

Botanical description: It is a fleshy herb that grows to 70 cm tall. The petiole is 15–80 cm long and form sheaths. The blade is sagittate, glossy, 15–55 cm × 8–15 cm, and with 6–12 lateral nervations. The inflorescence peduncle is 9–20 cm long. The spathe is 7–10 cm long, the limb lanceolate and 5.5–7.5 cm long. The spadix is 6–9.5 cm long. The female zone is 1–1.5 cm long. The fruits are 0.7 cm in diameter and red.

Medicinal use: counterirritant

Pharmacology: Unknown.

11.4 *Alocasia cucullata* (Lour.) G. Don.

Synonyms: *Alocasia cucullata* (Lour.) Schott; *Arum cucullatum* Lour.

Local names: Bilae kochu; lipkai

Common names: Chinese taro; Buddha's palm

Botanical description: It is a fleshy herb that grows to 1 m tall. The petiole is 25–80 cm long and form sheaths. The blade is broadly lanceolate-cordate and about 10–40 cm × 7–25 cm with 8 lateral nervations. The inflorescence peduncle is 20–30 cm long. The spathe is 9–15 cm long, green, and develops a limb that is cymbiform. The spadix is 8–15 cm long. The staminate zone is 3.5 cm long. The female zone is 1.5–2.5 cm long. The fruits are subglobose berries that are about 1 cm in diameter.

Medicinal uses: snake bites, tumors

Pharmacology: Extracts of the plant exhibited anti-tumor effects.[134,135]

11.5 *Alocasia macrorrhizos* (L.) G. Don

Synonyms: *Alocasia indica* (Lour.) Schott; *Colocasia indica* (Lour.) Hassk.; *Alocasia indica* (Lour.) Spach; *Arum indicum* Roxb.; *Caladium indica* (Lour.) K. Koch; *Colocasia indica* (Lour.) Kunth; *Colocasia indica* (Lour.) Kunth

Local name: Man kachu

Common name: Giant taro

Botanical description: It is a massive fleshy herb that grows to 3 m tall. The petiole is stout, up to 1.5 m, and form sheaths. The blade is sagittate, about 120 cm × 50 cm with about 18 lateral nervations. The flower peduncle is about 70 cm tall. The spathe is 15–35 cm long, ovoid. The limb spoon-shaped, and 10.5–30 cm long. The spadix comprises a staminate zone which is 3–7 cm long. The female zone is 1–2 cm long. The ovary develops a 3–5-lobed stigma. The fruits are red berries, which are olive-shaped and about 1 cm long.

Medicinal uses: anti-cancer, anti-inflammatory, laxative, edema

Pharmacology: The plant contains anti-inflammatory and cytotoxic indole alkaloids.[136,137] Extracts of the plant demonstrated anti-hyperglycemic effects.[138]

11.6 *Amorphophallus bulbifer* (Roxb.) Bl.

Synonyms: *Arum bulbiferum* Roxb.; *Conophallus bulbifer* (Sims) Schott; *Pythonium bulbiferum* (Roxb.) Schott

Local name: Amla bela

Common name: Voodoo lily

Botanical description: It is a fleshy aroid herb that grows up to 1.5 m tall. The petiole is 60–80 cm long and variegated. The blade is divided into lobes which are 5–15 cm × 2–4 cm and ovate-oblong. The inflorescence peduncle is about 70 cm tall. The spathe is ovate, 15–20 cm long, and light pink and eerie. The spadix is conspicuous, lanceolate, 20–25 cm long, and pale cream-colored. The staminate zone is 4–4.5 cm long. The female zone is 2.5–3.0 cm long. The flowers are minute. The fruits are ovoid, 1.0–1.5 cm long, and red.

Medicinal use: wounds

Pharmacology: The plant has not been much studied for its pharmacology. Extracts elicited anti-inflammatory effects.[139]

11.7 *Amorphophallus konjac* K. Koch

Synonyms: *Amorphophallus mairei* H. Lév.; *Amorphophallus nanus* H. Li & C.L. Long; *Amorphophallus rivieri* Durieu ex Rivière; *Brachyspatha konjac* (K. Koch) K. Koch; *Hydrosme rivieri* (Durieu ex Rivière) Engl.; *Proteinophallus rivieri* (Durieu ex Rivière) Hook. f.

Local name: Jongli kachu

Common name: Devil's tongue

Botanical description: It is an aroid that grows to 2.5 m tall. The petiole is up to 1 m tall, stout, and variegated. The blade is dissected, up to 1.5 m in diameter, the folioles elliptic, 2.5–10 cm × 2–6 cm, and acuminate. The inflorescence peduncle is up to 1 m tall. The spathe is purple, broadly lanceolate, and 10–60 cm × 10–50 cm. The spadix has an offensive odor and is up to 1 m tall. The female zone is 2–11 cm long. The male zone is 2–12.5 cm long. The fruits are olive-shaped, orange-red, and about 2 cm long.

Medicinal uses: tonic, cancer, antidote, itchiness, insecticide, carminative

Pharmacology: The plant is well known to produce glucomannan.[140] Extracts of the plant exhibited anti-tumor activity.[141]

11.8 *Amorphophallus paeoniifolius* (Dennst.) Nicolson

Synonyms: *Amorphophallus bangkokensis* Gagnep.; *Amorphophallus campanulatus* Blume ex Decne.; *Amorphophallus campanulatus* Decne.; *Amorphophallus dubius* Blume; *Amorphophallus gigantiflorus* Hayata; *Amorphophallus microappendiculatus* Engl.; *Amorphophallus paeoniifolius* var. *campanulatus* (Decne.) Sivad.; *Amorphophallus rex*

Prain ex Hook. f.; *Amorphophallus sativus* Blume; *Amorphophallus virosus* N.E. Br.; *Arum campanulatum* Roxb.; *Arum campanulatum* Roxb.; *Arum decurrens* Blanco; *Arum rumphii* Gaudich.; *Candarum rumphii* (Gaudich.) Schott; *Dracontium paeoniifolium* Dennst.; *Hydrosme gigantiflora* (Hayata) S.S. Ying; *Hydrosme gigantiflorus* (Hayata) S.S. Ying; *Plesmonium nobile* Schott

Local name: Ol

Common name: Elephant-foot yam

Botanical description: It is an odd-looking aroid, which grows 2.5 m tall from a massive discoid corm. The petiole is variegated and up to 1.5 m tall. The blade is dissected, about 1.5 m long, with elliptic-lanceolate leaflets, which are 10–20 cm × 2.5–12.5 cm, and acuminate at the apex. The flower peduncle is stout and up to 20 cm tall. The spathe is showy, campanulate, 10–45 cm × 15–60 cm, spreading, purplish, and glossy. The spadix has an offensive odor, 12.5–70 cm long, purplish and many folded. The male zone is obconic and up to 45 cm long. The female zone is up to 25 cm long. The fruiting zone is 10–50 cm × 3–10 cm with numerous reddish-yellow olive-shaped berries, which are up to 2 cm long.

Medicinal uses: asthma, dysentery, rheumatism, counterirritant, bronchitis

Pharmacology: Extracts demonstrated cytotoxic[142] and analgesic[143] effects. The plant produces amblyone, which is antibacterial and cytotoxic.[144]

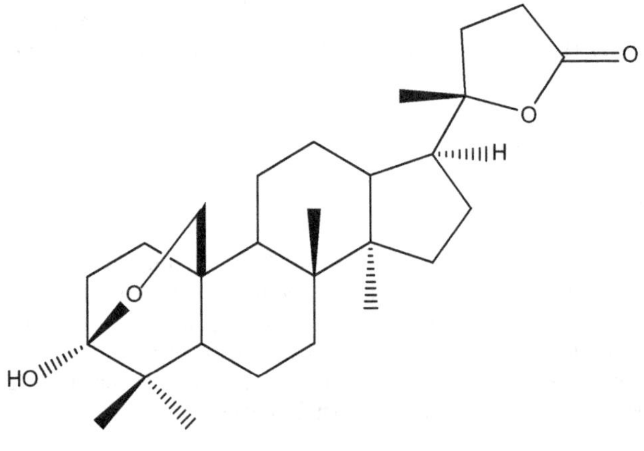

Structure of amblyone

11.9 *Anthurium andraeanum* Lind.

Synonyms: *Arum sylvaticum* Roxb.; *Brachyspatha sylvatica* (Roxb.) Schott; *Pythonium sylvaticum* (Roxb.) Wight; *Synantherias sylvatica* (Roxb.) Schott

Common name: Wild suran

Botanical description: It is an aroid herb that grows up to 1.5 m tall from a massive discoid tuber. The petiole is 30–60 cm long and variegated. The blade is dissected, up to 60 cm in diameter, the folioles 4–7.5 cm × 2.5–4.5 cm, ovate-elliptic, and acute at the apex. The flower peduncle is 40–50 cm long and variegated. The spathe is 3–5.5 cm × 4–6.5 cm, greenish-pink to purple, or pure white. The spadix protrudes from the spathe conspicuously, 15–25 cm long, and linear-lanceolate. The staminate zone is about 1 cm long. The female zone is 0.5–1.5 cm long. The ovary is bottle-shaped and minute. The fruits are red berries.

Medicinal uses: elephantiasis, piles

Pharmacology: Unknown

11.10 *Colocasia esculenta* (L.) Schott

Synonyms: *Alocasia dussii* Dammer; *Alocasia illustris* W. Bull; *Arum chinense* L.; *Arum colocasia* L.; *Arum nymphaeifolium* (Vent.) Roxb.; *Arum peltatum* Lam.; *Caladium acre* R. Br.; *Caladium esculentum* (L.) Vent.; *Caladium nymphaeaefolium* Vent. *Caladium violaceum* Engl.; *Calla gaby* Blanco; *Colocasia acris* (R. Br.) Schott; *Colocasia antiquorum* Schott; *Colocasia euchlora* K. Koch & Linden; *Colocasia fontanesii* Schott; *Colocasia formosana* Hayata; *Colocasia himalensis* Royle; *Colocasia konishii* Hayata; *Colocasia neocaledonica* Van Houtte; *Colocasia nymphaeifolia* (Vent.) Kunth; *Colocasia peregrina* (L.) Raf.; *Colocasia vulgaris* Raf.

Local name: Kachu

Common name: Taro

Botanical description: It is a massive fleshy herb which grows up to 3 m tall from an edible corm. The petiole is stout, 25–100 cm long, and form sheaths. The blade is suborbicular, 13–45 cm × 10–35 cm, cordate at base, wavy at margin, and acute at apex. The inflorescence peduncle is 15–25 cm long. The spathe is 3.5–5 cm × 1.2–1.5 cm and develops a limb that is yellow, lanceolate, and 10–20 cm × 2–5 cm. The spadix consists of a male zone that is 4–6.5 cm long. The female zone is 3 cm long. The fruits are berries that are green, and about 5 mm in diameter.

Medicinal uses: alopecia, skin diseases

Pharmacology: Extract of the plant demonstrated antibacterial activities.[145]

11.11 *Homalomena aromatica* (Spreng.) Schott

Synonyms: *Calla aromatoca* (Spreng.) Roxb.; *Zantedeschia aromatica* Spreng.

Local names: Gondhi kochu; barodaga; gandubi kachu

Botanical description: It is an aromatic aroid herb that grows to 70 cm in height. The petiole is 35 cm long and form sheaths. The blade is sagittate, glossy, green, fleshy, smooth, 20–30 cm × 10–15 cm, and with 8–12 primary lateral nervations. The flower peduncle is 10–20 cm long. The spathe is oblong and 8–10 cm long. The spadix is long as the spathe. The male zone is 5.5–6 cm long. The female zone is 1.5–3 cm long. The ovary is ovoid-with a capitate stigma. The fruits are berries ripening to orange yellow.

Medicinal use: tonic

Pharmacology: The plant produces anti-fungal[146] and anti-inflammatory[147] sesquiterpenes.

11.12 *Lasia spinosa* (L.) Thw.

Synonyms: *Dracontium spinosum L.*; *Lasia aculeata* Lour.; *Lasia crassifolia* Engl.; *Lasia desciscens* Schott; *Lasia hermannii* Schott; *Lasia heterophylla* (Roxb.) Schott; *Lasia jenkinsii* Schott; *Lasia loureirii* Schott; *Lasia roxburghii* Griff.; *Lasia zollingeri* Schott; *Pothos heterophyllus* Roxb.; *Pothos heterophyllus* Roxb.; *Pothos lasia* Roxb.; *Pothos spinosus* (L.) Buch.-Ham. ex Wall.

Botanical description: It is an aroid herb that grows to 1.5 m tall. The stem is creeping, stoloniferous, and spiny. The petiole is 30–100 cm long and spiny. The blade is sagittate-hastate or deeply dissected, 35–65 cm × 20–55 cm, and with 4–8 lateral nervations. The inflorescence peduncle is about 60 cm tall and spiny. The spathe is 18–35 cm, orangish-yellow, and develops a slender limb that is about 30 cm long. The spadix is 3–5 cm long. The ovary is tiny and ovoid. The fruits are packed into an oblong mass. The fruits are polygonal and about 1 cm in diameter.

Medicinal uses: stomach ache, sore throat

Pharmacology: Extracts of the plant exhibited anthelminthic properties.[148]

11.13 *Remusatia vivipara* Schott

Synonyms: *Arum viviparum* Roxb.; *Caladium viviparum* (Roxb.) Lodd.; *Caladium viviparum* (Roxb.) Nees; *Colocasia vivipara* (Roxb.) Thwaites; *Remusatia bulbifera* Vilm.; *Remusatia formosana* Hayata

Botanical description: It is a fleshy herb that grows to 1.5 m tall. The petiole is up to 30 cm long and form sheaths at the base. The blade is peltate, 12–40 cm × 7.5–30 cm, acuminate, cordate, with 3–4 nervations. The inflorescence peduncle is 6–20 cm long. The spathe is 15 cm long, green, 3–5 cm long, with a limb at first erect, later reflexed and yellow, apiculate, and 5.5–12.5 cm × 9.5 cm. The staminate and female portions are separated by a 1.5–2 cm long Asexual zone. The staminate zone is 1–1.5 cm long. The female zone is 2 cm long. The ovary is ovoid, unilocular with discoid stigma.

Medicinal uses: gonorrhea, infertility

Pharmacology: Extracts of the plant demonstrated anti-inflammatory effects.[149]

12 Family Arecaceae Schultz-Schultzenstein 1832

12.1 *Areca catechu* L.

Synonyms: *Areca catechu* Willd.; *Areca faufel* Gaertn.; *Areca himalayana* Griff. ex H. Wendl.; *Areca hortensis* Lour.; *Areca nigra* Giseke ex H. Wendl.; *Sublimia areca* Comm. ex Mart.

Local name: Gooa

Common names: Areca palm; betel-nut palm

Botanical description: It is a palm tree that grows up to 20m tall. The bole is straight, slender, 10–20cm in diameter, greyish, and with conspicuous marks. The leaves sheath forms a crownshaft that is up to 1 m long. The petiole is up to 5cm long. The rachis grows up to 2m, and supports 20–30 pinnae per side which are 30–60cm × 3–7cm. The inflorescences are up to 25cm long. The flowers comprise 3 sepals, 3 petals, and 6 stamens. The fruits are drupes which are yellow, orange, or red in color, ovoid, woody, and about 7cm × 4cm with persistent calyx.

Medicinal use: diarrhea

Pharmacology: The plant has been well studied for pharmacology.[150] Extracts of the plant promoted platelet aggregation and exhibited anti-inflammatory and analgesic effects.[151,152] Extracts of the plant demonstrated antibacterial and anti-fungal effects.[153,154]

12.2 *Borassus flabellifer* L.

Synonyms: *Borassus flabelliformis* Murray; *Borassus flabelliformis* Roxb.

Local names: Tar; taledare

Common names: Palmyra palm; toddy palm

Botanical description: It is a palm tree that grows up to 20m tall. The bole is rough. The petiole is 60–120cm long. The blade is 60–120cm long, with 60–80 linear-lanceolate, and induplicate segments. The male inflorescence is 30–150cm long. The calyx comprises 3 sepals. The corolla includes 3 petals. The androecium includes 6 stamens. The ovary is globose and minute. The fruits are globose, purple in color, 15–20cm in diameter, and contain 3 seeds.

Medicinal uses: cuts, abscesses

Pharmacology: The plant produces anti-diabetic and antibacterial saponins[155,156] and anti-inflammatory triterpenes.[157] The resveratrol polymer trans-scirpusin A isolated from the plant demonstrated anti-tumor property.[158]

13 Family Aristolochiaceae A.L. de Jussieu 1789

13.1 *Aristolochia indica* L.

Local name: Isharmul

Common name: Indian birthwort

Botanical description: It is a slender climber that grows up to 5 m long. The stem is angled and glabrous. The leaves are simple, alternate, and exstipulate. The petiole is 1–1.5 cm long. The blade is oblong, cordate at the base, acuminate at the apex, 5–10 cm × 2–4 cm, and fleshy. The inflorescences are axillary racemes. The perianth is tubular, greenish, curved, globose at the base and about 1 cm in diameter, and develops a fine tube, which reaches up to 2.5 cm long and terminates into a purplish throat and limb. The androecium includes 6 stamens in a column. The ovary is cylindrical and ribbed. The fruits are pendulous and dehiscent capsules, which are valved and contain deltoid seeds.

Medicinal uses: emmenagogue, rheumatism, snake bite, leprosy, fever

Pharmacology: Extracts of the plant protected rodents against viper venom.[159] Aristolacatam I and (–)-hinokinin from this plant demonstrated anti-inflammatory activities.[160] Essential oil and extract of the plant exhibited antibacterial properties.[161,162]

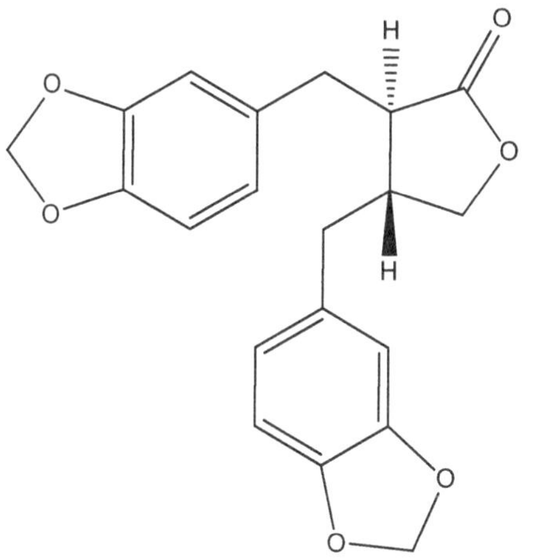

Structure of hinokinin

14 Family Asphodelaceae A.L. de Jussieu 1789

14.1 *Aloe vera* (L.) Burm.f.

Synonyms: *Aloe barbadensis* Mill.; *Aloe barbadensis* var. *chinensis* Haw.; *Aloe chinensis* (Haw.) Baker; *Aloe perfoliata* var. *vera* L.; *Aloe vera* var. *chinensis* (Haw.) A. Berger; *Aloe vulgaris* Lam.

Local name: Ghrita kumari

Common name: Barbados aloe

Botanical description: It is a fleshy herb forming a rosette that grows up to 80cm tall. The leaves are linear-lanceolate, spiny at the margin, green, fleshy, 15–35 cm × 4–5 cm, produce a yellow gum when incised, and accumulate an abundant translucent aqueous gel. The inflorescence is an erect raceme, which is 60–90cm tall. The flowers are showy. The perianth is 2.5–3 cm long, light-yellow, with 6 petals. The ovary is ovoid and minute, the style is elongated, and the stigma 3-lobed. The fruits are capsules that are 1.5 cm long.

Medicinal uses: inflammation, sore eyes, health drink

Pharmacology: The plant yields an aqueous gel and anthraquinones, which account for its medicinal properties. The pharmacological properties of this plant have been well studied.[163–165]

15 Family Asteraceae Martynov 1820

15.1 *Ageratum conyzoides* L.

Synonyms: *Ageratum album* Willd. ex Steud.; *Ageratum arsenei* B.L. Rob.; *Ageratum ciliare* L.; *Ageratum cordifolium* Roxb.; *Ageratum hirsutum* Poir.; *Ageratum hirtum* Lam.; *Ageratum humile* Salisb.; *Ageratum latifolium* Cav.; *Ageratum latifolium* var. *galapageium* B.L. Rob.; *Ageratum microcarpum* (Benth.) Hemsl.; *Ageratum pinetorum* (L.O. Williams) R.M. King & H. Rob.; *Ageratum suffruticosum* Regel; *Alomia microcarpa* (Benth.) B.L. Rob. *Alomia pinetorum* L.O. Williams; *Cacalia mentrasto* Vell.; *Caelestina microcarpa* Benth.; *Carelia conyzoides* (L.) Kuntze; *Coelestina microcarpa* Benth.; *Eupatorium conyzoides* (L.) E.H.L. Krause

Local name: Uchunti

Common names: Billy goat weed; tropical white weed

Ageratum conyzoides

Botanical description: It is an herb that grows up to 50 cm tall. The stem is hairy. The leaves are simple, opposite, and exstipulate. The petiole is 1–3 cm long. The blade is hairy, broadly lanceolate, 3–8 cm × 2–5 cm, cuneate or acute at base, serrate at margin, and acute at apex. The inflorescence is a small capitulum, which is globose and about 0.5 cm in diameter and somewhat bluish and almost fluorescent especially at dawn. The calyx is minutely 5-lobed. The corolla is tubular, 5-lobed, minute, and purplish. The fruits are achenes, which are 5-angled and minute.

Medicinal uses: fever, cut, diarrhea, intestinal worms, boils, wounds, leprosy

Pharmacology: Extracts displayed anti-inflammatory[166] and antibacterial[167] activities. The essential oil from this plant showed schistosomicidal effects.[168] The plant produces flavonoids with anti-trypanosomal activity.[169]

15.2 *Chromolaena odorata* (L.) R.M. King & H. Rob.

Synonyms: *Eupatorium conyzoides* Vahl; *Eupatorium odoratum* L.; *Osmia odorata* (L.) Sch. Bip.

Local name: Assam lata

Common name: Butterfly-weed

Botanical description: It is a climber that grows up to 2 m long. The stems are terete and striated. The leaves are simple, opposite, and exstipulate. The petiole is 1–2.5 cm long. The blade is cordate, subdeltoid, 7–15 cm × 3.5–8 cm, acute at apex, wedge-shaped or acute at base, with 3 pairs of secondary nerves, and serrate at margin. The inflorescences are lax corymbs of capitula that are about 1 cm long. The florets are light bluish-white, tubular, 5-lobed, and with conspicuous protruding stigmas. The fruits are achenes, which are 5-ribbed and with a pappus.

Medicinal use: snake bites

Pharmacology: Extracts of the plant displayed anti-inflammatory, anti-pyretic, and anti-spasmodic properties.[170] The plant produces a pro-coagulant flavonoid.[171]

15.3 *Eclipta alba* (L.) Hassk.

Synonyms: *Eclipta prostrata* L.; *Eclipta brachypoda* Michx.; *Eclipta erecta* L.; *Eclipta erecta* Sessé & Moc.; *Eclipta prostrata* (L.) L.; *Eclipta punctata* L.; *Verbesina alba* L.; *Verbesina prostrata* L.

Local name: Kalachona

Common names: Trailing eclipta plant; false daisy

Botanical description: It is a water-loving prostrate herb that grows up to 30 cm long from fibrous roots. The stem is fleshy, hairy, and purplish. The leaves are simple, opposite, and exstipulate. The blade is subsessile, lanceolate, 3–10 cm × 0.5–2.5 cm, serrate hairy, dark green, acute at base and apex. The inflorescences are terminal or axillary capitula, which are about

5 mm across on about 3 cm long and slender peduncles. The ray florets have a bifid lamina and are minute. Disc florets have a 4-lobed corolla. The fruits are achenes, which are minute.

Medicinal uses: liver diseases, alopecia, boils

Pharmacology: The plant produces wedelolactone that has hepatoprotective,[172] anti-cancer,[173] anti-HCV,[174] and anti-HIV-1[175] properties.

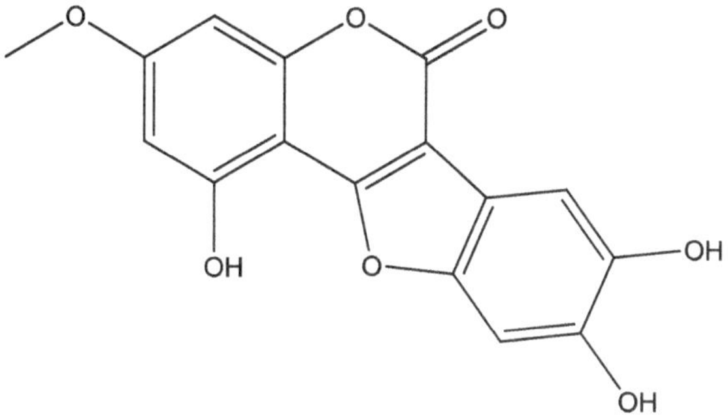

Chemical structure wedelolactone

15.4 *Enydra fluctuans* Lour.

Synonyms: *Cryphiospermum repens* P. Beauv.; *Enydra anagallis* Gardner; *Enydra heloncha* DC.; *Enydra longifolia* (Blume) DC.; *Enydra paludosa* (Reinw.) DC.; *Enydra woollsii* F. Muell.; *Hingtsha repens* Roxb.; *Meyera fluctuans* (Lour.) Spreng.; *Meyera guineensis* Spreng.; *Tetraotis longifolia* Blume; *Tetraotis paludosa* Reinw.; *Wahlenbergia globularis* Schumach.

Local name: Hencha

Common name: Marsh herb

Botanical description: It is a marsh herb that grows up to 80 cm long. The stem is stout, terete, fleshy, and basally prostrate and rooting. The leaves are simple, decussate, sessile, and exstipulate. The blade is oblong, 2–6 cm × 4–15 mm, amplexicaul serrate at margin, obtuse at apex, and somewhat fleshy. The inflorescences are terminal capitula that are about 1 cm in diameter. The ray florets are minute and 3- or 4-lobed. The disc florets are white and 5-lobed. The fruits are achenes, which are cylindrical and minute.

Medicinal uses: laxative, anemia

Pharmacology: Extracts of the plant demonstrated hepatoprotective[176] properties. The plant produces cytotoxic[177] and antibacterial[178] flavonoids.

15.5 *Pseudognaphalium luteo album* (L.) Hilliard & B.L. Burtt

Synonyms: *Dasyanthus conglobatus* Bubani; *Gnaphalium helichrysoides* Ball; *Gnaphalium luteoalbum* L.; *Gnaphalium nanum* Kunth; *Gnaphalium nanum* Willd.;

Gnaphalium pallidum Lam.; *Helichrysum luteoalbum* (L.) Rchb.; *Laphangium luteoalbum* (L.) Tzvelev

Local name: Jabra

Common names: Everlasting cudweed; fragrant everlasting; weedy cudweed

Botanical description: It is a light glaucous erect herb, which grows up to 30 cm in dry areas. The stem is hairy. The leaves are simple, spiral, sessile, and exstipulate. The blade is oblong, 1–3 cm × 2–5 mm hairy, with a revolute margin. The inflorescences are terminal cymes of ovoid capitula. Florets are minute. The fruits are achenes, which are ribbed and present a pappus.

Medicinal use: cuts

Pharmacology: Extracts of the plant demonstrated cytotoxic[179] and anti-fungal[180] effects.

15.6 *Spilanthes acmella* (L.) L.

Synonyms: *Acmella calva* (DC.) R.K. Jansen; *Bidens acmella* (L.) Lam.; *Bidens ocymifolia* Lam.; *Pyrethrum acmella* (L.) Medik.; *Spilanthes ocymifolia* (Lam.) A.H. Moore; Spilanthes paniculata wall. ex Dc. *Verbesina acmella* L.

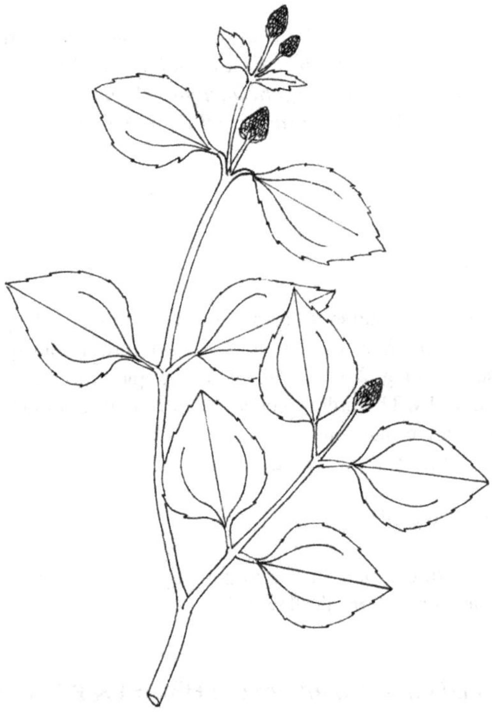

Spilanthes acmella (L.) L

Local name: Mariccha

Common name: Toothache plant

Botanical description: It is an herb that grows up to 30 cm tall. The leaves are simple, spiral, alternate, subopposite, and exstipulate. The petiole is 1–2 cm long. The blade is broadly lanceolate, fleshy, glossy, 2–4 cm × 1–2.5 cm, cordate or wedge-shaped at the base, apex acute, 3-nerved, and serrate at margin. The inflorescences are showy, yellow and dark red, ovoid capitula, which are up to about 1 cm long and sustained by axillary or terminal 2.5–10-cm long peduncles. The disc florets are tubular, minute, and 4- or 5-lobed. The fruits are achenes, which are trigonal, minute, with a pappus of 2 bristles.

Medicinal uses: toothache, infection of throat and gums, tongue paralysis, dysentery

Pharmacology: The plant produces the anti-inflammatory alkyl amide spilanthol.[181] Extracts of the plant displayed antibacterial and anti-fungal effects.[182]

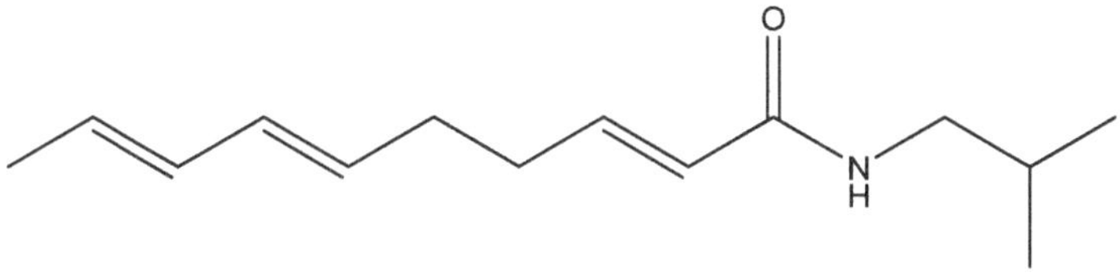

Chemical structure of spilanthol

15.7 *Wedelia chinensis* (Osbeck) Merr.

Synonyms: *Complaya chinensis* (Osbeck) Strother; *Solidago chinensis* Osbeck; *Thelechitonia chinensis* (Osbeck) H. Rob. & Cuatrec.; *Verbesina calendulacea* L.; *Wedelia calendulacea* (L.) Less.

Local name: Kechuria

Common name: Chinese Wedelia

Botanical name: It is a prostrate herb that grows up to 30 cm long. The stems have ascending apex, rooting at nodes, and are hairy. The leaves are simple, sessile, opposite, and exstipulate. The blade is lanceolate, glossy, fleshy, 2–10 cm × 0.6–2 cm, acute at the apex, with about 3–4 pairs of secondary nerves, and serrate at margin. The inflorescences are capitula, which are showy, 2–2.5 cm in diameter, solitary on 6–12 cm long peduncles. The ray florets are yellow, with a corolla which is 2- to 3-dentate and about 1 cm long. Disc flowers are minute, and 5-lobed. The fruits are achenes, which are obovoid, minute, and with a pappus.

Medicinal use: removes obstructions

Pharmacology: Extracts of the plant demonstrated anti-cancer,[183] cytotoxic, antibacterial,[184] and hepatoprotective[185] activities. The plant produces anti-cancer and anti-inflammatory triterpenes.[186]

15.8 *Xanthium strumarium* L.

Synonyms: *Xanthium americanum* Walter; *Xanthium cavanillesii* Schouw; *Xanthium chasei* Fernald; *Xanthium chinense* Mill.; *Xanthium curvescens* Millsp. & Sherff; *Xanthium cylindricum* Millsp. & Sherff; *Xanthium echinatum* Murray; *Xanthium echinellum* Greene ex Rydb.; *Xanthium inaequilaterum* DC.; *Xanthium indicum* Koenig. ex Roxb.; *Xanthium italicum* Moretti; *Xanthium mongolicum* Kitag.; *Xanthium natalense* Widder; *Xanthium orientale* L.; *Xanthium oviforme* Wallr.; *Xanthium pensylvanicum* Wallr.; *Xanthium pungens* Wallr.; *Xanthium sibiricum* Patrin ex Widder; *Xanthium speciosum* Kearney; *Xanthium varians* Greene; *Xanthium wootonii* Cockerell

Local name: Lengra

Common name: Cocklebur

Botanical description: It is an herb that grows up to 50cm tall. The stem is fleshy and somewhat purplish. The leaves are simple, spiral, and exstipulate. The petiole is 3.5–10cm long and purplish. The blade is ovate-deltoid, 9–15cm long, papery, cordate at base, margin irregularly dentate, and acute at apex. The inflorescence is a capitulum, which is globose, whitish-green, and about 0.5cm in diameter. Male capitula are arranged in terminal umbels and comprise minute flowers with white and 5-lobate corolla. The female capitula are axillary. The fruits sessile, oblong, ovoid, about 1cm long and spiny.

Medicinal use: infections of the skin

Pharmacology: Extracts of the plant demonstrated anti-hyperglycemic and anti-nociceptive activities.[187]

16 Family Balsaminaceae Berchtold and J. Presl 1820

16.1 *Impatiens balsamina* L.

Synonyms: *Balsamina hortensis* Desp.; *Balsamina hortensis* Desp.; *Impatiens eriocarpa* Launert; *Impatiens stapfiana* Gilg

Local name: Dopati

Common name: Garden balsam

Botanical description: It is an elegant fleshy herb that grows to about 1 m tall. The stem is light green, and longitudinally striated. The leaves are simple, alternate or opposite, and exstipulate. The petiole is 1–3 cm long and channeled. The blade is elliptical to lanceolate, 4–12 cm × 1.5–3 cm, with 4–7 pairs of secondary nerves, base cuneate, margin serrate, and acuminate apex. The inflorescences are axillary fascicles of showy and whitish-pink flowers. The flower peduncles are 2–2.5 cm long and pubescent. There are 2 lateral sepals that are ovate and minute. Lower sepal is navicular in shape, up to 2 cm long, and narrowed into an incurved spur that is 1–2.5 cm long. The upper petal is orbicular. Lateral united petals are clawed, 2-lobed and up to 2.5 cm long. The androecium consists of 5 stamens with linear filaments. The ovary is minute, fusiform, and pubescent. The fruit is a fusiform capsule, 1–2 cm long, hairy, and containing numerous tiny seeds.

Medicinal uses: burns, diuretic

Pharmacology: Extracts of the plant demonstrated anti-nociceptive effects.[188] It produces neuroprotective flavonoids[189] and antimicrobial naphthoquinones.[190]

17 Family Bignoniaceae A.L. de Jussieu 1789

17.1 *Oroxylum indicum* (L.) Vent.

Synonyms: *Bignonia indica* L.; *Bignonia pentandra* Lour.; *Calosanthes indica* (L.) Blume; *Spathodea indica* (L.) Pers.

Local names: Sona; taita

Common names: Broken bones tree; tree of Damocles; Indian trumpet flower

Botanical description: It is a strange-looking tree that grows up to 10 m tall. The bole is straight. The bark is smooth and the wood is yellow. The leaves are opposite, exstipulate at the apex of the bole, and pinnately compound. The blade is 60–130 cm. The rachis is stout, and swollen at base. The folioles are broadly elliptic, 5–12.5 cm × 3–10 cm, glabrous, round at base, acuminate apex, and with 5–6 pairs of inconspicuous secondary nerves. The inflorescences are gathered at the apex of stems and comprise showy and ephemerous, somewhat smelly flowers that open at night. The flower peduncle is stout and 3–7 cm long. The calyx is purple, campanulate, and 2.5–4.5 cm × 2–3 cm. The corolla is purplish-red, bell-shaped, fleshy, 3–9.5 cm × 1–1.5 cm, the upper lip bilobed, the lower lip trilobed, and reflexed. The androecium includes 5 stamens inserted at the middle of corolla tube on 4 cm long filaments. A 1.5 cm in diameter 5-lobed disc is present. The ovary is cylindrical and develops a style, which is about 6 cm long. The fruit is a sword-shaped capsule, which is 40–120 cm × 5–9 cm, 1 cm thick, and contains numerous seeds, which are about 5 cm across and equipped with membranous wings. Among the trees of Asia, the fruits of this plant are among the largest.

Medicinal uses: jaundice, asthma, bronchitis, piles, festering sores

Pharmacology: Extracts of the plant exhibited antibacterial[191] and anti-diabetic[192] properties. The plant produces oroxilin that has anti-diabetic properties.[193]

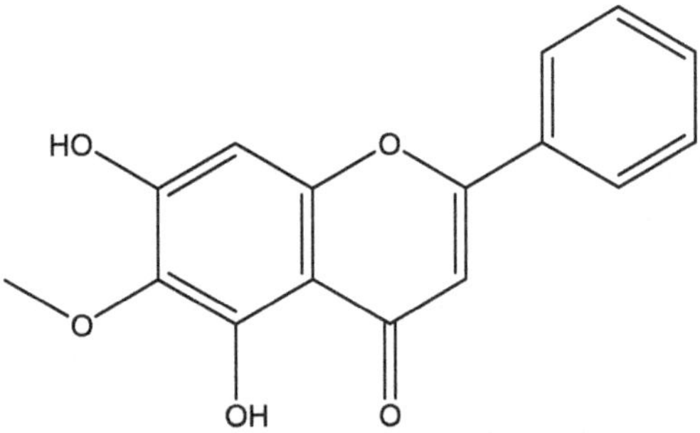

Structure of oroxilin

18 Family Bombacaceae Kunth 1822

18.1 *Adansonia digitata* L.

Synonyms: *Adansonia bahobab* L.; *Adansonia scutula* Steud.; *Adansonia integrifolia* Raf.; *Adansonia situla* (Lour.) Spreng.; *Adansonia sphaerocarpa* A. Chev.; *Adansonia sulcata* A. Chev.; *Baobabus digitata* (L.) Kuntze; *Ophelus sitularius* Lour.

Local names: Baobab; kattio-daghor

Common names: Baobab tree; monkey bread-tree

Botanical description. It is a gigantic, strange-looking tree that can grow up to 40 m in height. The bole can reach 10 m across and looks like a huge smooth cylinder. The leaves have caducous tiny stipules and are compound. The petiole is slender and about 10 cm long. The blade is palmately compound, and 5–7 foliates. The folioles are elliptic and about 5–15 cm × 3–7 cm, and hairy below. The flowers are massive, white, and pendulous. The corolla is membranaceous and develops 5 lobes, which are orbiculate and are about 7 cm long. The androecium produces numerous slender stamens that are white and about 10 cm long. The fruits are ovoid, up to 25 cm × 11 cm, hairy and contain numerous seeds, which are black and kidney-shaped.

Medicinal uses: fever, diarrhea, dysentery

Pharmacology: Extracts of the plant demonstrated analgesic[194] and chemopreventive effects in rodents.[195]

18.2 *Bombax ceiba* L.

Synonyms: *Bombax heptaphyllum* L.; *Bombax malabaricum* DC.; *Gossampinus malabarica* (DC.) Merr.; *Salmalia malabarica* (DC.) Schott & Endl.

Local name: Simul

Common name: Hill glory bower

Botanical description: It is a massive tree that grows up to 25 m tall. The bole is buttressed, and spiny. The leaves are spiral, palmate, and stipulate. The stipules are minute. The petiole is slender and up to 20 cm long. The blade comprises 5–7 leaflets on petiolules, which are up to 4 cm long. The folioles are oblong, 10–15 cm × 3.5–5.5 cm, with 15–17 pairs of secondary nerves, and acuminate at the apex. The flowers are solitary and terminal. The calyx is cup-shaped, fleshy, glossy, 2–3 cm long with 5 lobes that are triangular. The corolla comprises 5 oblong, 8–10 cm × 3–4 cm, red petals. The androecium includes numerous stamens, which are showy. The fruit is a capsule, which is oblong, 10–15 cm × 4.5–5 cm and open to release a cottonous mass and numerous seeds.

Medicinal uses: fever, rheumatism, diuretic, diarrhea

Pharmacology: The plant has been the subject of numerous pharmacological studies.[196] It produces an anti-HBV lignan.[197] Extracts exhibited diuretic,[198] male aphrodisiac,[199] hypotensive, and hypoglycemic[200] effects.

19 Family Brassicaceae Burnett 1835

19.1 *Brassica alba* L.

Synonym: *Sinapis alba* L.

Local name: Sada sorse

Common name: White mustard

Botanical description: It is an erect herb that grows up to 80cm tall. The leaves are simple, spiral, and exstipulate. The petiole is up to 6cm long. The blade is lyrate, serrate, deeply incised, and 5–14cm × 2–8cm. The flowers are terminal. The calyx includes 4 sepals, which are narrow and up to about 1.5cm long. The corolla includes 4 petals, which are yellowish, obovate, and 0.8–1.4cm × 4–7mm. The androecium consists of 4 stamens. The fruits are fusiform capsules, which are up to 5cm long and contain numerous minute globose seeds.

Medicinal uses: fever, bronchitis

Pharmacology: Extracts of the plant exhibited anti-inflammatory effects.[201] The plant contains antibacterial isothiocyanates.[202]

19.2 *Brassica campestris* L.

Synonyms: *Brassica rapa* L.; *Gorinkia campestris* (L.) J. Presl & C. Presl; *Napus campestris* (L.) Schimp. & Spenn.; *Raphanus campestris* (L.) Crantz

Local names: Sharisha shak; sada sharisha

Common name: Bird rape

Botanical description: It is an erect herb that grows up to 1 m tall. The leaves are simple, spiral, and exstipulate. The petiole is up to 15cm long. The blade is ovate, 10–50cm × 3–15cm, with 2–6 pairs of lateral lobes plus a terminal lobe. The upper leaves are entire and amplexicaul. The flowers are terminal. The calyx includes 4 sepals, which are oblong and about 1cm long. The corolla includes 4 petals, which are yellow, about 1cm long, obovate, and round at apex. The androecium includes 4 stamens. The fruits are linear capsules which are up to 8cm long and containing numerous globose and minute seeds.

Medicinal uses: abdominal pain, rheumatism

Pharmacology: The plant produces glucosinolates with antibacterial[203] and anti-cancer[204] properties.

19.3 *Brassica juncea* (L.) Czern.

Synonyms: *Brassica besseriana* Andrz. ex Trautv.; *Brassica cernua* (Thunb.) F.B. Forbes & Hemsl.; *Brassica chenopodiifolia* Sennen & Pau; *Brassica integrifolia* (H. West) O.E. Schulz; *Brassica integrifolia* (H. West) Rupr.; *Brassica japonica* (Thunb.) Siebold ex Miq.; *Brassica juncea* (L.) Coss.; *Brassica timoriana* F. Muell.; *Raphanus junceus* (L.) Crantz; *Sinapis abyssinica* A. Braun; *Sinapis cernua* Thunb.; *Sinapis integrifolia* H. West; *Sinapis japonica* Thunb.; *Sinapis juncea* L.; *Sinapis rugosa* Roxb.; *Sinapis timoriana* DC.

Local names: Rai; rai sharisha

Common names: Brown mustard; Indian mustard

Botanical description: It is an erect herb that grows up to 1 m tall. The leaves are simple, spiral, and exstipulate. The petiole is 1–15 cm long. The blade of basal leaves is lyrate, pinnatifid or pinnatisect, 6–30 cm × 1.5–15 cm, incised, with 2–6 lateral lobes. The flowers are gathered at the apex of the stems. The calyx comprises 4 sepals that are oblong and about 1.5 cm long. The corolla includes 4 petals, which are yellow, up to 1.3 cm long, and obovate. The androecium includes 4 stamens. The fruit is a linear capsule that is up to 6 cm long, and containing numerous minute seeds, which are globose.

Medicinal uses: diabetes, counterirritant, indigestion

Pharmacology: The plant has been the subject of numerous pharmacological studies demonstrating notably anti-diabetic and anti-tumor effects.[205]

19.4 *Brassica napus* L.

Synonyms: *Brassica napobrassica* (L.) Mill.; *Brassica rugosa* (Roxb.) L.H. Bailey

Local name: Sharisha

Common names: Indian rape; colza

Botanical description: It is an erect herb that grows up to 1 m tall. The leaves are simple, spiral, and exstipulate. The petiole is up to 15 cm long. The blade is oblong, 5–20 cm × 2–7 cm, lyrate, dentate, with 2–12 laterally lobed. The upper cauline leaves are sessile and amplexicaul. The flowers are terminal. The calyx consists of 4 sepals, which are oblong and 5–10 mm × 1.5–2.5 mm. The corolla comprises 5 yellow petals, which are up to about 2 cm long and broadly obovate. The androecium includes 4 stamens which are up to 1 cm long. The fruits are linear capsules, which are about 1 cm long, and contain numerous seeds that are minute and globose.

Medicinal use: diuretic

Pharmacology: The plant has been studied for pharmacology.[206] It contains glucosinolates[207] that are chemoprotective.[208]

19.5 *Brassica sinapis* Visiani

Synonym: *Sinapis arvensis* L.

Local name: Sada sharshe

Common name: White mustard

Botanical description: It is an herb that grows up to 1.5 m tall. The leaves are simple, spiral, and exstipulate. The petiole is 1–3 cm long. The blade is deeply incised, 5–15 cm × 2–5 cm, and with 2–6 lateral lobes. The flowers are terminal. The calyx includes 4 sepals, which are narrowly oblong and about 1.5 cm long. The corolla consists of 5 petals, which are pale yellow, obovate, and about 0.8–1.2 cm × 4–6 mm. The androecium includes 4 stamens. The fruit is a fusiform capsule that is up to 5 cm long and contains numerous minute seeds that are globose.

Medicinal use: counterirritant

Pharmacology: The plant produces glucosinolates.[209]

20 Family Bromeliaceae A.L. de Jussieu 1789

20.1 *Ananas comosus* (L.) Merr.

Synonyms: *Ananas ananas* (L.) Voss; *Ananas ananas* Ker Gawl.; *Ananas domestica* Rumph.; *Ananas parguazensis* L.A. Camargo & L.B. Sm.; *Ananas sativa* Lindl.; *Ananas sativus* Schult. & Schult. f.; *Ananassa sativa* Lindl.; *Ananassa sativa* Lindl. ex Beer; *Bromelia ananas* L.; *Bromelia ananas* Willd.; *Bromelia comosa* L.

Local name: Anarash

Common name: Pineapple

Botanical description: It is an herb with leaves in a rosette that grows up to 1.50m. The leaves are triangular, narrow, coriaceous, and serrate. The scape is short. The inflorescence is a conical spike which is about 20cm long. The calyx includes 3 sepals. The corolla is purplish and consists of 6 petals. The androecium is made of 6 stamens. The ovary is globose and develops a slender style. The infructescence is a comose syncarp, which is globose, ovoid, up to 30cm long fleshy, fragrant, and delicious.

Medicinal uses: intestinal worms, diuretic, liver diseases

Pharmacology: The plant produces the proteolytic and anti-helminthic enzyme bromelaine.[210,211]

21 Family Burseraceae Kunth 1824

21.1 *Garuga pinnata* Roxb.

Local name: Jum

Common name: Garuga

Botanical description: It is a resinous tree that grows up to 10 m tall. The bark is grey and rough. The leaves are imparipinnate and spiral. The blade comprises 4–11 pairs of folioles. The rachis is hairy. The folioles are oblong-lanceolate, 5–11 cm × 2–3 cm, base rounded, and shortly acuminate at apex, and present 6–15 pairs of secondary nerves. The inflorescences are axillary panicles that are up to about 20 cm long and hairy. The calyx is urceolate and consists of 5 deltoid sepals. The corolla is white and produces 5 petals, which are oblong and about 0.5 cm long, and hairy. The androecium includes 5 stamens. A disc is present. The ovary is oblong and develops a style with a 5-lobed stigma. The fruits are globose, dull light green, and about 1.5 cm in diameter.

Medicinal uses: stomach ache, asthma, conjunctivitis

Pharmacology: Extracts of the plant demonstrated anti-diabetic properties.[212] The plant produces antibacterial phenolic compounds.[213]

22 Family Caprifoliaceae A.L. de Jussieu 1789

22.1 *Abelia chinensis* R. Br.

Synonyms: *Abelia aschersoniana* (Graebn.) Rehder; *Abelia cavaleriei* H. Lév.; *Abelia hanceana* Mart. ex Hance; *Abelia ionandra* Hayata; *Abelia lipoensis* M.T. An & G.Q. Gou; *Abelia rupestris* Lindl.; *Linnaea aschersoniana* Graebn.; *Linnaea chinensis* (R. Br.) A. Braun & Vatke; *Linnaea rupestris* (Lindl.) A. Braun & Vatke

Local name: Gunal

Common name: Chinese abelia

Botanical description: It is a much-branched fragrant shrub, which grows to 2 m tall. The leaves are opposite or whorled, simple and exstipulate. The blade is lanceolate, 2–5 cm × 1–3.5 cm, rounded at base, and acute at apex. The inflorescences are terminal panicles. The calyx includes 5 sepals, which are elliptic, and 5 mm long. The corolla is tubular, 5-lobed, whitish-pink, an up to 1.2 cm long. The androecium includes 4 stamens inserted to the corolla tube. The ovary is cylindrical and develops a capitate stigma.

Medicinal use: blood circulation

Pharmacology: Unknown; the plant contains iridoids.[214]

23 Family Clusiaceae Lindley 1836

23.1 *Mesua ferrea* L.

Synonyms: *Calophyllum nagassarium* Burm. f.; *Mesua nagassarium* (Burm. f.) Kosterm.

Local name: Nageswar champa

Common name: Ironwood

Botanical description: It is a slow-growing, upright, and beautiful tree, which grows up to 10 m in height. The bole is erect and the wood is extremely hard. The bark is greyish brown, scaly, and yields upon incision a yellow latex. The leaves are simple, opposite, and exstipulate. The petiole is 5–8 mm long. The blade is light red at first, dark green and glossy at maturity, glaucous below, linear-lanceolate, coriaceous, 6–10 cm × 2–4 cm, cuneate base, acuminate apex, and presents numerous inconspicuous secondary nerves. The flowers are showy, ephemeral, fragrant, solitary, axillary, or terminal. The calyx comprises 4 sepals, which are somewhat ovate, fleshy, and coriaceous. The corolla comprises 5 petals which are membranous, obovate-cuneate, and 3–3.5 cm long. The androecium is showy, golden yellow, and comprises numerous stamens with filiform and 1.5–2 cm long anthers. The ovary is tear-shaped, 1.5 cm long, and develops a style which is up to 1.5 cm long. The fruits are globose, up to about 2.5 cm in diameter, light dull green and contain 1–4 seeds.

Medicinal uses: tonic, rheumatism

Pharmacology: Extracts of the plant demonstrated immunomodulatory,[215] anti-arthritic,[216] and antibacterial[217] effects.

24 Family Combretaceae R. Brown 1810

24.1 *Anogeissus acuminata* (Roxb. ex DC.) Guill., Perr. & A. Rich.

Synonyms: *Anogeissus harmandii* Pierre; *Anogeissus lanceolata* (Wall. ex C.B. Clarke) Wall. ex Prain; *Anogeissus pierrei* Gagnep.; *Anogeissus tonkinensis* Gagnep.; *Conocarpus acuminatus* Roxb. ex DC.

Local names: Itchri; chakwa

Common name: Axlewood

Botanical description: It is a resinous tree that grows up to 20 m tall with pendant stems. The leaves are simple, opposite, and exstipulate. The petiole is about 0.5 cm long. The blade is elliptic, 4–8 cm × 1–3 cm, the base and apex are acute, and 5–7 pairs of secondary nerves are present. The inflorescences are globose and about 1.5 cm in diameter heads which are axillary or terminal. The hypanthium, 5 mm long, and expands into 5 campanulate lobes. The androecium includes 10 stamens, which are up to about 1 cm long. The fruit is globose, winged, and about 5 mm in diameter.

Medicinal uses: wounds, ulcers, diarrhea

Pharmacology: Extracts of the plant demonstrated anti-inflammatory, antimicrobial, and anti-diabetic activities.[218,219] The plant produces anolignan A that inhibits HIV-1.[220]

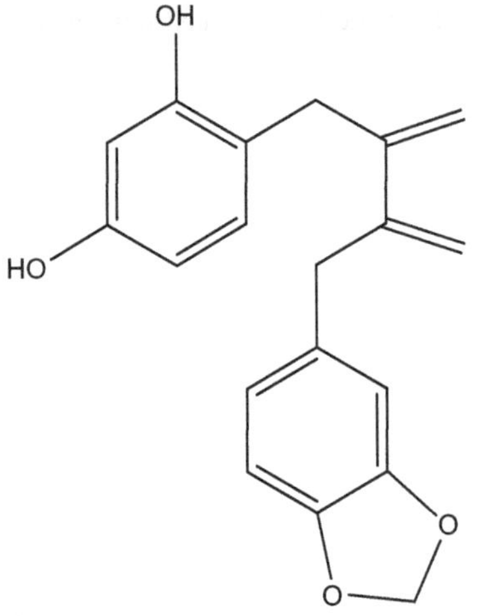

Structure of anolignan A

24.2 *Anogeissus latifolia* (Roxb. ex DC.) Wall. ex Bedd.

Synonym: *Conocarpus latifolius* Roxb. ex DC.

Local name: Doya

Common names: Button tree; ghatti tree

Botanical description: It is a resinous tree that grows up to 10 m tall. The leaves are alternate or sub-opposite, simple, and exstipulate. The petiole is up to 1.5 cm long. The blade is elliptic, 5–10 cm × 3–5 cm, round or acute at the base and apex, coriaceous, and with 8–14 pairs of secondary nerves. The inflorescences are globose heads that are about 1 cm in diameter. The hypanthium, expands at apex into 5 campanulate lobes. The androecium includes 10 stamens. The fruits are about 5 mm long and laterally winged.

Medicinal uses: diarrhea, ulcers, tonic

Pharmacology: Extracts of the plant exhibited anti-ulcer[221] and antibacterial[222] effects.

24.3 *Terminalia arjuna* (Roxb. ex DC.) Wight & Arn.

Synonyms: *Pentaptera arjuna* Roxb.; *Pentaptera arjuna* Roxb. ex DC.; *Terminalia glabra* Wt. & Arn.; *Terminalia urjan* Royle.

Local name: Arjoon

Common name: Arjuna

Botanical description: It is an elegant tree that grows up to 15 m tall. The bole is straight and the bark is somewhat smooth. The leaves are simple, sub-opposite, oblong-elliptic, and exstipulate. The petiole is about 1 cm long. The blade is oblong, 7–18 cm × 4–6 cm, with rounded or cordate base, and acute at apex. The inflorescences are axillary or terminal paniculate brush bottle-like, 3–6 cm long spikes of yellow flowers. The hypanthium is campanulate, 4–5 mm long, with 5 triangular lobes. The androecium includes 10 stamens. A barbate disc is present. The fruit is ovoid-oblong, 2.5–5 cm long, yellowish green, and with 5 wings.

Medicinal uses: tonic, earache, liver troubles, bone fractures, tachycardia

Pharmacology: The plant has been the subject of numerous pharmacological investigations for its cardiovascular effects.[223] It produces anti-cancer and antibacterial flavonoids.[224] Extracts of the plant exhibited gastroprotective[225] and analgesic[226] effects.

24.4 *Terminalia bellirica* (Gaertn.) Roxb.

Synonyms: *Myrobalanus bellirica* Gaertn.; *Myrobalanus laurinoides* (Teijsm. & Binn.) Kuntze; *Terminalia attenuata* Edgew.; *Terminalia eglandulosa* Roxb. ex C.B. Clarke; *Terminalia gella* Dalzell; *Terminalia laurinoides* Teijsm. & Binn.; *Terminalia punctata* Roth.

Local name: Bahura

Common name: Belliric myrobalan

Botanical description: It is a tree that grows up to 10 m tall. The leaves are simple, spiral, and exstipulate. The petiole is about 2 cm long. The blade is coriaceous, elliptic or obovate, 8–20 cm × 7.5–15 cm, acute at base, and round at apex, and with 3–7 pairs of secondary nerves. The inflorescences are slender, 5–15 cm long, and pendulous axillary spikes of whitish flowers. The hypanthium is minute, cup-shaped, hairy, and produces 5 triangular lobes. The androecium includes 10 stamens. A densely hairy disc is present. The fruits are tear-shaped, 1.5–2.5 cm in diameter, velvety and contain a 5-angled endocarp.

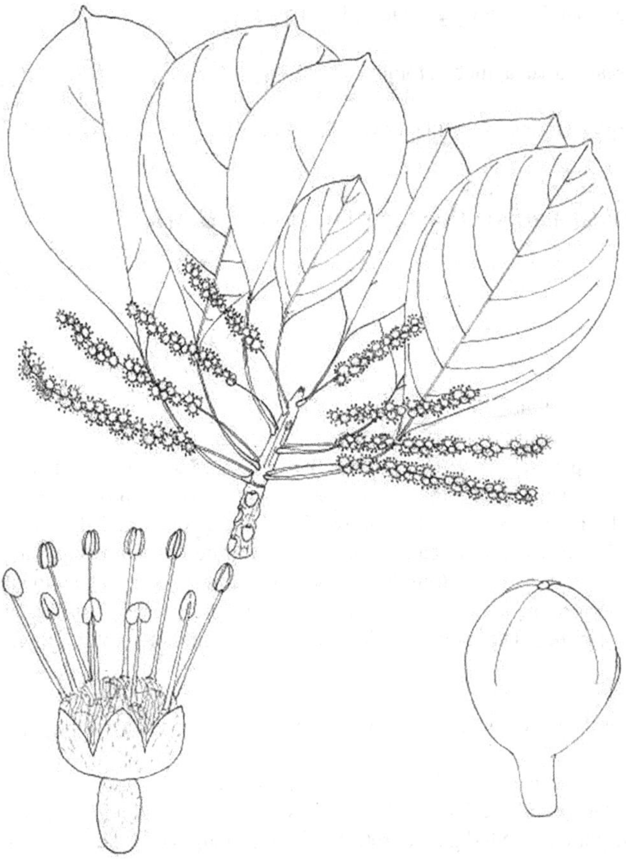

Terminalia bellirica (Gaertn.) Roxb

Medicinal uses: leprosy, diarrhea, cough, piles, eye infection, heart diseases, cholera

Pharmacology: The plant has been the subject of numerous pharmacological studies.[227] Extracts of the plant exhibited cardioprotective,[228] anti-diabetic,[229] wound healing,[230] analgesic,[231] and hepatoprotective[232] effects.

24.5 *Terminalia chebula* Retz

Synonyms: *Myrobalanifera citrina* Houtt.; *Myrobalanus chebula* (Retz.) Gaertn.; *Myrobalanus gangetica* (Roxb.) Kostel.; *Terminalia gangetica* Roxb.; *Terminalia reticulata* Roth.

Local name: Hari tuki

Common names: Black myrobalan; Indian myrobalan; Chebulic myrobalan

Botanical description: It is a tree which grows up to 10 m tall. The leaves are alternate or sub-opposite, simple, and exstipulate. The petiole is 1.25–3.5 cm long. The blade is broadly elliptic, coriaceous, 10–15 cm × 3.5–7.5 cm, acute at the base and apex, and with 7–11 pairs of secondary nerves. The inflorescences are terminal spikes, which are bottle brush-shaped and 15 cm long. The hypanthium is about 5 mm long and produces 5 lobes. The androecium includes 10 stamens. The fruit is ellipsoid, dull green, smooth, 2–3.5 cm long, and shelters a somewhat 5-ribbed stone.

Medicinal uses: diarrhea, aphthae, purgative, pimple, dysentery

Pharmacology: The plant has been the subject of numerous pharmacological studies.[233] It produces chebulagic acid, which is anti-inflammatory and cytotoxic activities.[234,235] Extract of the plant exhibited anti-diabetic and renoprotective activities.[236,237]

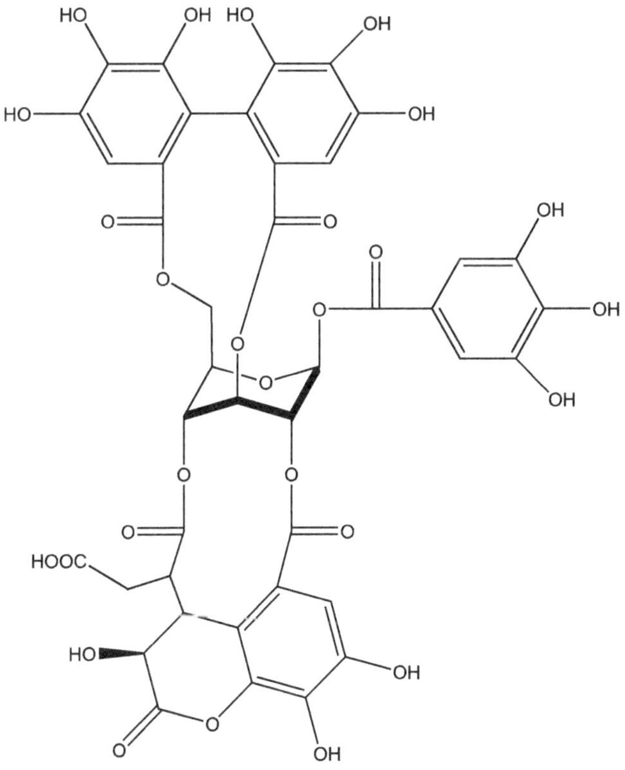

Structure of chebulagic acid

25 Family Convolvulaceae A.L. de Jussieu 1789

25.1 *Evolvulus nummularius* L.

Synonyms: *Convolvulus nummularius* L.; *Evolvulus capreolatus* Mart. ex Choisy; *Evolvulus dichondroides* Oliv.; *Evolvulus domingensis* Spreng. ex Choisy; *Evolvulus reniformis* Salzm. ex Choisy; *Evolvulus repens* D. Parodi; *Evolvulus veronicaeifolius* Kunth; *Evolvulus yunnanensis* S.H. Huang; *Volvulopsis nummularium* (L.) Roberty

Local name: Sada sankhapushpi

Common name: Roundleaf bindweed

Botanical description: It is a prostrate and covering herb that grows up to 40 cm in length. The stems are terete, rooting at nodes, and woody. The leaves are simple, regularly alternate and exstipulate. The petiole is minute. The leaves are round, coriaceous, 1.3–1.7 cm × 1.2–1.4 cm, deeply cordate at base, and round at apex. The flowers are axillary. The calyx comprises 5 sepals, which are oblong and minute. The corolla is pure white, campanulate, about 5 mm long, and 5-lobed. The androecium includes 5 stamens, which are inserted at the middle of the corolla tube. The ovary develops a capitate stigma. The fruit is an ovoid capsule containing few minute seeds.

Medicinal use: dysentery

Pharmacology: Extracts of the plant exhibited antibacterial effects.[238]

25.2 *Ipomoea aquatica* Forssk.

Synonyms: *Convolvulus repens* Vahl; *Convolvulus reptans* L.; *Ipomoea natans* Dinter & Suess.; *Ipomoea repens* Roth; *Ipomoea reptans* Poir.; *Ipomoea sagittifolia* Hochr.; *Ipomoea subdentata* Miq.

Local name: Kolmi shaak

Common names: Kangkong; water spinach; swamp morning-glory

Botanical description: It is a fleshy, aquatic, soft, herb that grows to 1.5 m tall. The stems are terete and rooting at the nodes. The leaves are edible, simple, floating, alternate, and exstipulate. The petiole is 3–14 cm long. The blade is fleshy, dull green, oblong to sagittate, 3.5–17 cm × 0.9–8.5 cm, cordate at the base and acute at the apex. The inflorescences are axillary, 1–3 flowered cymes on 1.5–10 cm long peduncles. The calyx comprises 5 sepals, which are lanceolate and about 1 cm long. The corolla is tubular, funnel-like, membranous, light purplish, with a darker center, 3.5–5 cm long and inconspicuously 5 lobed. The androecium includes 5 stamens. The ovary is conical and develops a bifid stigma. The fruit is a capsule, which is ovoid, dehiscent, and contains pubescent seeds.

Medicinal uses: ringworm, liver complaints

Pharmacology: Extracts of the plant elicited anti-diabetic properties.[239–241]

25.3 *Ipomoea fistulosa* Mart. ex Choisy

Synonyms: *Batatas crassicaulis* Benth.; *Ipomoea crassicaulis* (Benth.) B.L. Rob.; *Ipomoea gossypioides* Hort. ex Dammann; *Ipomoea gossypioides* Parodi; *Ipomoea texana* J.M. Coult.

Local name: Donkalos

Common name: Bush morning glory

Botanical description: It is a shrub that grows up to 2 m tall. The stem is woody and pubescent. The leaves are simple, alternate, and exstipulate. The petiole is slender and up to 20 cm long. The blade is cordate, 10–25 cm long, with about 7–10 pairs of secondary nerves, and acuminate at apex. The inflorescences are axillary cymes on a 7–15 cm long peduncle. The calyx comprises 5 sepals, which are suborbicular, and 5–6 mm long. The corolla is tubular, infundibuliform, light pink with a dark throat, membranous, obscurely 5-lobed, and 5–8 cm long. The fruit is an ovoid capsule, which is 2 cm long, dehiscent, and contains hairy seeds.

Medicinal uses: cold, fever

Pharmacology: Extract of the plant exhibited neuromuscular blocking activity.[242]

25.4 *Ipomoea mauritiana* Jacq.

Synonyms: *Batatas paniculata* (L.) Choisy; *Convolvulus paniculatus* L.; *Ipomoea digitata* L.; *Ipomoea paniculata* (L.) R. Br.; *Ipomoea pedata* G. Don

Local name: Bhoomi koomra

Common names: Giant potato; milk yam

Botanical description: It is a climber that grows from a large tuber up to 10 m long. The leaves are simple, spiral, and exstipulate. The petiole is slender and 3–11 cm long. The blade is 7–18 cm × 7–22 cm, palmately lobed, the lobes lanceolate, acuminate, or acute at the apex. The inflorescences are axillary cymes on 2.5–20 cm long peduncles. The calyx comprises 5 sepals, which are oblong and up to 1 cm long. The corolla is tubular, infundibuliform, membranous, pink or reddish purple, with a darker center, 5-lobed, and 5–6 cm long. The androecium consists of 5 stamens. The ovary produces a bifid stigma. The fruit is a dehiscent capsule which is ovoid, 1.2–1.4 cm long, and contains numerous hairy and globose seeds, which are about 5 mm long.

Medicinal uses: constipation, sexual impotence, galactagogue, tonic

Pharmacology: The plant produces scopoletin and sitosterol[243] as well as the cardiotonic glycoside paniculatin.[244] The plant exhibited anti-diabetic effects.[245]

Ipomoea mauritiana

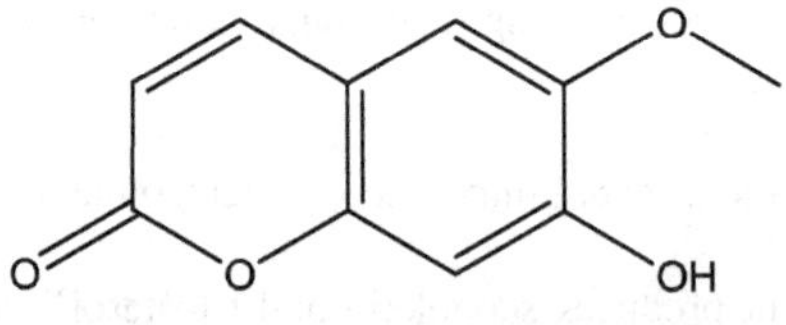

Structure of scopoletin

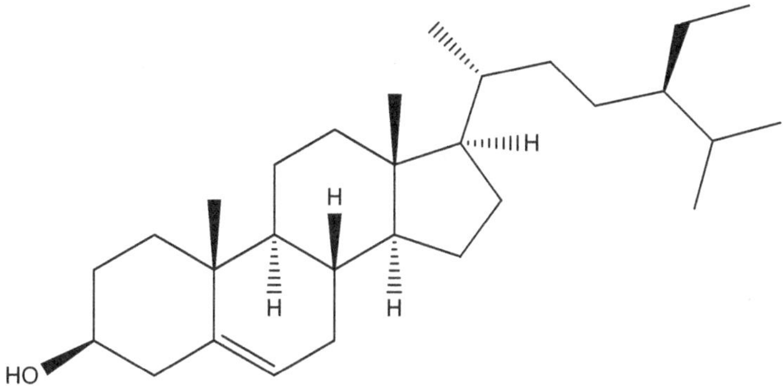

Structure of sitosterol

26 Family Crassulaceae J. Saint-Hilaire 1805

26.1 *Kalanchoe pinnata* (Lam.) Pers.

Synonyms: *Bryophyllum calycinum* Salisb.; *Bryophyllum germinans* Blanco; *Bryophyllum pinnatum* (Lam.) Asch. & Schweinf.; *Bryophyllum pinnatum* (Lam.) Kurz; *Bryophyllum pinnatum* (Lam.) Oken; *Calanchoe pinnata* Pers.; *Cotyledon calycina* Roth; *Cotyledon calyculata* Solander; *Cotyledon pinnata* Lam.; *Cotyledon rhizophilla* Roxb.; *Crassula pinnata* L. f.; *Crassuvia floripendia* Comm. ex Lam.; *Crassuvia floripenula* Comm.; *Kalanchoe calycina* Salisb.; *Sedum madagascariense* Clus.; *Verea pinnata* (Lam.) Spreng.

Local names: Gios; Patharkuchi

Common name: Air plant

Botanical description: It is a fleshy, glossy, and erect herb that grows up to 1.5 m tall. The stems are terete, fleshy, and purplish. The leaves are imparipinnate, decussate, and exstipulate. The blade is 10–30 cm long, and comprises 3–5 folioles. The folioles are oblong to elliptic, 5–20 cm × 2.5–12 cm, crenate, rounded at base, fleshy, dull green, and obtuse at the apex. The inflorescences are axillary or terminal panicles, which are 10–40 cm long and many flowered. The flowers are pendulous and showy. The calyx is tubular, cylindrical, purplish or light green, 2–4 cm long, and produces 4 lobes that are triangular. The corolla is reddish to purple, up to 5 cm long and presents 4 lobes, which are lanceolate. The androecium consists of 8 stamens inserted basally on the corolla. Nectar scales are present. The ovary is ovoid, minute, and develops into a slender style. The fruits are follicles, which are about 1.5 cm long, included in the calyx and corolla tube and containing numerous minute seeds.

Medicinal uses: inflammation, diabetes, cold, cough, tumors

Pharmacology: Extracts of the plant exhibited anti-lithiatic,[246] gastroprotective,[247] anti-convulsant,[248] and anti-hypertensive[249] effects.

27 Family Cucurbitaceae A.L. de Jussieu 1789

27.1 *Bryonopsis laciniosa* (L.) Naudin

Synonym: *Diplocyclos palmatus* (L.) C. Jeffrey

Local name: Mala

Common name: Lollipop climber

Botanical description: It is a climber with tendrils that are bifid. The leaves are simple, alternate, and exstipulate. The petiole is 4–6 cm long. The blade is palmately 5-lobed, 8–10 cm × 2–3.5 cm, the lobes are linear-lanceolate to elliptic, and acute or acuminate at the apex. The inflorescences are axillary clusters of greenish-yellow flowers. The calyx tube is 5-lobed and minute. The corolla is tubular, with 5 triangular lobes, and about 1 cm in diameter. The androecium comprises 3 stamens. The ovary is ovoid and 5 mm long. The fruits are spherical, dark green to reddish-purple with white lines, 1.5–2.5 cm across, and contain a few seeds.

Medicinal uses: venereal diseases, inflammation

Pharmacology: Antibacterial, cytotoxic, and larvicidal activities have been recorded.[250,251]

27.2 *Coccinia grandis* (L.) J. Voigt

Synonyms: *Bryonia grandis* L.; *Cephalandra indica* Naudin; *Coccinia cordifolia* (L.) Cogn; *Coccinia indica* Wight & Arn.

Local name: Telakucha

Common names: Ivy gourd; scarlet gourd

Botanical description: It is a climber that grows up to 5 m in length. The stems are terete. The leaves are simple, spiral, and exstipulate. The petiole is slender, striate, and 2–5 cm long. The blade is 5–10 cm long, cordate, 5-angular and serrate. The flowers are large, pure white, on 4 cm long peduncles, and solitary. The corolla is membranous, funnel-like, and 5-lobed. The fruits are pendulous, cylindrical, 2.5–5 cm × 1.2–2.5 cm, red, and contain numerous reniform seeds that are 5 mm long.

Medicinal uses: cold, fever, asthma, diabetes

Pharmacology: Extracts of the plant demonstrated anti-inflammatory,[252] anti-diabetic,[253] antibacterial,[254] and hepatoprotective[255] effects.

27.3 *Momordica charantia* L.

Synonyms: *Momordica indica* L.; *Momordica muricata* Willd.; *Momordica ceylanicum* Mill.

Local name: Karala

Common name: Bitter gourd

Botanical description: It is a climber with pubescent stems. The tendrils are up to 20 cm long. The leaves are simple, spiral, and exstipulate. The petiole is slender and 4–6 cm long. The blade is 5-lobed, 4–12 cm × 4–12 cm, the lobes ovate-oblong, and the margin crenate. The flowers are axillary on 3–12 cm long peduncles. The calyx comprises 5 sepals. The corolla is yellow, membranous, 5-lobed the lobes up to 2 cm long. The androecium includes 3 stamens. The ovary is fusiform, verrucose, and expands in a 2-lobed stigma. The fruit is pendulous, elliptical, 10–20 cm long, verrucose, dehiscent, edible but very bitter and contains numerous seeds embedded in a bright red aril.

Medicinal uses: ulcers, leprosy, intestinal worms, diabetes

Pharmacology: The plant has been the subject of numerous pharmacological investigations.[256] Extracts of the plant exhibit hypoglycemic[257] and anti-tumor[258] activities.

27.4 *Trichosanthes anguina* L.

Synonym: *Trichosanthes cucumerina* L. var. *anguina* (L.) Haines

Local name: Chichinga

Common name: Snake gourd

Botanical description: It is a climber with slender and angular stems. The leaves are simple, spiral, and exstipulate. The petiole is 3–7 cm long. The blade is 5-lobed, 8–16 cm × 6–18 cm, the lobes obovate, and the margin finely denticulate. The flowers are axillary on 1–18 cm long peduncles. The calyx tube is about 5 mm long with 5 segments. The androecium includes 3 stamens. The ovary is fusiform. The corolla is funnel-like, pure white, 5-lobed and laciniate. The fruit is cylindrical, up to about 1.5 m long, smooth, and contorted. The seeds are oblong and up to 1.7 cm long.

Medicinal uses: fever, intestinal worms

Pharmacology: Extracts of the plant exhibited anti-hyperglycemic and anti-nociceptive activities.[259]

28 Family Cyperaceae A.L. de Jussieu 1789

28.1 *Cyperus rotundus* L.

Synonyms: *Chlorocyperus rotundus* (L.) Palla; *Cyperus agrestis* Willd. ex Spreng. & Link; *Cyperus bicolor* Vahl; *Cyperus hexastachyus* Rottb.; *Cyperus hydra* Michx.; *Cyperus rubicundus* Vahl; *Cyperus tetrastachyos* Desf.

Local name: Mutha

Common name: Nut-grass; purple nut-sedge

Botanical description: It is an invasive sedge that grows up to 60 cm tall from fragrant dark brown tubers that are about 2 cm long. The stem is trigonous. The leaves are linear, with a 10 cm sheath, and a blade that reaches 30 cm long. The inflorescence is an anthelodium, which grows up to 10 cm in length. The spikes are brownish-red, fusiform and about 1.5 cm long.

Medicinal uses: cholera, fever, dysentery, diarrhea, diuretic, vermifuge, leprosy, ulcers

Pharmacology: The plant has been the subject of numerous pharmacological studies.[260] It contains sesquiterpenes such as α-cyperone that is anti-inflammatory,[261] the cytotoxic 6-acetoxy cyperene,[262] and sesquiterpenes with antiviral effects.[263]

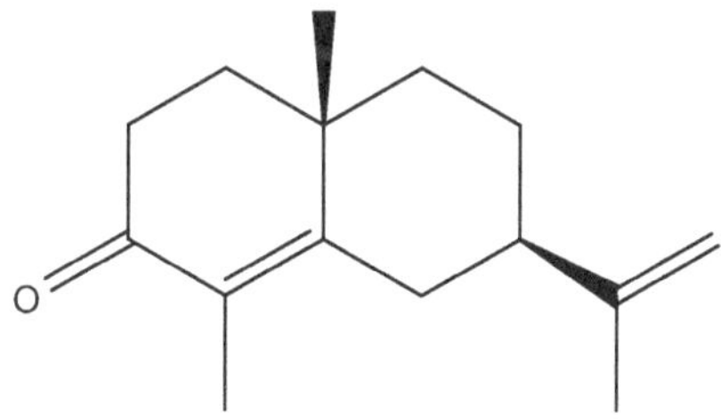

Structure of α-cyperone

29 Family Dilleniaceae R.A. Salisbury 1807

29.1 *Dillenia indica* L.

Synonyms: *Dillenia speciosa* Thunb.; *Dillenia speciosa* Thunb.

Local names: Chalta; chalita

Common name: Elephant apple

Botanical description: It is a handsome tree that grows up to 30 m tall. The bark is reddish brown and scaly. The leaves are simple, spiral, and exstipulate. The petiole is narrowly winged and 2–4 cm long. The blade is obovate-oblong, coriaceous, 15–40 cm × 7–14 cm, with 30–40 pairs of secondary nerves, and serrate at margin. The flowers are magnificent, showy, 12–20 cm in diameter, terminal, and solitary. The calyx comprises 5 sepals that are round and 4–6 cm in diameter, fleshy, dull green, and coriaceous. The corolla comprises 5 petals that are pure white, obovate, membranous, and about 7–9 cm long. The androecium comprises numerous stamens. The gynoecium includes up to 20 carpels. The fruit is globose, 10–15 cm in diameter, heavy with persistent sepals and is dull green on color.

Medicinal uses: fever, constipation

Pharmacology: Extracts of the plant evoked anti-inflammatory and analgesic effects.[264] The plant produces an anti-diabetic chromane.[265]

30 Family Dioscoreaceae R. Brown 1810

30.1 *Dioscorea bulbifera* L.

Synonyms: *Dioscorea anthropophagorum* A. Chev. ex Jum.; *Dioscorea hoffa* Cordem.; *Dioscorea hofika* Jum. & H. Perrier; *Dioscorea latifolia* Benth.; *Dioscorea longipetiolata* Baudon; *Dioscorea perrieri* R. Knuth; *Dioscorea sativa* L.; *Dioscorea sativa* Thunb.; *Dioscorea violacea* Baudon; *Helmia bulbifera* (L.) Kunth

Local name: Kukuralu

Common names: Air potato; aerial yam

Botanical description: It is a climber that grows from a large tuber. The stems are glabrous, twining, and smooth. The leaves are simple, alternate, and exstipulate. The petiole is 2.5–5.5 cm long. The blade is broadly and deeply cordate, glossy, dark green, 8–15 cm × 2–14 cm, glabrous, and with about 6–10 paralleled secondary nerves. The inflorescences are axillary, slender spikes, which are pendulous and 6.5–25 cm long. The perianth comprises 6 lobes The androecium includes 6 stamens. The fruits are about 2 cm long, oblate to spheroid, smooth, dark brown, somewhat glossy and contain winged seeds.

Medicinal uses: boils, ulcers, sores

Pharmacology: Extracts of the plant displayed analgesic, anti-inflammatory,[266] and antibacterial[267] properties. The plant produces anti-tumor flavonoids and saponins,[268,269] an antibacterial diterpene and phenolic compounds[270–273]

31 Family Dipterocarpaceae Blume 1825

31.1 *Anisoptera scaphula* (Roxb.) Kurz

Synonym: *Hopea scaphula* Roxb.

Local name: Boilam

Botanical description: It is a buttressed timber tree that grows up to 45 m in height. The bark is greyish. The bole is straight and the wood is very hard. The leaves are simple and alternate. The blade is oblong, coriaceous, shortly acuminated, with about 17 pairs of secondary nerves as well as scalariform tertiary nerves. The inflorescences are axillary racemes. The flowers are white. The androecium includes 15 stamens. The gynoecium develops a trifid stigma. The fruit is a nut with woody, narrowly oblong-elliptic and glossy wings which are about 15 cm long.

Medicinal use: asthma

Pharmacology: Unknown.

31.2 *Dipterocarpus turbinatus* Gaertn.

Synonyms: *Dipterocarpus jourdainii* Pierre; *Dipterocarpus laevis* Buch.-Ham.

Local name: Gurjun

Common names: Gurjun tree; wood oil tree

Botanical description: It is a majestic timber tree that grows up to 30 m in height. An aromatic exudate can be obtained by cutting and burning parts of the bark. The bark is brown, fissured, and flaky. The leaves are simple, alternate, and stipulate. The stipules are 2–6 cm long and puberulous. The petiole is 2–3 cm long. The blade is ovate-oblong, 20–30 cm × 8–13 cm, coriaceous, with 15–20 pairs of secondary nerves, base rounded and acuminated or acute apex. The inflorescences are axillary and 3–6-flowered racemes. The calyx comprises 5 sepals. The corolla produces 5 lobes, which are linear and pinkish-white. The androecium includes 30 stamens. The ovary is pubescent and develops a terete style. The fruit is a nut with 2 oblong wings that are purplish-brown, 12–15 cm × 3 cm, and reflexed.

Medicinal uses: diuretic, ringworm, skin affections, ulcers, rheumatism

Pharmacology: The plant has been little studied for its pharmacology. It produces borneol.[274]

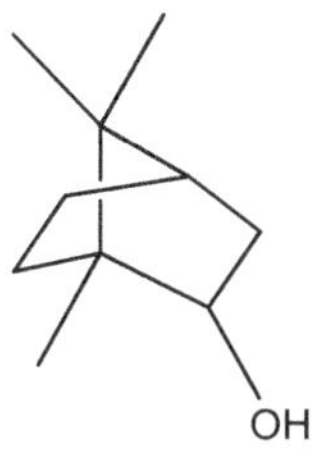

Structure of borneol

31.3 *Shorea robusta* Gaertn.

Local name: Shal

Common name: Red Balau

Botanical description: It is a timber tree which grows to 40m in height. The bark is grey-brown and fissured. The trunk exudes a resin. The wood is very hard and heavy. The leaves are simple, alternate, and stipulate. The stipules are small and lanceolate. The petiole is 2–2.5cm long. The blade is 10–40cm × 5–24cm, ovate to oblong, with about 12 pairs of secondary nerves glossy, base obtuse to cordate, and acuminate at apex. The inflorescences are axillary racemes, which grow up to 20cm long. The calyx includes 5 sepals, which are ovate, subequal, and pubescent. The corolla produces 5 lobes, which are contorted, creamy-yellow to pinkish, 1–1.5cm × 5mm, and linear. The androecium includes numerous anthers. The ovary is ovoid and pubescent. The fruit is an ovoid, about 1cm long nut with 5 wings, which are up to about 8cm long.

Medicinal use: ulcers

Pharmacology: Extracts of the plant demonstrated wound healing,[275] anti-ulcerogenic,[276] and antibacterial[277] properties.

32 Family Dracaenaceae Salisbury 1866

32.1 *Dracaena spicata* Roxb.

Synonyms: *Draco spicata* (Roxb.) Kuntze; *Pleomele spicata* (Roxb.) N.E. Br.

Local name: Kadorateng

Common name: Dragon tree

Botanical description: It is a shrubby plant that grows up to about 1.2 m tall. The stem is terete. The leaves are gathered at the apex of stems. The blades are lanceolate, dull dark green, without conspicuous longitudinal nervations, and about 15–30 cm long. The flowers are arranged in fascicles. The perianth is tubular, 6-lobed, the lobes linear, and greenish-yellow. The androecium includes 6 stamens inserted at the throat of the perianth. The ovary produces a 3-lobed style. The fruits are round, irregular berries, which are reddish orange and contain 3 seeds.

Medicinal use: measles

Pharmacology: Unknown.

33 Family Ebenaceae Gürke 1891

33.1 *Diospyros peregrina* (Gaertn.) Gürke

Synonyms: *Embryopteris peregrina* Gaertn.; *Diospyros embryopteris* Pers.; *Diospyros malabarica* (Desr.) Kostel.

Local name: Gap

Common name: Gaub persimmon

Botanical description: It is a tree that can grow up to 15 m tall. The bark is dark grey. The stem is hairy at apex. The leaves are simple, alternate, and exstipulate. The petiole is up to 1.2 cm long. The blade is oblong, coriaceous, 3–6 cm × 10–20 cm, with 6–8 inconspicuous pairs of secondary nerves, round at base and acute apex. The inflorescences are axillary cymes. The calyx comprises 4 ovate lobes and is about 5 mm long. The corolla is light yellow, 1–1.2 cm long, with 4 lobes, and campanulate. The androecium includes about 40 stamens. The ovary is globose, minute, with 4 styles. The fruit is an ovoid berry, yellow, with persistent calyx lobes, 5–6 cm in diameter, velvety, and containing 4–8 seeds.

Medicinal uses: wounds, ulcers, sore throats, fever, diarrhea

Pharmacology: Extracts of the plant exhibited anti-diabetic[278] and antibacterial[279] properties.

34 Family Euphorbiaceae A.L. de Jussieu 1789

34.1 *Acalypha hispida* Burm.f.

Synonyms: *Caturus spiciflorus* L.; *Ricinocarpus hispidus* (Burm. f.) Kuntze

Local name: Bara hatisur

Common name: Chenille plant

Botanical description: It is a shrub that grows up to about 2.5 m tall. The leaves are simple, spiral, and stipulate. The stipules are up to 1 cm long and hairy. The petiole is 4–8 cm long. The blade is ovate, 8–20 cm × 5–14 cm, cuneate at the base, serrate at margin, acute or acuminate at the apex, and with 3–8 pairs of secondary nerves. The inflorescence is a purplish-red pendulous somewhat furry or caterpillar-like spike of minute flowers that is 15–30 cm long and axillary. The calyx comprises 4 sepals. The ovary is globose and develops 3 styles, which are laciniate. The fruits are seldom seen, and are capsular and 3-lobed.

Medicinal uses: diarrhea, ulcers

Pharmacology: Extracts of the plant exhibited anti-inflammatory properties.[280] The plant contains phenolic compounds such as gallic acid.[281]

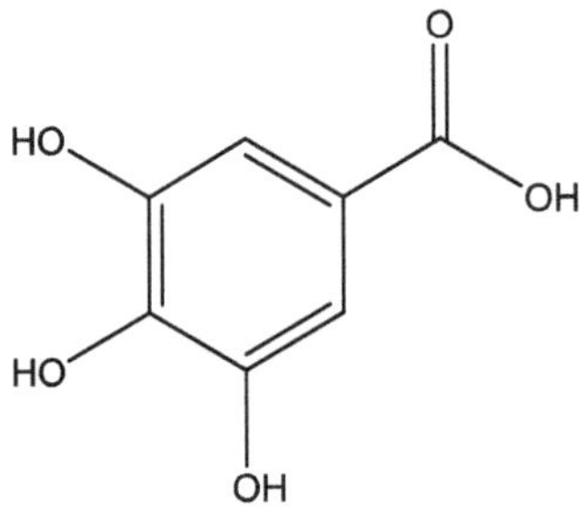

Structure of gallic acid

34.2 *Acalypha indica* L.

Synonyms: *Acalypha ciliata* Wall.; *Acalypha canescens* Wall.; *Acalypha spicata* Forsk.

Local name: Muka jhuri

Common names: Indian Acalypha; Copperleaf

Botanical description: It is an erect herb that grows to about 50 cm tall. The plant looks like some sort of strange dwarf tree. The leaves are simple, spiral, and stipulate. The stipules are minute. The petiole is slender, straight, and up to 3.5 cm long. The blade is rhombic-ovate, dull green, 2–3.5 cm × 1.5–2.5 cm, membranous, the base cuneate, serrate margin, acute apex, and presents about 5 pairs of secondary nerves. The inflorescences are axillary spikes that are 2–7 cm long. The flowers are minute. The calyx comprises 4 sepals. The androecium includes 8 stamens. The ovary is pilose and develops 3 styles that are laciniate. The fruit is a capsule that is trilocular, and minute containing few tiny seeds.

Medicinal uses: bronchitis, pneumonia, asthma, tuberculosis, intestinal worms, sores, colds, infections, bronchitis, ringworm

Pharmacology: The plant has been the subject of numerous pharmacological studies.[282] Extracts of the plant promoted wound healing in rodents[283] and exhibited antibacterial effects.[284,285]

34.3 *Euphorbia hirta* L.

Synonyms: *Anisophyllum piluliferum* (L.) Haw.; *Chamaesyce gemella* (Lag.) Small; *Chamaesyce hirta* (L.) Millsp.; *Chamaesyce hirta* (L.) Small; *Chamaesyce karwinskyi* (Boiss.) Millsp.; *Chamaesyce microcephala* (Boiss.) Croizat; *Chamaesyce pilulifera* (L.) Small; *Chamaesyce rosei* Millsp.; *Euphorbia bancana* Miq.; *Euphorbia capitata* Lam.; *Euphorbia chrysochaeta* W. Fitzg.; *Euphorbia gemella* Lag.; *Euphorbia globulifera* Kunth; *Euphorbia karwinskyi* Boiss.; *Euphorbia microcephala* Boiss.; *Euphorbia nodiflora* Steud.; *Euphorbia obliterata* Jacq.; *Euphorbia pilulifera* A. Chev.; *Euphorbia pilulifera* Jacq.; *Euphorbia pilulifera* L.; *Euphorbia verticillata* Vell.; *Tithymalus pilulifer* (L.) Moench; *Tithymalus piluliferus* (L.) Moench

Local names: Dudhia; lalkeru

Common name: Asthma weed

Botanical description: It is a laticiferous, bent, herb that grows up to about 30 cm tall. The stem is terete, pilose, somewhat reddish-brown and zigzag-shaped. The leaves are simple, opposite, and stipulate. The stipule is minute. The petiole is minute. The blade is lanceolate, asymmetrical, 1–5 cm × 3 mm–1.5 cm, pilose below, round at base, acute at apex, serrate, and presents 1–2 pairs of secondary nerves. The inflorescence is a globular cyathium of minute flowers on a 2.5-cm long peduncle. The calyx comprises 4 sepals that are white. The androecium includes 1 anther that is red. The ovary is triangular and pilose. The fruits are minute capsules, which are pilose and contain trigonous seeds.

Medicinal uses: dysentery, diarrhea, cold, cough, ringworms, gonorrhea, sores, galactagogue

Pharmacology: Extracts of the plant exhibited cytotoxic[286] and anti-allergic[287] effects. It produces antibacterial phenolics.[288]

34.4 *Euphorbia nivulia* Buch.-Ham.

Synonym: *Euphorbia neriifolia* sensu Roxb.

Local name: Sij

Common name: Leafy milk hedge

Botanical description: It is a succulent, laticiferous, spiny tree that grows up to about 5 m tall. The leaves are simple, sessile, alternate, and stipulate. The stipules are spines. The blade is spathulate, 10–25 cm × 3–8 cm, rounded at apex, tapered at base, with 6–8 pairs of secondary nerves, and fleshy. The inflorescences are cymes of cyathia. The flowers are minute. The fruits are 6 mm × 13 mm, smooth, and contain quadrangular, 4 mm long, smooth seeds.

Medicinal uses: rabies, rheumatism, constipation

Pharmacology: Extracts of the plant displayed antibacterial effects.[289] The plant produces diterpenes that are cytotoxic.[290]

34.5 *Jatropha curcas* L.

Synonyms: *Castiglionia lobata* Ruiz & Pav.; *Curcas adansonii* Endl.; *Curcas curcas* (L.) Britton & Millsp.; *Curcas drastica* Mart.; *Curcas indica* A. Rich.; *Curcas purgans* Medic.; *Curcas purgans* Medik.; *Jatropha acerifolia* Salisb.; *Jatropha afrocurcas* Pax; *Jatropha condor* Wall.; *Jatropha edulis* Cerv.; *Jatropha moluccana* Wall.; *Jatropha tuberosa* Elliot; *Jatropha yucatanensis* Briq.; *Manihot curcas* (L.) Crantz; *Ricinus americanus* Mill.; *Ricinus jarak* Thunb.

Local names: Kuruzdare; bagh dharanda

Common names: Barbados nut; angular-leaved physic nut

Botanical description: It is a shrub that grows about 3 m tall. The stem is terete, glaucous, smooth, and yield a latex upon incision. The leaves are simple, spiral, and stipulate. The stipules are minute. The petiole is straight and about 5–15 cm long. The blade is fleshy, glossy, wavy, somewhat 3-lobed, 7–14 cm × 6–12 cm, cordate at base, acute at the apex, and 5–7 pairs of secondary nerves. The inflorescences are cymose, axillary, and up to 10 cm long. The calyx comprises 5 sepals that are about 5 mm long. The corolla includes 5 petals that are about 5 mm long. The androecium comprises 10 stamens. The gynoecium develops a bifid stigma. The fruits are smooth, ovoid, 2–3 cm long, and contain a few black seeds.

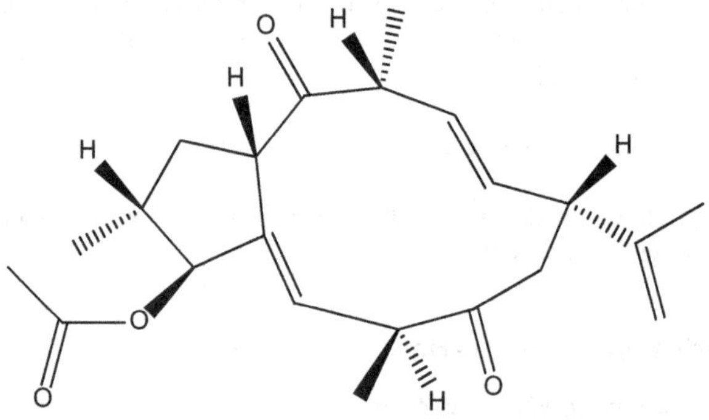

Structure of jatrophenone

Medicinal uses: wounds, ringworms, rheumatisms, constipation, syphilis

Pharmacology: Extracts of the plant exhibit anti-inflammatory[291] and anti-diarrheal[292] properties. The plant produces phorbol esters with antimicrobial effects.[293]

34.6 *Jatropha gossypifolia* L.

Local names: Beddha; laljeol

Common name: Bellyache bush

Botanical description: It is a laticiferous shrub that grows about 3 m tall. The stem is hirsute and glandular. The leaves are simple, spiral, and stipulate. The stipules are up to 1.2 cm long, filiform and divided. The petiole is straight, 3–12 cm, reddish, glossy, and hirsute-glandular. The blade is trifoliolate, 4–18.2 cm × 4.2–13.4 cm, base cordate, margin glandular-ciliate, apex acuminate, dark reddish and glossy. The inflorescences are terminal cymes on 2.5–10.5 cm long peduncles. The calyx comprises 5 sepals that are hirsute-glandular, minute, and elliptic. The corolla includes 5 petals that are obovate and glossy. The androecium includes 8 stamens in 2 whorls and arranged in a column. A disc is present. The ovary is minute and hirsute. The fruits are dull green, 3-lobed, dehiscent capsules growing up to 1.2 cm long. The seeds are ovoid and up to 7 mm long.

Medicinal uses: dysentery, boils, carbuncles, headache

Pharmacology: The plant has been the subject of numerous pharmacological investigations.[294] It produces the antibacterial diterpene jatrophenone.[295] Extracts of the plant exhibited anti-inflammatory effects.[296]

34.7 *Macaranga peltata* (Roxb.) Mull-Arg

Synonyms: *Osyris peltata* Roxb.; *Macaranga roxburghii* Wight; *Macaranga tomentosa* Wight; *Macaranga flexuosa* Wight; *Macaranga wightiana* Baill.

Local name: Nainna bichi gash

Common name: Indian lotus croton

Botanical description: It is a shrub that grows up to 4 m tall. The stems are green, articulated, smooth, and hollow. The leaves are simple, spiral, and stipulate. The stipules are ovate and up to 1 cm long. The petiole is slender, straight, and up to 19 cm long. The blade is soft, peltate, dull green, 13–25 cm × 12–21 cm, with about 6 pairs of secondary nerves, round at the base, with prominent tertiary scalariform nervation below, and acute to acuminate at the apex. The inflorescences are cauliflorous, about 6 cm long racemes of minute flowers. The calyx comprises 3 sepals. The androecium includes 3 stamens. The ovary is minute and glandular. The fruits are about 5 mm in diameter, hairy, and 3-lobed.

Medicinal uses: boils, piles

Pharmacological activity: Extracts of the plant demonstrated antibacterial and anti-fungal activities.[297,298]

34.8 *Mallotus philippensis* (Lam.) Müll. Arg.

Synonyms: *Echinus philippinensis* Baill.; *Rottlera tinctoria* Roxb.

Local names: Kuruar gash; ruda; toong

Common name: Kamala tree

Botanical description: It is a tree that grows up to 15 m tall. The stems are stellate-tomentose. The leaves are simple, spiral, and stipulate. The petiole is stellate-tomentose and 2–9 cm long. The blade is lanceolate, 5–20 cm × 3–6 cm, leathery, stellate-tomentose below, the apex acute or acuminate, and glossy. The inflorescences are axillary fascicles, which are up to 10 cm long. The calyx produces 3 or 5 lobes. The androecium includes 15–30 stamens. The ovary is stellate-tomentose and produces 3 styles. The fruits are capsules, which are 3-lobed, dehiscent, reddish, stellate-tomentose, up to 1.2 cm across, and contain seeds that are 4 mm long.

Medicinal uses: wounds, piles, ringworm, intestinal worms, herpes

Pharmacology: The plant produces the anti-fungal kamalachalcone E,[299] anti-allergic phloroglucinol derivatives[300] as well as anti-tumor-promoting 3α-hydroxy-D: A-friedooleanan-2-one.[301]

34.9 *Pedilanthus tithymaloides* (L.) Poit.

Synonym: *Euphorbia tithymaloides* L.

Local names: Rang chita; patabahar

Common names: Devil's backbone; Japanese poinsettia

Botanical description: It is a fleshy, laticiferous shrub that grows to about 1.5 m tall. The stems are terete, green, fleshy, and zigzag-shaped. The leaves are simple, sessile, alternate, and stipulate. The stipule is minute. The blade is lanceolate, fleshy, dull green, somewhat wavy at the margin, the base attenuate, the apex obtuse, and with about 9 pairs of secondary nerves that are inconspicuous. The inflorescences are cyathia with involucres, which are pink and about. 1.5 cm long. The flowers are minute and much reduced. The fruits are 3-lobed capsules, which are about 5 mm across and contain minute seeds. The plant is particularly visited by olive-backed sunbirds early morning.

Medicinal uses: warts, syphilis

Pharmacology: Extracts of the plant demonstrated anti-inflammatory[302] and wound healing[303] activities. It produces anti-plasmodial and antibacterial jatrophane diterpenoids.[304]

34.10 *Ricinus communis* L.

Synonyms: *Ricinus africanus* Mill.; *Ricinus angulatus* Thunb.; *Ricinus armatus* Andrews; *Ricinus atropurpureus* Pax & K. Hoffm.; *Ricinus badius* Rchb.; *Ricinus borboniensis* Pax & K. Hoffm.; *Ricinus cambodgensis* Benary; *Ricinus digitatus* Noronha; *Ricinus europaeus* T. Nees; *Ricinus giganteus* Pax & K. Hoffm.; *Ricinus glaucus* Hoffmanns.; *Ricinus hybridus* Besser; *Ricinus inermis* Mill.; *Ricinus japonicus* Thunb.; *Ricinus krappa* Steud.; *Ricinus*

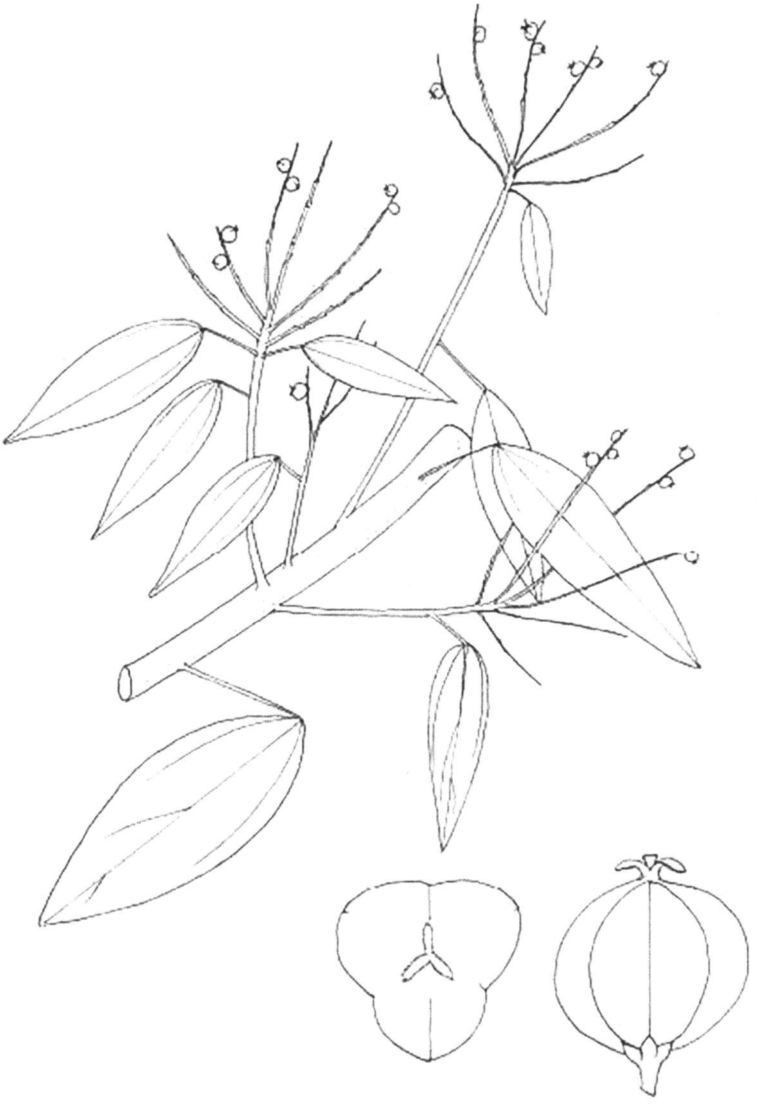

Mallotus philippensis (Lam.) Müll. Arg

laevis DC.; *Ricinus leucocarpus* Bertol.; *Ricinus lividus* Jacq.; *Ricinus macrocarpus* Popova; *Ricinus macrophyllus* Bertol.; *Ricinus medicus* Forssk.; *Ricinus medius* J.F. Gmel.; *Ricinus megalospermus* Delile; *Ricinus messeniacus* Heldr.; *Ricinus metallicus* Pax & K. Hoffm.; *Ricinus microcarpus* Popova; *Ricinus minor* Mill.; *Ricinus nanus* Bald.; *Ricinus obermannii* Groenl.; *Ricinus peltatus* Noronha; *Ricinus perennis* Steud. *Ricinus persicus* Popova *Ricinus purpurascens* Bertol. *Ricinus ruber* Miq. *Ricinus rugosus* Mill. *Ricinus rutilans* Müll. Arg. *Ricinus sanguineus* Groenl. *Ricinus scaber* Bertol. ex Moris; *Ricinus speciosus* Burm. f. *Ricinus spectabilis* Blume *Ricinus tunisensis* Desf. *Ricinus undulatus* Besser *Ricinus urens* Mill. *Ricinus viridis* Willd. *Ricinus vulgaris* Mill. *Ricinus zanzibarinus* Popova

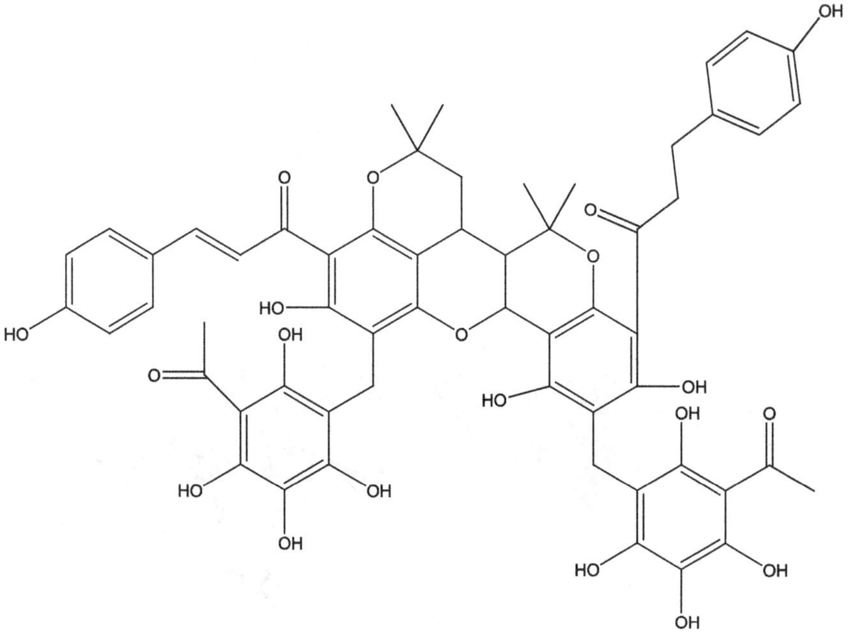

Structure of kamalachalcone E

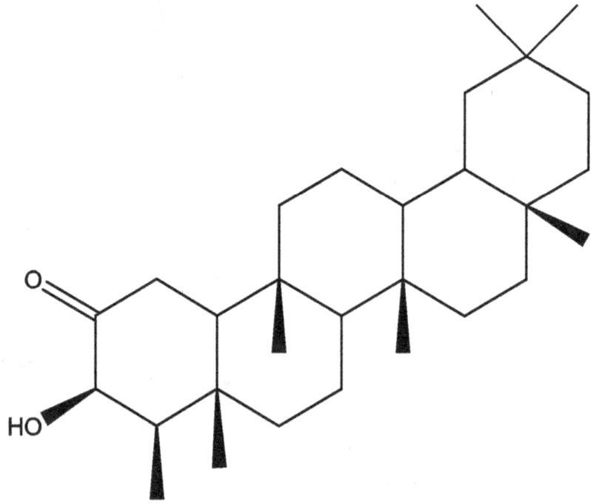

Structure of 3α-hydroxy-D: A-friedooleanan-2-one

Local names: Lal bherol; bherenda; veron; araddom

Common name: Castor oil tree

Botanical description: It is a shrub that grows up to about 5 m. The stems are terete, hollow, articulate, somewhat glaucous. The leaves are simple, spiral, and stipulate. The stipules are up to 2 cm long and form a sheath. The petiole is straight, slender and 5–20 cm long. The blade is 7–9-lobed, glossy, the median lobe up to 20 cm × 5 cm, the lobes lanceolate, acuminate, and serrate with 10–20 pairs of secondary nerves. The inflorescences are terminal racemes, which are 10–25 cm long. The calyx comprises 3 or 5 sepals that are elliptic-ovate, 6–8 mm × 3–4 mm, and yellowish-green. The androecium includes numerous stamens. The ovary is trilobate, echinate, and produces 3 bifid styles, which are red. The fruits are echinate, trilobate, 1–1.8 cm × 1–1.5 cm, green and dehiscent capsules containing a few seeds, which are glossy, ellipsoid, marbled, and about 5 mm long.

Medicinal uses: constipation, induces lactation, eye treatment, enlarged spleen, rheumatism, herpes, pruritus, wounds, ringworm, boil, toothache, fever, diarrhea, stomach ache, cough, abdominal pain during childbirth, breathing pain after childbirth, scorpion stings, sore, wounds

Pharmacology: The plant has been the subject of numerous pharmacological investigations.[305] It produces commercial Castor oil.[306] Extracts of the plant exhibit anti-inflammatory[307] and antibacterial[308] effects that contain the lectin ricin, which is a dreadful poison.

34.11 *Suregada multiflora* (A. Juss.) Baill.

Synonym: *Gelonium multiflorum* A. Juss.

Local name: Charchu

Common name: False lime

Botanical description: It is a tree that grows up to about 8 m in height. The leaves are simple, alternate, and stipulate. The stipules are connate and deciduous. The petiole 3–12 mm. The blade is elliptic, 5–15 cm × 3–8 cm, base cuneate, margin entire, apex acute and presents 5–7 pairs of secondary nerves. The inflorescences are short cymes on peduncles that are up to 1 cm long. The flowers are 8–12 mm in diameter. The calyx comprises 5 sepals. The androecium includes 30–60 stamens. A disc is present. The ovary is globose, and produces 3 styles that are 2-lobed. The fruits are 3-lobed capsules, which are up to 2.5 cm in diameter, fleshy, orangish, dehiscent, and containing 3 seeds.

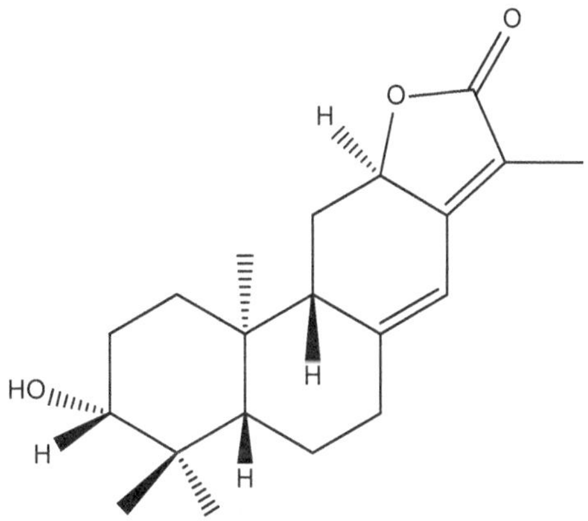

Structure of helioscopinolide A

Medicinal use: kill fish

Pharmacology: The plant produces helioscopinolide A that is anti-inflammatory[309] and anti-inflammatory kaurene diterpenes.[310]

34.12 *Tragia involucrata* L

Synonym: *Tragia hispida* Willd.

Local names: Bichuti; sengal sing

Common name: Indian stinging nettle

Botanical description: It is an erect herb that grows up to about 1 m tall. The stem is terete and hairy. The leaves are simple, spiral, and stipulate. The stipules are lanceolate, hairy, and up to about 5 mm long. The petiole grows up to 4 cm long. The blade is lanceolate 5–16 cm × 1.5–7.5 cm, hairy, serrate, and with 3–7 per pairs of secondary nerves. The inflorescences are axillary racemes that are 1.5–4 cm long. The calyx comprises 3 sepals. The androecium includes 3 stamens. The ovary is hispid and produces trifid styles. The fruits are 3-lobed, up to 1 cm long, hairy and greenish capsules.

Medicinal uses: alopecia, scorpion stings, diuretic, syphilis, fever

Pharmacology: Extracts of the plant exhibited anti-inflammatory, analgesic,[311] sedative,[312] anti-diabetic, hypolipidemic,[313] and wound healing[314] properties.

35 Family Fabaceae Lindley 1836

35.1 *Abrus precatorius* L.

Synonyms: *Abrus abrus* (L.) W. Wight; *Abrus maculatus* Noronha; *Abrus minor* Desv.; *Abrus pauciflorus* Desv.; *Abrus tunguensis* P. Lima; *Abrus wittei* Baker f.; *Glycine abrus* L.

Local names: Kunch; gunch; kawet

Common names: Crab's eye; Indian liquorice

Botanical description: It is a climber that grows up to about 4 m in length. The stems are terete. The leaves are paripinnate, spiral, and exstipulate. The stipules are about 5 mm long and linear. The blade comprises 8–20 pairs of folioles that are oblong, membranous, and about 1–2 cm × 0.4–0.8 cm. The inflorescences are axillary racemes that are 3–8 cm long. The calyx is tubular and 5-lobed. The corolla grows up to 1.5 cm long and consists of 5 petals which are purple. The androecium includes 9 stamens. The fruits are pods which are 2–3.5 cm × 0.5–1.5 cm, dehiscent, with 2–6 seeds, which are glossy, globose, reddish brown, and black.

Medicinal uses: contraceptive, sprain, gonorrhea, inflammations

Pharmacology: The plant contains saponins and flavonoids.[315] Extracts of the plant exhibited antimicrobial, insecticidal, and anti-protozoal properties.[316–318] The seeds contain the protein abrin, which is a dreadful poison.[319]

35.2 *Acacia arabica* var *indica* Benth

Local name: Blabla

Common names: Indian gum; Arabic tree

Botanical description: It is a gummiferous tree that grows up to about 10 m tall. The leaves are bipinnate, spiral, and stipulate. The stipules are spinescent and up to 8 cm long. The blade presents 2–11 pairs of pinnae with 7–25 pairs of folioles, which are about 5 mm long. The inflorescences are axillary pedunculated, heads of minute flowers which are 6–15 mm in diameter. The calyx is tubular, 5-lobed, and minute. The corolla comprises 5 petals. The fruits are pods constricted between the seeds, which are 4–22 cm × 1.3–2.2 cm. The seeds are blackish-brown and up to about 8 mm long.

Medicinal uses: astringent, anti-inflammatory, abortion

Pharmacology: The gum has beneficial effects against periodontopathic bacteria and plaque.[320,321] The plant abounds with tannins.[322]

35.3 *Acacia catechu* (L. f.) Willd.

Synonyms: *Acacia wallichiana* DC.; *Mimosa catechu* L. f.

Local name: Khayer

Common names: Cutch tree; catechu

Botanical description: It is a tree which grows to 6 m tall. The bark is fissured. The stem are pubescent, often with a pair of flat, brown, hooked spines below the stipules. Leaf glands near petiolar base and between several upper leaflets of rachis are present. The rachis is villous and supports 10–30 pairs of pinnae. The pinnae include 20–50 pairs of leaflets which are linear, 2–6 mm × 1–1.5 mm, and ciliate. The inflorescence is a spike which is about 2.5–10 cm long. The calyx is campanulate and about 1.2–1.5 cm long with tiny deltoid lobes. The corolla includes 5 petals which are lanceolate or oblanceolate, 2.5 cm, long and sparsely pubescent. The androecium includes numerous stamens. The gynaecium comprises an ovary which is minute and glabrous. The fruit is brown, straight, strap-shaped, 12–15 cm × 1–1.8 cm, dehiscent pod containing 3–10 seeds.

Medicinal uses: diarrhea, leprosy, bleeding, fever, ringworm, indigestion, rheumatism

Pharmacology: Extract of the plant exhibited anti-diarrheal,[323] chemoprotective,[324] and hypotensive[325] activities. The plant abounds with tannins and phenolic compounds, which are astringent and anti-inflammatory.[326] The anti-leprosy property of the plant has been clinically investigated.[327]

35.4 *Acacia concinna* (Willd.) DC.

Synonyms: *Acacia sinuata* (Lour.) Merr.; *Mimosa concinna* Willd.; *Mimosa sinuata* Lour.

Local names: Shikakai

Common name: Soap pod

Botanical description: It is a tree which grows to 5 m tall. The stems are hairy and minutely prickly. The leaves are bipinnate, alternate, and stipulate. The stipules are ovate-cordate and about 3–8 mm × 1.5–6 mm. The blade is 10–20 cm long and presents 6–18 pairs of pinnae that are 8–12 cm long. The pinnae bear 15–25 pairs of folioles, which are glaucous below, linear-oblong, asymmetrical, 8–12 mm × 2–3 mm, and membranous. The inflorescences are axillary panicles of globose flower heads which are 9–12 mm in diameter. The calyx is tubular, 5-lobed, and minute. The corolla is tubular, 5-lobed and whitish-yellow. The ovary is stipitate. The fruits are pods, which are blackish, glossy, wrinkled, strap-shaped, 8–15 cm × 2–3 cm, and fleshy.

Medicinal use: hair wash

Pharmacology: The plant produces saponins[328] with immunomodulatory activity.[329] Extracts of the plant exhibited anti-helminthic properties.[330]

35.5 *Acacia farnesiana* (L.) Willd.

Synonyms: *Acacia acicularis* Humb. & Bonpl. ex Willd.; *Acacia acicularis* R. Br.; *Acacia densiflora* (Alexander ex Small) Cory; *Acacia edulis* Humb. & Bonpl. ex Willd.; *Acacia ferox* M. Martens & Galeotti; *Acacia lenticellata* F. Muell.; *Acacia leptophylla* DC.; *Acacia pedunculata* Willd.; *Acacia smallii* Isely; *Farnesia odora* Gasp.; *Mimosa arcuata* M. Martens & Galeotti; *Mimosa farnesiana* L.; *Mimosa scorpioides* Forssk.; *Pithecellobium minutum* M.E. Jones; *Poponax farnesiana* (L.) Raf.; *Vachellia farnesiana* (L.) Wight & Arn.; *Vachellia guanacastensis* (H.D. Clarke, Seigler & Ebinger) Seigler & Ebinger

Local names: Guya babula; kanta naksha

Common name: West Indian blackthorn

Botanical description: It is a tree that grows up to 4m tall. The stems are zigzag-shaped. The leaves are bipinnate, alternate, and stipulate. The stipules are spines, which are up to 2cm long. The rachis is 1.2–2.5cm long and bears 2–8 pairs of pinnae that are 1.5–3.5cm long. The pinnae include 10–20 pairs of folioles which are 2.5–5.5mm × 1–1.5mm, linear-oblong, and asymmetrical. The inflorescences are axillary fascicles of flowering heads on 1.2–2.5cm long, slender, and pubescent peduncles. The calyx is campanulate, minute, and 5-lobed. The corolla is tubular, 5-lobed, yellow, and minute. The androecium is showy and consists of numerous stamens, which are about twice as long as the corolla. The fruits are pods, which are 4.5–7.5cm × 1.2cm, dark brown and contain numerous seeds, which are brown, ovoid, and about 5mm across.

Medicinal uses: headache, leucorrhea, infection of the gums, gonorrhea

Pharmacology: The plant contains tannins with antibacterial activity.[331] Methyl gallate from the plant elicited antibacterial effects against *Vibrio cholerae*.[332]

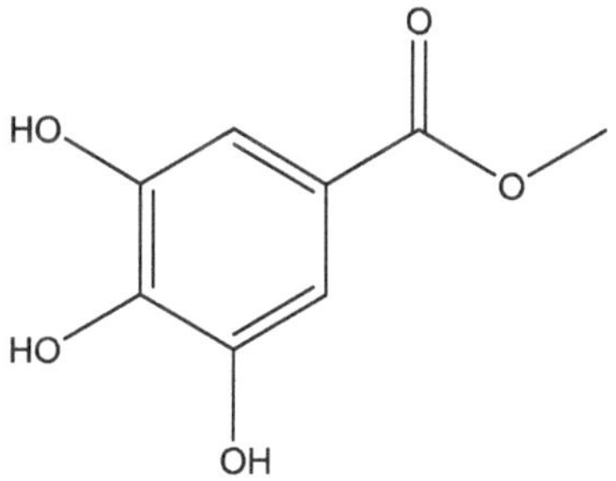

Structure of methyl gallate

35.6 *Adenanthera pavonina* L.

Local name: Raktochandan

Common name: Coral pea

Botanical description: It is a tree that grows up to 10m tall. The leaves are bipinnate, alternate, and stipulate. The rachis is up to 40cm long, and presents 2–6 pairs of 7–12cm long pinnae bearing 7–15 pairs of folioles, which are alternate, obovate, 1.5–4.0 cm × 5–25 mm broad, dull green, and membranous. The inflorescences are axillary or terminal racemes that are 7.5–15 cm long. The flower peduncles are 3.5–4 cm long. The calyx is tubular, 5-lobed and minute. The corolla consists of 5 petals that are pale yellow and minute. The androecium includes 10 stamens. The ovary is sessile and develops a capitate stigma. The fruits are dehiscent and spiral pods, which are 10–22.5 cm and containing numerous beautiful bright red and glossy ellipsoid seeds that are about 5 mm across.

Medicinal uses: boils, abscesses

Pharmacology: Not much is known about the pharmacological properties of this common plant. It contains antibacterial flavonoids.[333] Extracts demonstrated anti-inflammatory activity.[334]

35.7 *Albizia lebbeck* (L.) Benth.

Synonyms: *Acacia lebbeck* (L.) Willd.; *Albizzia lebbeck* Benth.; *Feuilleea lebbeck* (L.) Kuntze; *Mimosa lebbeck* L.; *Mimosa lebbek* Forssk.; *Mimosa sirissa* Roxb.; *Mimosa speciosa* Jacq.

Local names: Sherisha; siris

Common name: Woman's tongue

Botanical description: It is a tree that grows up to about 10 m tall. The bole is straight and the plant has a strange architecture. The stems are hairy when young and very hard. The leaves are bipinnate, alternate, and stipulate. The stipule is about 5 mm long, linear, caducous, and hairy. The rachis is 7.5–15 cm long and presents 1–4 pairs of pinnae, which are 5–20 cm long as well as minute elliptic glands. The pinnae comprise 3–9 pairs of folioles, which are oblong and 2–4.5 cm × 1.3–2 cm. The inflorescences are heads of fragrant flowers that are solitary or fasciculated on 3.5–10 cm long peduncles. The calyx is campanulate, about 5 mm long and with 5 minute lobes. The corolla is 7–8 mm long, funnel-shaped, with 5 deltoid-ovate lobes. The androecium is showy and includes numerous stamens, which are 2.5–3.5 cm long. The ovary is glabrous and sessile. The fruits are pods, which are 15–30 cm × 2.5–5 cm, flat, and contain 5–12 seeds.

Medicinal uses: rat bites, piles, body pains

Pharmacology: Extract of the plant demonstrated anti-inflammatory activity.[335]

35.8 *Albizia odoratissima* (L.f.) Benth.

Synonyms: *Acacia odoratissima* Willd.; *Albizia odoratissima* (L. f.) Kuntze; *Mimosa odoratissima* L.f.

Local name: Koroi

Common name: Ceylon rosewood

Botanical description: It is a tree that grows about 10 m tall. The stems are hairy. The leaves are bipinnate, alternate, and stipulate. The stipules are filiform and minute. The rachis is 10–20 cm long. The blade presents 3–8 pairs of pinnae which are 4.5–9 cm long, and with

Albizia lebbeck

8–20 pairs of folioles. The folioles are 1.7–2.5 mm × 5–10 mm, asymmetrical, oblong, and hairy below. The inflorescences are hairy panicles of fragrant flower heads. The calyx is cup-shaped, hairy, and minute. The corolla is pale yellow, tubular, and develops 5 lobes. The androecium includes 20 stamens, which are about 1.5 cm long. The ovary is hairy and develops a filiform style. The fruits are pods, which are 10–30 cm × 1.5–4 cm, hairy, reddish brown, and contain 6–12 seeds.

Medicinal uses: leprosy, ulcers

Pharmacology: Extract of the plant demonstrated anti-diabetic properties.[336]

35.9 *Albizia procera* (Roxb.) Benth.

Synonyms: *Acacia procera* (Roxb.) Willd.; *Feuilleea procera* (Roxb.) Kuntze; *Mimosa procera* Roxb.

Local names: Shil-koroi; kori gach; bai-keowra; sada sirish

Common names: Albizia; black siris; false lebbeck; forest siris; silver bark rain tree; tall albizia

Botanical description: It is a tree that grows up to 10 m in height. The leaves are bipinnate, alternate, and stipulate. The rachis is 30–45 cm long and presents 2–6 pairs of pinnae, which are up to 15 cm long. The pinnae include 4–16 pairs of folioles, which are oblong-ovate, hairy below, obtuse, asymmetrical, with about 7 pairs of secondary nerves, and round at the apex. The inflorescences are axillary or terminal panicles of globose heads of about 20 flowers. The calyx is minute, tubular and 5-lobed. The corolla is tubular, 5-lobed, yellow-white, and about 5 mm long. The androecium includes numerous stamens, which protrude out of the corolla. The ovary is glabrous and develops a slender style. The fruits are pods, which are flat, 10–15 cm × 1.5–2.5 cm, and containing 8–12 seeds.

Medicinal use: ulcers

Pharmacology: The plant contains (+)-catechin and protocatechuic acid which inhibited HIV-1 integrase.[337] Extracts of the plant exhibited analgesic, antibacterial, and central nervous system-depressant activities.[338]

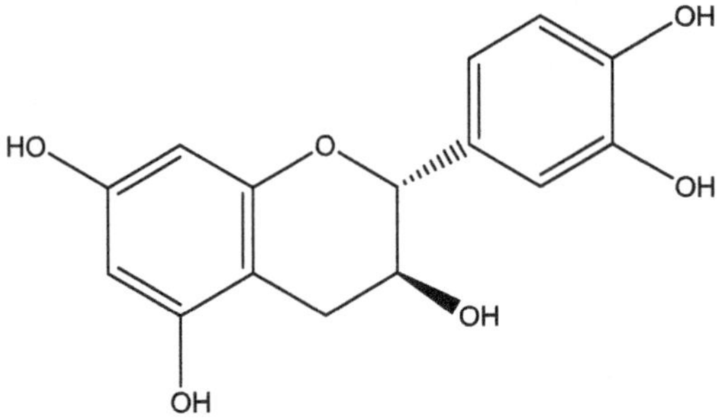

Structure of (+)-catechin

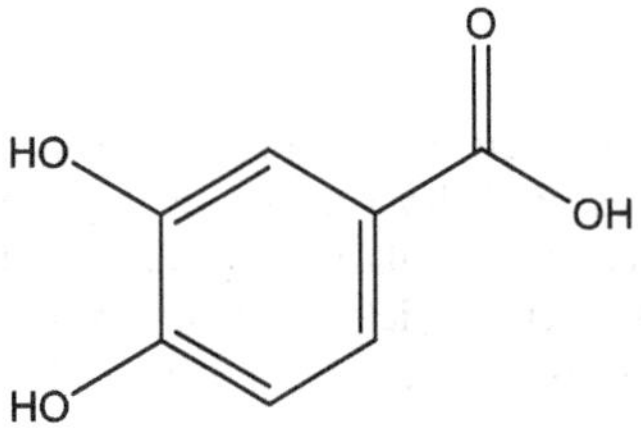

Structure of protocatechuic acid

35.10 *Butea monosperma* (Lam.) Taub.

Synonyms: *Butea frondosa* K.D. Koenig ex Roxb.; *Butea frondosa* Roxb. exWilld.; *Butea monosperma* Kuntze; *Erythrina monosperma* Lam.; *Plaso monosperma* (Lam.) Kuntze

Local names: Palas; murut

Common name: Bastard teak

Botanical description: It is a tree that grows up to 10 m tall. The bark exudes a red sap upon incision. The stems are hairy. The leaves are imparipinnate, alternate, and stipulate. The stipules are minute. The rachis is 10–20 cm long. The blade presents 3 folioles, which are coriaceous, about 12.5–20 cm × 11–17.5 cm, broadly ovate, and with about 5 pairs of secondary nerves. The inflorescences are axillary or terminal racemes, which are up to 17.5 cm long. The calyx is 1.2 cm long, campanulate, and hairy. The corolla is papilionaceous, bright red, with a recurved standard that reaches 4.5 cm long. The androecium includes 10 stamens. The ovary is hairy. The fruits are flat pods, which are 10–20 cm × 2.5–5 cm, and coriaceous. The seeds are reddish brown, reniform, flat, and about 3 cm long.

Medicinal uses: sores, diarrhea, urinary infection, leprosy, ringworm, intestinal worms, diarrhea, leucorrhea

Pharmacology: Extracts of the plant exhibited anti-diabetic[339] and anthelminthic[340] properties. The plant produces flavonoids with hepatoprotective,[341] estrogenic,[342] antiviral,[343] antibacterial,[344] and anti-fungal[345] properties.

35.11 *Caesalpinia bonduc* (L) Roxb.

Synonyms: *Caesalpinia bonducella* (L.) Fleming; *Caesalpinia crista* L.; *Caesalpinia crista* Thunb.; *Guilandina bonduc* Aiton; *Guilandina bonduc* Griseb.; *Guilandina bonduc* L.; *Guilandina bonducella* L.; *Guilandina crista* (L.) Small

Local names: Natakaranj; nata; koranju; karanj

Common names: Yellow nicker; nicker nut; molucca bean

Botanical description: It is a woody climber that grows up to about 10 m in length. The stems are prickly and hairy. The prickles are straight. The leaves are bipinnate, alternate, and stipulate. The stipules are deciduous, large, and about 2 cm long. The blade is 30–45 cm long with 6–9 pairs of pinnae bearing 6–12 pairs of folioles, which are oblong, 1.5–4 cm × 1.2–2 cm, membranous, oblique at base rounded to acute at apex, and glossy. The inflorescences are axillary racemes, which grow up to 30 cm long. The calyx includes 5 sepals, which are 8 mm long, and hairy. The corolla includes 5 petals, which are up to 1.5 cm, long yellow, the standard tinged with red spots, and clawed. The androecium includes 10 stamens. The ovary is hairy. The fruits are oblong, 5–7 cm × 4–5 cm, coriaceous, dehiscent pods beaked at apex, spiny, the spines up to 1 cm long and contain 2–3 seeds, which are globose, smooth, and light grey.

Medicinal uses: tonic, placental expulsion, headache, wounds, boils

Pharmacology: The plant produce anti-fungal phenolic compounds.[346] Extracts demonstrated antibacterial,[347,348] anti-inflammatory,[349] anti-diabetic,[350] and anti-pyretic[351] properties.

35.12 *Cajanus cajan* (L.) Huth

Synonyms: *Cajan cajan* (L.) Huth; *Cajan inodorum* Medik.; *Cajan inodorum* Medik.; *Cajanum thora* Raf.; *Cajanus bicolor* DC.; *Cajanus cajan* (L.) Druce; *Cajanus cajan* (L.) Merr. *Cajanus cajan* (L.) Millsp. *Cajanus flavus* DC.; *Cajanus indicus* Spreng.; *Cajanus luteus* Bello; *Cajanus obcordifolia* Singh; *Cajanus pseudocajan* (Jacq.) Schinz & Guillaumin; *Cajanus striatus* Bojer; *Cytisus cajan* L.; *Cytisus cajan* L.; *Cytisus cajan* L.; *Cytisus cajan* L.; *Cytisus cajan* L.; *Cytisus guineensis* Schumach. & Thonn.; *Cytisus pseudocajan* Jacq.

Local names: Arahor; baredare

Common name: Pigeon pea

Botanical description. It is a shrub that grows up to 2m tall. The stems are hairy and pubescent. The leaves are imparipinnate, alternate, and stipulate. The stipules are minute and ovate-lanceolate. The blade is 3-foliolate. The folioles are spathulate to elliptic, dull green, 2.8–10cm × 0.5–3.5cm, papery, hairy below, acute or acuminate at apex, and with about 8 pairs of secondary nerves. The inflorescence is a terminal or axillary raceme, which is 3–7cm long. The calyx is campanulate, hairy, 5–7mm long, and 5-lobed. The corolla is papilionaceous, yellow to reddish brown, and about 1.5cm long. The ovary is hairy and develops a long style and a capitate stigma. The fruits are pods, which are linear-oblong, 4–8.5cm × 0.6–1.2cm, pubescent, beaked at apex, and contain 3–6 grey, subspherical, 5-mm long seeds with brown spots.

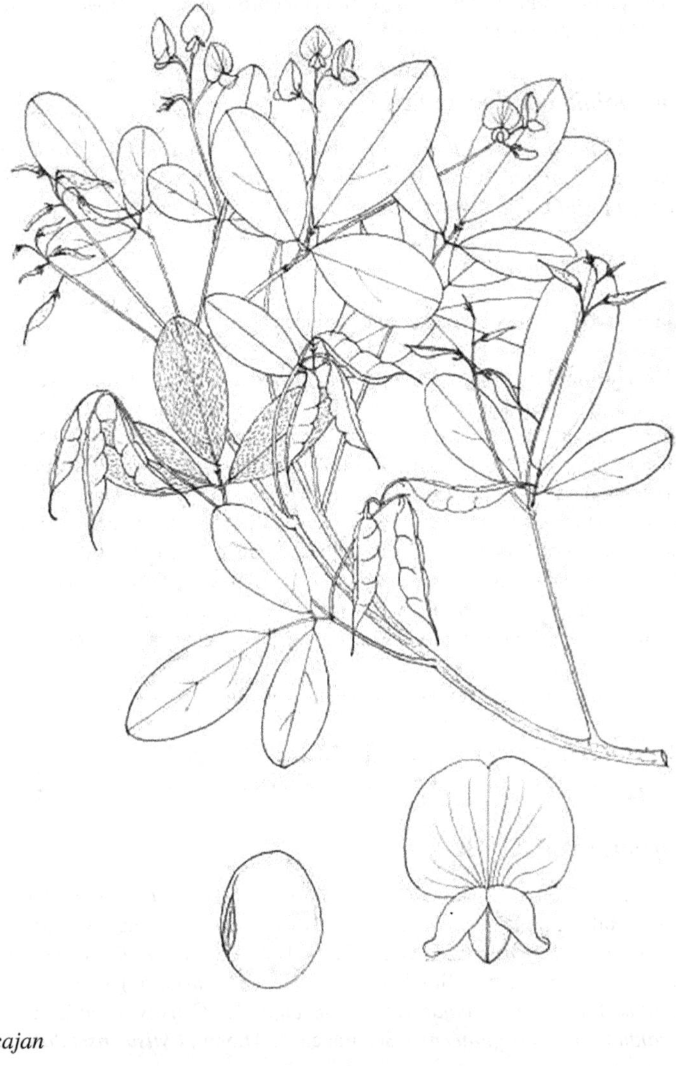

Cajanus cajan

Medicinal use: jaundice

Pharmacology: Extract of the plant demonstrated hepatoprotective effects.[352]

35.13 *Cassia fistula* L.

Synonyms: *Bactyrilobium fistula* Willd.; *Cassia bonplandiana* DC.; *Cassia excels* H.B. K.; *Cassia fistuloides* Colladon; *Cassia rhombifolia* Roxb.; *Cathartocarpus excelsus* G. Don; *Cathartocarpus fistula* Pers.; *Cathartocarpus fistuloides* (Colladon) G. Don; *Cathartocarpus rhombifolius* G. Don

Local name: Bandar lathi

Common name: Golden shower tree

Botanical description: It is a magnificent tree that grows up to about 10 m in height. The bole is straight, pale greyism to almost white, and smooth. The leaves are a paripinnate, alternate, and stipulate. The stipule is deltoid and minute. The blade is 30–40 cm long, with 3 or 8 pairs of folioles, which are broadly lanceolate, dull green, 8–13 cm × 4–8 cm, cuneate at base, and acute at apex. The inflorescences are showy axillary racemes, which are 20–40 cm long, lax, pendant, and many flowered. The corolla is 3.5–4 cm in diameter. The calyx comprises 5 sepals, which are narrowly ovate and 1–1.5 cm long. The corolla includes 5 petals that are golden yellow, broadly ovate, subequal, 2.5–3.5 cm, and shortly clawed. The androecium includes 10 stamens up to 4 cm long. The ovary develops a small stigma. The fruit is a cylindrical, blackish-brown, terete pendulous, sausage-like, indehiscent, 30–60 cm × 2–2.5 cm pod containing numerous seeds in a brown edible pulp.

Medicinal uses: constipation, leprosy, ringworm

Pharmacology: The plant produces anthraquinones, which are laxatives and are used therapeutically.[353] Extracts of the plant displayed anti-helminthic[354] and anti-fungal[355–357] properties.

35.14 *Dalbergia sissoo* Roxb. ex DC.

Synonym: *Dalbergia pseudo-sissoo* Miq.

Local name: Sissoo

Common name: Indian rosewood

Botanical description: It is a tree that grows up to 10 m. The stems are hairy. The leaves are imparipinnate, alternate, and stipulate. The stipules are 5 mm long and lanceolate. The rachis is 3.7–7.5 cm long. The blade is 12–15 cm long and comprises 3–5 folioles, which are rhombic-obovate, 3.5–6.5 cm, the apex rounded and shortly caudate. The inflorescences are axillary panicles which are about 7 cm long and hairy. The calyx is campanulate, 6–7 mm long, and 5-toothed. The corolla is papilionaceous, yellowish white. The standard is broadly obovate and emarginate. The androecium includes 9 stamens. The ovary oblong, hairy, and with a capitate stigma. The fruit is a pod, which is linear-oblong, 3–10 cm long with 1–3 reniform and compressed seeds.

Medicinal uses: gonorrhea, wound, itches, abscess

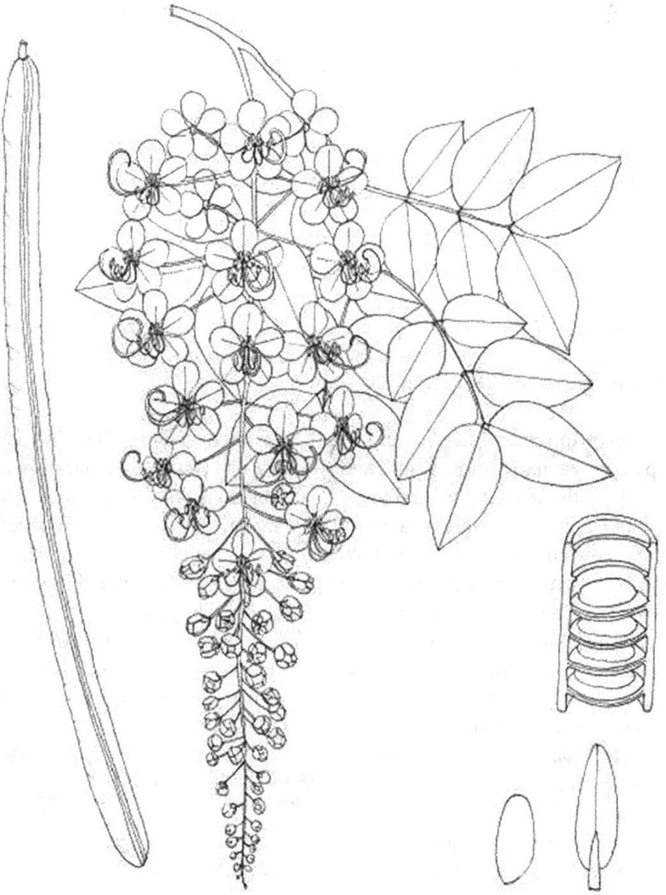

Cassia fistula

Pharmacology: Extracts of the plant showed anti-spermatogenic[358] and fracture healing[359] properties. The plant contains estrogenic neoflavonoids.[360]

35.15 *Desmodium motorium (Houtt.)* Merr.

Synonyms: *Codariocalyx gyrans* (L.f.) Hassk.; *Codariocalyx motorius* (Houtt.) Ohashi; *Hedysarum gyrans* L.f.; *Hedysarum motorium* Houtt.

Local name: Turi-chombhol

Common name: Semaphore plant

Botanical description: It is an erect herb that grows up to about 1 m in height. The stems are hairy. The leaves are simple or imparipinnate, alternate, and stipulate. The stipules are about 5 mm long. The petiole is 1–2.5 cm long. The lateral folioles are 1–2 cm × 3.5–4.5 mm and the terminal foliole is 2.5–7 cm × 6.5–13 mm oblong-lanceolate, hairy below. The folioles have the extraordinary ability to move very slowly and elliptically. The inflorescences are terminal and axillary racemes. The calyx is campanulate, minute, and bilabiate. The corolla is papilionaceous, 7.5–8.5 mm long, and pink. The androecium includes 10 stamens. The fruits are pods, which are 3–4.5 cm long, and slightly falcate.

Medicinal use: rheumatism

Pharmacology: Extracts of the plant exhibited anti-inflammatory[361] and anti-diabetic[362] effects.

35.16 *Erythrina variegata* L.

Synonyms: *Erythrina indica* Lam.; Erythrina orientalis (L.) Merr.

Local names: Madar; maharbaha; palita mundar

Common name: Indian coral tree

Botanical description: It is a tree that grows up to about 10 m tall. The bark is dark brown. The stems are minutely prickly. The leaves are crowded at apex of stems, imparipinnate, and stipulate. The stipules are lanceolate. The blade presents 3 folioles. The folioles are broadly ovate or rhomboid-ovate, 15–30 cm × 15–30 cm, membranous, with 3–5 pairs of secondary nerves, base broadly cuneate, and apex acuminate to obtuse. The inflorescences are terminal racemes which are 10–15 cm long and woody. The calyx is spathaceous and 2–3 cm long. The corolla is papilionaceous, intensely red, 6–7 cm. The standard is elliptic and about 5–6 cm × 2.5 cm. The androecium includes 10 stamens. The ovary is stipitate. The fruits are straight, up to about 15 cm long pods, which are constricted between the seeds.

Medicinal uses: fever, waist pain

Pharmacology: Extracts of the plant exhibited coagulant activity.[363] The plant produces antibacterial flavonoids.[364]

35.17 *Indigofera tinctoria* L.

Synonyms: *Indigofera anil* L. var. *orthocarpa* DC; *Indigofera argentea* L.; non Burm.f.; *Indigofera. Bergii* Vatke; *Indigofera cinerascens* DC; *Indigofera houer* Forssk.; *Indigofera indica* Lam.; *Indigofera sumatrana* Gaertn.; *Indigofera ornithopodioides* Schum.; *Indigofera orthocarpa* (DC.) Berg.

Local name: Neel

Common name: Common indigo

Botanical description: It is an herb that grows up to about 1 m in height. The stems are hairy. The leaves are imparipinnate, alternate, and stipulate. The stipules are minute and triangular. The blade is up to about 10 cm long, comprises 4–6 pairs of folioles plus a terminal one. The folioles are obovate-oblong, 1.5–3 cm × 0.5–1.5 cm, broadly cuneate to rounded at base, rounded to emarginated at apex, dull green, and membranous. The inflorescences are axillary racemes that grow up to about 10 cm long. The calyx is minute, campanulate, cup-shaped and 5-lobed. The corolla is papilionaceous, reddish to pink, the standard broadly obovate and 5 mm long. The androecium includes 10 stamens. The ovary develops a capitate stigma. The fruits are linear pods which are 2.5–3 cm long and contain 5–12 cubic and minute seeds.

Medicinal uses: diuretic, antidote

Pharmacology: The plant is used as a source of blue dye.[365] Extracts of the plant exhibited anthelminthic,[366] anti-seizure[367] and anti-inflammatory[368] properties.

35.18 *Lablab purpureus* (L.) Sweet

Synonyms: *Dolichos lablab* L.; *Dolichos benghalensis* Jacq.; *Dolichos purpureus* L.; *Lablab niger* Medikus; *Lablab purpurea* (L.) Sweet; *Lablab vulgaris* (L.) Savi; *Vigna aristate* Piper

Local name: Bun shim

Common name: Bonavista bean

Botanical description: It is a climber that grows up to about 5 m in length. The stems are purplish. The leaves are imparipinnate, alternate, and stipulate. The stipules are lanceolate. The folioles are deltoid-ovate, 6–10 cm × 6–10 cm, membranous, asymmetrical, acute or wedge-shaped at base, and acute or acuminate at apex. The inflorescences are axillary racemes which are erect and 15–25 cm long. The calyx is bilabiate and 6 mm long. The corolla is papilionaceous, about 1.5 cm long, white or purple, and the standard is orbicular. The ovary is linear. The fruits are oblong-falcate, 5–7 cm × 1.4–1.8 cm, glossy, purple, beaked pods which contain 3–5 ovoid and compressed seeds.

Medicinal uses: nausea, vomiting, abdominal pains

Pharmacology: Extracts of the plant demonstrated hepatoprotective[369] and anti-tumor[370] activities.

35.19 *Mucuna pruriens* (L.) DC.

Synonyms: *Carpopogon niveus* Roxb.; *Carpopogon pruriens* (L.) Roxb.; *Dolichos pruriens* L.; *Dolichos pruriens* L.; *Mucuna aterrima* (Piper & Tracy) Merr.; *Mucuna esquirolii* H. Lév.; *Mucuna prurita* Hook.; *Mucuna prurita* Wight; *Mucuna prurita* Wight; *Stizolobium niveum* (Roxb.) Kuntze; *Stizolobium pruriens* (L.) Medik.; *Stizolobium pruritum* (Wight) Piper

Local name: Alkushi

Common names: Velvet bean; cowitch

Botanical description: It is a somewhat sinister climber that grows up to 5 m long. The stems are hairy. The leaves are imparipinnate, alternate, and stipulate. The stipules are 5 mm long. The petiole is 2–4 cm long. The blade presents a pair of folioles plus a terminal one. The folioles are 4.8–19 cm × 3.5–16.5 cm, membranous, asymmetrical, elliptic, acute, or acuminate at apex and hairy. The inflorescences are axillary racemes that grow up to 30 cm long. The calyx is 7.5–9 mm long, hairy, campanulate and 5-lobed. The corolla is papilionaceous, with a beautiful deep purple coloration, and up to 3.8 cm long. The fruits are 5–6.3 cm long, somewhat S-shaped, furry, and contain 5–6-dark, glossy, elliptic seeds. The hairs provoke an intense and extremely unpleasant sensation of itchiness.

Mucuna pruriens

Medicinal uses: spermatorrhea, post-partum vaginal pain, cholera, intestinal worms, diuretic, diarrhea, leucorrhea

Pharmacology: Extracts of the plant increased male fertility,[371] and displayed anti-epileptic,[372] and hypoglycemic[373] effects.

35.20 *Pterocarpus santalinus* L.f.

Synonym: *Lingoum santalinum* (L.f.) Kuntze

Local name: Rukta-chandana

Common name: Red sandalwood

Botanical description: It is a tree that grows up to about 10 m tall. The bark is blackish-brown that yields a red sap upon incision. The wood is very hard. The leaves are imparipinnate, alternate, and stipulate. The blade comprises 3 folioles and is10–18 cm long. The folioles are 3.8–7.6 cm long, broadly ovate coriaceous, base round, apex emarginated, coriaceous, and glossy. The inflorescences are racemes, which are axillary or terminal. The calyx is about 5 mm long and 5-lobed. The corolla is papilionaceous, about 1 cm long, yellow, made of 5 petals and a standard which is ovate. The fruits are flat pods, which are 3.8–5 cm across, winged, and contain1 or 2 seeds.

Medicinal uses: hemorrhages, scabies, skin diseases, infected eyes

Pharmacology: The plant produces savinin, which is anti-inflammatory.[374] Extracts of the plant demonstrated anti-diabetic[375,376] and cytotoxic[377] effects.

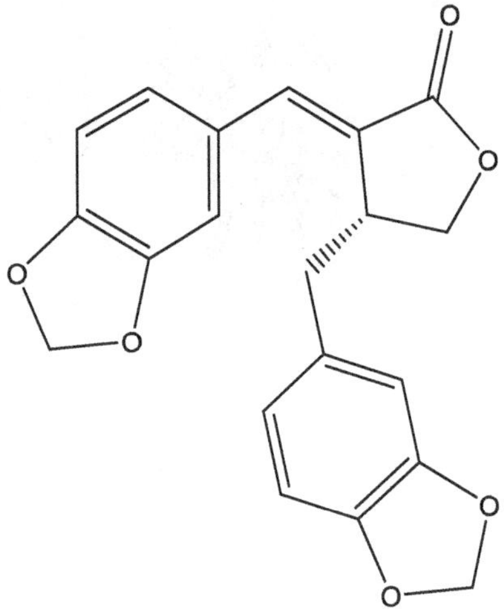

Structure of savinin

35.21 *Saraca asoca* (Roxb.) De Wilde

Synonyms: *Jonesia asoca* Roxb.; *Saraca indica* L.

Local names: Asoke; ashoka

Common name: Asoka tree

Botanical description: It is a tree that grows to about 8 m tall. The leaves are paripinnate, alternate, and stipulate. The stipules are small and caduceus. The blade comprises 4–6 pairs of folioles, which are 10–20 cm long, often pendulous, oblong, lanceolate, glossy, and coriaceous. The inflorescences are axillary corymbs. The calyx is tubular, up to about 1.5 cm long, orange, and produces 4 round lobes. The corolla is absent. The androecium includes 4–8 stamens, which are coming out from the calyx and up to about 2 cm long. The ovary is stipitate and forms a style, which is curved and up to 1.3 cm long. The fruits are pods, which are flat, dark green to black, 15–20 cm long, dehiscent with 4–8 seeds.

Medicinal uses: increases fertility, tonic, leucorrhea

Pharmacology: The plant has been the subject of numerous pharmacological studies.[378] Extracts of the plant exhibited cardioprotective[379] and antibacterial[380] properties.

35.22 *Sesbania sesban* (L.) Merr.

Synonyms: *Aeschynomene sesban* L.; *Emerus sesban* (L.) Kuntze; *Sesbania aegyptiaca* Pers; *Sesbania aegyptiaca* Poir.

Local name: Joyonti

Common name: Common sesban

Botanical description: It is a shrub that grows to about 2 m in height. The leaves are compound, alternate, and stipulate. The stipules are lanceolate and about 5 mm long. The rachis is 4–10 cm long. The blade comprises 10–20 pairs of folioles which are 1.3–2.5 cm × 3–4 mm, rounded at base and apex. The inflorescences are racemes, which are 8–10 cm long. The calyx is campanulate and develops 5 triangular teeth. The corolla is papilionaceous and develops 5 petals which are yellow. The standard is elliptic and 1.3 cm long. The androecium comprises 10 stamens jointed in a tube which is about 1 cm long. The ovary is about 5 mm long with a globose stigma. The fruits are linear and spiral pods that are 15–20 cm long and contain numerous minute seeds.

Medicinal use: boils

Pharmacology: Extract of the plant exhibited anti-cancer properties.[381]

35.23 *Tephrosia purpurea* (L.) Pers.

Synonyms: *Cracca leptostachya* (DC.) Rusby; *Cracca purpurea* L.; *Cracca wallichii* (Graham ex Fawcett & Rendle) Rydb.; *Galega diffusa* Roxb.; *Galega purpurea* (L.) L.; *Glycyrrhiza mairei* H. Lév.; *Tephrosia crassa* Bojer ex Baker; *Tephrosia diffusa* Wight & Arn.; *Tephrosia hamiltonii* J.R. Drummond; *Tephrosia indigofera* Bertol.; *Tephrosia ionophlebia* Hayata; *Tephrosia lanceaefolia* Link; *Tephrosia leptostachya* DC.; *Tephrosia pumila* (Lam.) Pers.; *Tephrosia tenella* A. Gray; *Tephrosia wallichii* Graham ex Fawcett & Rendle

Local name: Jongli-niil

Common name: Bastard indigo

Botanical description: It is an erect or spreading herb that grows about 50 cm tall. The leaves are imparipinnate, spiral, and stipulate. The blade includes 5–10 pairs of folioles. The rachis is 7–15 cm long. The folioles are oblong-elliptic, 1.5–3.5 cm × 0.4–1.5 cm, with 7–12 pairs of secondary nerves, acute at base, and obtuse or round at apex. The inflorescences are axillary or terminal racemes, which are 10–15 cm long. The flowers are 8 mm long. The calyx is about 5 mm long with 5 obscure lobes. The corolla is papilionaceous and comprises 5 petals, which are purple and includes an orbicular standard. The androecium includes 10 stamens. The ovary is hairy. The fruit is a pod, which is 3–5 cm × 3.5–4 mm, hairy, and contains minute ellipsoid seeds.

Medicinal uses: fever, gonorrhea, boils, intestinal worms

Pharmacology: Extracts of the plant exhibited hepatoprotective[382] and anti-carcinogenic[383] properties. The plant produces anti-diabetic flavonoids.[384]

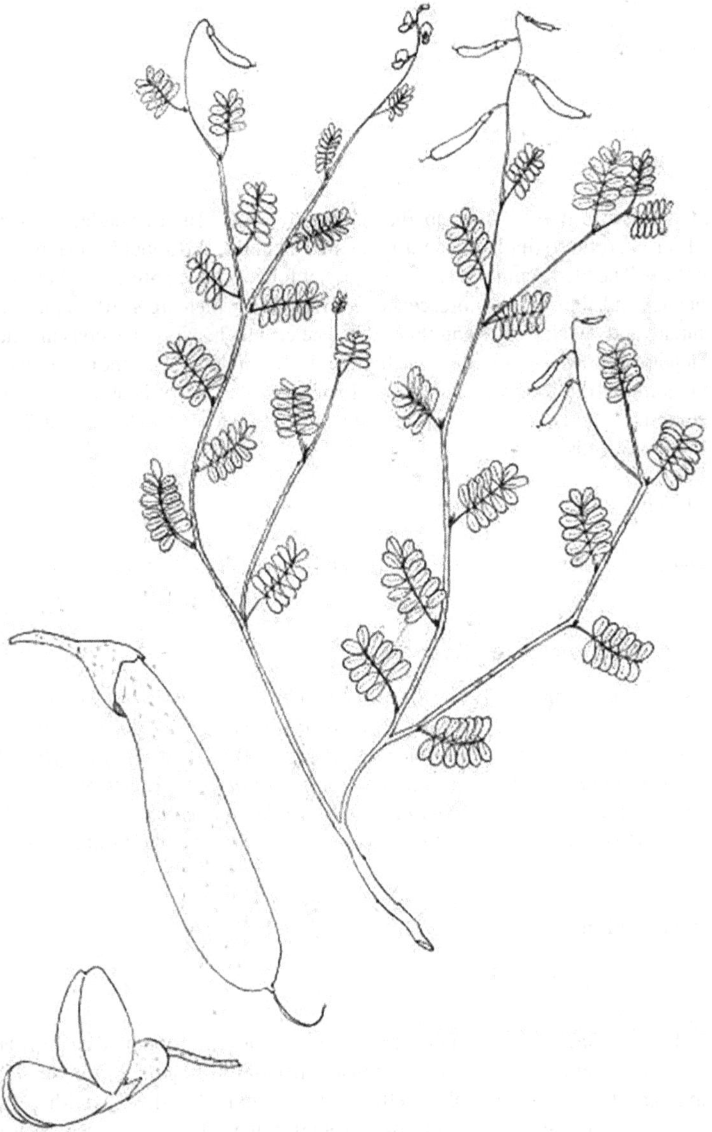

Tephrosia purpurea

36 Family Flacourtiaceae Richard ex A.P. de Candolle 1824

36.1 *Flacourtia indica* (Burm.f.) Merr.

Synonyms: *Flacourtia parvifolia* Merr.; *Flacourtia ramontchi* L'Hér.

Local name: Bainchi

Common name: Governor's plum

Botanical description: It is a treelet that grows up to 5 m tall. The stems present axillary spines and are hairy. The leaves are simple, alternate, and stipulate. The petiole is red, short, 3–5 mm, and hairy. The blade is ovate, oblong-obovate, 2–4 cm × 1.5–3 cm, with 5–7 pairs of secondary nerves, thickly papery, margin serrate, and the apex and base acute. The inflorescences are axillary and racemose. The calyx comprises 5–6 sepals, which are minute and hairy. The androecium includes numerous stamens. The corolla is absent. The ovary is minute, on a disc, glabrous, and produces 5–6 styles. The fruits are globose berries, which are dull to blackish red about 1 cm across and containing 5–6 seeds.

Medicinal uses: cholera, jaundice, wound

Pharmacology: Extract of the plant demonstrated cytotoxic[385] effects. The plant produces anti-malarial phenolics.[386]

36.2 *Flacourtia jangomas* (Lour.) Raeusch

Synonyms: *Flacourtia cataphracta* Roxb. ex Willd.; *Roumea jangomas* Spreng.; *Stigmarota jangomas* Lour.; *Xylosma borneense* Ridley

Local name: Bainchi

Common name: Puneala plum

Botanical description: It is a tree that grows to about 10 m in height. The stems are spiny. The leaves are simple, alternate, and stipulate. The stipules are minute. The petiole is up to 8 cm long. The blade is ovate-oblong, 7–14 cm × 2–5 cm, with 3–6 pairs of secondary nerves, base acute, margin serrate, and apex obtuse. The inflorescences are axillary, racemose, and up to 5 cm long. The calyx comprises 4 or 5 minute sepals. The corolla is absent. The androecium is showy and includes numerous stamens. The ovary is bottle-shaped, minute, and develops 4–6 styles. The fruits are brownish red or purple, blackish, globose, berries, which are 1.5–2.5 cm in diameter, and contain 4 or 5 seeds.

Medicinal uses: ringworm, diuretic, digestive, appetizer, spleen enlargement, clearing urine

Pharmacology: Extract of the plant demonstrated anti-diabetic activity.[387]

36.3 *Gynocardia odorata* R.Br.

Synonym: *Chaulmoogra odorata* (R.Br.) Roxb.

Local name: Chalmogra

Common names: Chaulmoogra; false gynocardia oil

Botanical description: It is a tree that grows up to 20 m tall. The leaves are simple, alternate, and stipulate. The stipule is minute and caducous. The petiole is 1–3 cm long. The blade is oblong-elliptic, 13–20 cm × 5–10 cm, coriaceous, with 4–8 pairs of secondary nerves , margin somewhat wavy, rounded at base, and acuminate at apex. The inflorescences are cauliflorous corymbs of fragrant flowers. The flower peduncles are 2.5–5 cm long. The calyx comprises 5 oblong sepals, which are 7 mm long. The corolla includes 5 petals, which are yellowish green, oblong, and 1.5–2 cm long. The androecium consists of numerous stamens, which are 1 cm long. The ovary is minute and develops 5 styles. The fruits are woody, globose, leprous-like, dull brown, 8–12 cm in diameter, and contain several seeds, which are ellipsoid and 2.5–3 cm long.

Medicinal uses: intestinal worms, fever, piles, ulcers, bronchitis, diabetes, skin disease, leprosy, rheumatism

Pharmacology: Unknown.

36.4 *Hydnocarpus kurzii* (King) Warb.

Synonym: *Taraktogenos kurzii* King

Local name: Gupto mul

Common name: Chaulmoogra tree

Botanical description: It is a tree that grows up to 25 m in height. The stems are hairy. The leaves simple, alternate, and stipulate. The stipules are triangular and minute. The petiole is 3 cm long. The blade is oblong, 15–28 cm × 4–9 cm, cuneate at base, margin entire, acuminate at apex, and with 5–7 pairs of secondary nerves. The inflorescences are axillary clusters, which are about 2 cm long and hairy. The calyx comprises 4 sepals, which are orbicular and about 7 mm long. The corolla includes 5–9 petals, which are greenish-white and 5 mm long. The androecium includes 15–30 stamens. The ovary is hairy. The fruit is globose, woody, 10 cm in diameter, hairy, and contain numerous seeds, which are about 1.5 cm long.

Medicinal uses: skin diseases, eczema, leprosy

Pharmacology: The oil expressed from the seeds has been used for the treatment of leprosy on account of chaulmoogric acid, which is active against *Mycobacterium leprae*.[388]

36.5 *Xylosma longifolia* Clos

Local name: Katari

Common name: Long-leaved xylosma

Botanical description: It is a tree that grows to about 7 m in height. The stems are spiny. The leaves are simple, alternate, and stipulate. The petiole is 5–8 mm long. The blade is narrowly elliptic, 4–15 cm × 2–7 cm, coriaceous with 7–11 pairs of secondary nerves, cuneate at base, margin serrate, and acuminate at apex. The inflorescences are short-branched clustered racemes. The calyx comprises 4 or 5 minute sepals. The androecium includes numerous stamens, which are showy. A disc is present. The ovary is ovoid, minute, and develops 2–3 styles. The fruit is a berry, which is red, globose, 4–6 mm in diameter, and containing 4 or 5 seeds.

Medicinal uses: gastritis, dysentery

Pharmacology: Extracts of the plant demonstrated anti-fungal effects.[389]

37 Family Gentianaceae A.L. de Jussieu 1789

37.1 *Swertia chirayita* H. Karst.

Local name: Chirota

Common name: Indian gentian

Botanical description: It is an erect herb that grows up to about 1 m tall. The stems are terete. The leaves are simple, sessile, opposite, and exstipulate. The blade is broadly lanceolate, 1.5–10 cm × 0.5–3.8 cm, amplexicaul, margin entire, and acute at apex. The inflorescences are axillary or terminal cymes. The calyx is tubular, minute, and 4-lobed. The corolla is tubular, 4-lobed, and yellowish tinged with purple. The androecium includes 4 stamens, which are about 5-mm long. The ovary is ovoid, minute and develop a capitate stigma. The fruits are about 5 mm long capsules.

Medicinal uses: blood purifier, tonic, gastric pain, diabetes, intestinal worms, liver disorder, fever, leprosy, leukoderma, bronchitis, gonorrhea

Pharmacology: The plant has been the subject of numerous pharmacological investigations.[390] Extracts of the plant exhibited hepatoprotective,[391] hypoglycemic,[392] anthelmintic,[393] and gastroprotective[394] effects.

38 Family Gesneriaceae Richard et A.L. de Jussieu ex A.P. de Candolle 1816

38.1 *Rhynchotechum ellipticum* (Wall. ex D. Dietr.) A. DC.

Synonyms: *Chiliandra obovata* Griff.; *Corysanthera elliptica* Wall. ex D. Dietr.; *Rhynchotechum latifolium* Hook. f. & Thomson ex C.B. Clarke; *Rhynchotechum obovatum* (Griff.) B.L. Burtt

Local names: Manipuri; yenbum

Common name: Taiwan Rhynchotechum

Botanical description: It is an herb that grows up to 2 m long. The stem is hairy. The leaves are simple, opposite, and exstipulate. The petiole is 0.8–5 cm long and hairy. The blade is oblanceolate, dark green, glossy, 9.5–32 cm × 3–10 cm, hairy below, attenuate at base, serrate at margin, and acute to acuminate at apex. The inflorescences are axillary cymes which are 15–70 flowered on a peduncle that is 0.9–4 cm long and hairy peduncles. The calyx is tubular, 5-lobed, hairy, and minute. The corolla is white or tinged pink, membranous, 5–6 mm long, and 5-lobed. The androecium includes 4 minute stamens attached to the corolla tube. The ovary is minute and develops a slender style. The fruits are about 5 mm long white berries.

Medicinal use: indigestion

Pharmacology: Unknown.

39 Family Hydrophyllaceae R. Brown 1817

39.1 *Hydrolea zeylanica* (L.) Vahl

Synonyms: *Beloanthera oppositifolia* Hassk.; *Hydrolea arayatensis* Blanco; *Hydrolea ceilonica* F. Muell.; *Hydrolea inermis* Lour.; *Hydrolea javanica* Blume; *Hydrolea prostrata* Exell; *Hydrolea sansibarica* Gilg; *Nama zeylanica* L.; *Steris aquatica* Burm. f.; *Steris javana* L.; *Steris javanica* L. & Christm.

Hydrolea zeylanica (L.) Vahl

Local names: Kasschra; isha-langula

Common name: Ceylon hydrolea

Botanical description: It is an erect to prostrate herb that grows in rice fields and other humid areas. The stem grows to 50 cm long and rooting at base. The leaves are simple, alternate, and exstipulate. The petiole is minute. The blade is lanceolate, 2–10 cm × 0.5–2.5 cm, acute at base, margin entire, acute at apex, and presents about 7 pairs of secondary nerves. The inflorescences are axillary and terminal clusters. The calyx is tubular, 5-lobed, the lobes are lanceolate, hairy, 3-nerved, and up to 8 mm long. The corolla is tubular, about 5 mm long, 5-lobed, dark dull blue (a blue typical of members of the order Boraginales) and the lobes are ovate. The androecium includes 5 stamens adnate to the corolla tube. The ovary is ovoid and develops 2 styles. The fruit is a capsule, which is ovoid, 5 mm long, dehiscent, and contains numerous minute ovoid seeds.

Medicinal uses: leprosy, wounds, ulcers, diabetes

Pharmacology: Extracts of the plant exhibited anti-ulcer[395] and anthelminthic[396] effects.

40 Family Lamiaceae Martynov 1820

40.1 *Ajuga macrosperma* Wall. ex Benth.

Synonyms: *Ajuga macrosperma* Kudô; *Bulga macrosperma* (Wall. ex Benth.) Kuntze

Local names: Athing-phang; shi-priee; charan tulshi

Botanical description: It is an erect or prostrate herb that grows up to about 50 cm tall. The stems are hairy and quadrangular. The leaves are simple, opposite, and exstipulate. The petiole is 2–5 cm long and hairy. The blade is oblanceolate, 4–10 cm × 1.8–4.5 cm, hairy, base cuneate-decurrent, margin undulate-crenate, apex obtuse to acute, and with about 5 pairs of secondary nerves. The inflorescences are axillary and terminal verticils. The calyx is tubular, 5–6 mm long, and produces 5 triangular lobes. The corolla is blue to purple, tubular, 7–9 mm long, and bilabiate. The androecium includes 4 stamens. The fruits are nutlets.

Medicinal uses: asthma, pneumonia, breathing difficulty, respiratory

Pharmacology: Unknown. The plant contains neo-clerodane diterpenes[397,398]

40.2 *Anisomeles heyneana* Benth.

Local name: Gobhura

Botanical description: It is an erect herb that grows up to about 1 m in height. The stems are quadrangular. The leaves are simple, opposite, and exstipulate. The petiole is 1.5–4.5 cm long. The inflorescences are verticils that are axillary with 4–8 flowers. The calyx is campanulate, minute, and 5-lobed. The corolla is tubular, pinkish to purplish, bilabiate, upper lip elliptical and erect, lower lip about 5 mm long, broad, emarginated with darker strikes. The androecium includes 4 stamens, which are up to 1.2 cm long. The ovary develops a style, which is up to 1.3 cm long. The fruits are minute nutlets.

Pharmacology: The plant produces ovatodiolide that inhibited the growth of *Mycobacterium tuberculosis*.[399]

40.3 *Anisomeles indica* (L.) Kuntze

Synonyms: *Anisomeles ovata* R.Br.; *Epimeredi indicus* (L.) Rothm.; *Marrubium indicum* (L.) Burm. f.; *Nepeta indica* L.

Local names: Harinchi; kukumurta

Common name: Indian catmint

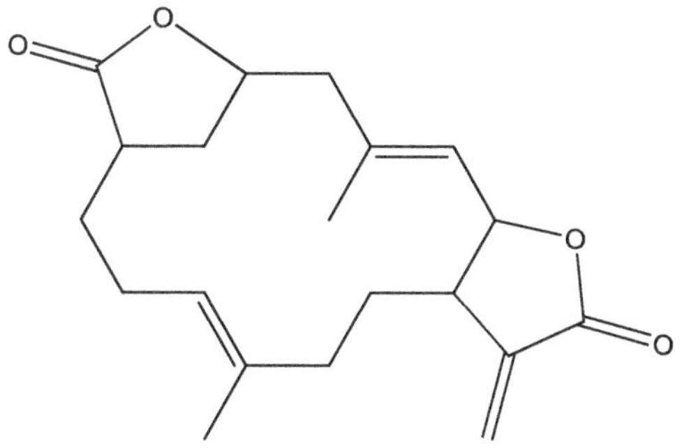

Ovatodiolide

Botanical description: It is an erect herb that grows up to about 1 m in height. The stem is quadrangular and hairy. The leaves are simple, opposite, and exstipulate. The petiole is 1–4.5 cm long. The blade is broadly triangular, 4–9 cm × 2.5–6.5 cm, dull light green, hairy below, broadly truncate-cuneate at base, margin serrate, acute at apex, and with about 5 pairs of secondary nerves. The inflorescences are many-flowered verticils. The calyx is 5-lobed, about 5 mm long, and hairy. The corolla is tubular, purplish, greenish-white, about 1.5 cm long, upper lip oblong, lower lip spreading and emarginated. The androecium includes 4 stamens. The ovary is glabrous and develops a slender style. The fruits are minute, glossy, and black nutlets.

Medicinal uses: fever, sexual impotence

Pharmacology: The plant produces cytotoxic and anti-inflammatory diterpenes[400,401]

40.4 *Hyptis suaveolens* (L.) Poit.

Synonyms: *Ballota suaveolens* L.; *Bystropogon suaveolens* (L.) L'Hér.; *Hyptis congesta* Leonard; *Hyptis ebracteata* R. Br.; *Hypti plumieri* Poit.; *Mesosphaerum suaveolens* (L.) Kuntze; *Schaueria suaveolens* (L.) Hassk.

Local names: Jangulijungol; changa dana

Common names: Horehound; pignut; wild spikenard

Botanical description: It is an aromatic erect herb that grows up to about 1 m in height. The stems are slender, hairy, and quadrangular. The leaves are simple, opposite, and exstipulate. The petiole is up to 3 cm long and hairy. The blade is broadly ovate, hairy 3–7 cm × 2–4 cm, rounded at the base, serrate at margin, acute at the apex, and presents about 5 pairs of secondary nerves. The inflorescences are axillary clusters. The calyx is campanulate about 5 mm long, 5-lobed and hairy. The corolla is tubular, bilabiate, 7 mm long, blue, the upper lip 2-lobed and the lower 3 lobes, the lobes oblong. The androecium comprises 4 stamens. The fruits are minute nutlets.

Medicinal uses: boils, itchiness, fever, sexual impotence

Pharmacology: Extracts of the plant exhibited anti-fungal[402] and anti-nociceptive[403] effects. The plant produces gastroprotective[404] and anti-inflammatory[405] diterpenes.

Anisomeles indica

40.5 *Leonurus sibiricus* L.

Synonym: *Leonurus manshuricus* Yabe

Local names: Dondo-kolosh; dorpie; Raktodrone

Common name: Honeyweed

Botanical description: It is an erect herb that grows up to about 1 m tall. The stems are quadrangular and stiff. The leaves are simple, decussate, sessile and exstipulate. The blade is 3-palmatisect and up to 5 cm long. The inflorescences are verticils, which are many flowered. The calyx tube is tubular, 9 mm long, hairy, obscurely 2-lipped, upper lip straight, 3-toothed, lower lip 2-toothed. The corolla is tubular, subglabrous, bilabiate, purplish-red, about 1.8 cm long, the upper lip oblong, straight, concave, about 1 cm long, the lower lip about 7 mm long and emarginate. The androecium includes 4 stamens. The ovary develops a style that is bifid. The fruits are triquetrous and minute nutlets.

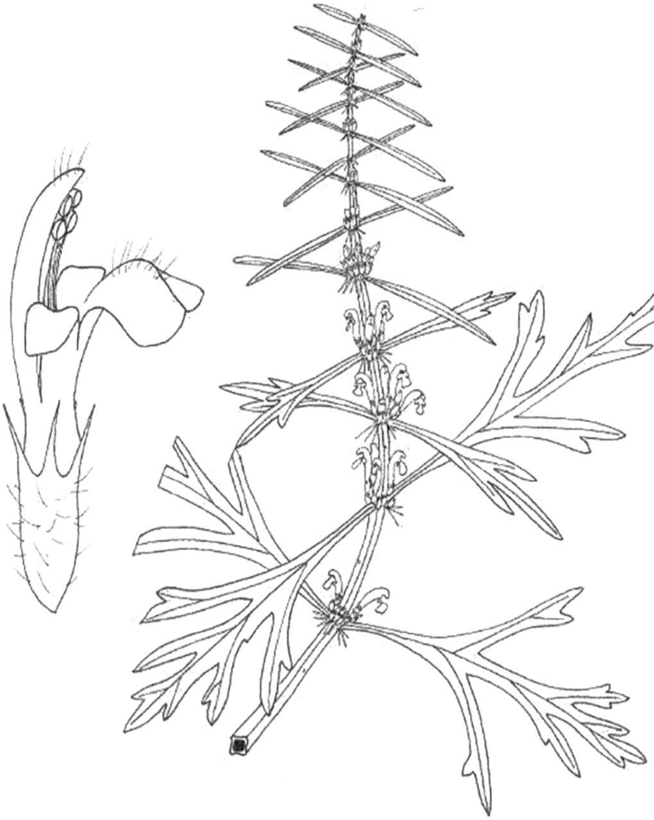

Leonurus sibiricus

Medicinal uses: Fatigue; fever

Pharmacology: Extracts of the plant exhibited anti-inflammatory activity.[406,407] Quercetin from this plant exhibited insulinotropic effects.[408]

40.6 *Leucas aspera* (Willd.) Link

Synonyms: *Leucas plukenetii* (Roth) Spreng.; *Phlomis aspera* Willd.; *Phlomis plukenetii* Roth

Local names: Swetad-rona; chota halkusa

Common name: Common leucas

Botanical description: It is an herb that grows up to 30cm tall. The stem is quadrangular, somewhat stiff, and hairy. The leaves are simple, opposite, and exstipulate. The petiole is minute. The blade is linear-lanceolate, hairy below, serrate, tapering at the base and acute at the apex and presents about 3–6 pairs of secondary nerves. The inflorescences are verticils, which are 16–20 flowered. The calyx is tubular, hairy, about 1 cm long and develops 8–10, irregular but slender lobes. The corolla is 8–10 mm long, pure white, bilabiate, upper lip short, densely hairy, lower lip longer, and deltoid. The fruits are trigonous nutlets which are minute.

Medicinal use: fever

Pharmacology: Extracts of the plant exhibited anti-venom,[409] anti-inflammatory,[410] and anti-nociceptive[411] effects.

40.7 *Ocimum tenuiflorum* L.

Synonyms: *Geniosporum tenuiflorum* Merr.; *Ocimum sanctum* L.

Local names: Tulshi; torshi; tulsi

Common name: Holy basil

Botanical description: It is an erect aromatic herb that grows up to about 1 m tall. The stems are woody at base and hairy at apex. The leaves are simple, opposite, and exstipulate. The petiole is 1–2.5 cm long and hairy. The blade is ovate, 2.5–5.5 cm × 1–3 cm, hairy, cuneate to rounded at base, margin serrate, obtuse at apex, and presents about 4–5 pairs of secondary nerves. The inflorescences are verticils, which are 6-flowered. The calyx is campanulate, minute, hairy, bilobed, upper lip 3-toothed, lower lip 2-toothed, the teeth spinescent. The corolla is tubular, pure white, upper lip oblong and entire, lower lip 3-lobed. The androecium includes 4 stamens, which protrude out of the corolla. The fruits are minute ovoid nutlets.

Medicinal uses: cold, ringworm, wounds, fever, bronchitis, asthma, cough, eczema, skin diseases, measles

Pharmacology: Extracts of the plant demonstrated antimicrobial,[412] angiotensin converting enzyme inhibitor,[413] antiviral,[414] and immunomodulator,[415] activities.

40.8 *Salvia plebeia* R.Br

Synonyms: *Lumnitzera fastigiata* (Roth) Spreng.; *Ocimum fastigiatum* Roth; *Ocimum virgatum* Thunb.; *Salvia brachiata* Roxb.; *Salvia minutiflora* Bunge

Local names: Hath mutha; buthulsi

Common name: Sage weed

Botanical description: It is an erect herb that grows up to 60 cm. The stems are quadrangular. The leaves are simple, opposite, and exstipulate. The petiole is narrowly winged and 5–30 mm long. The blade is oblong to ovate, 2.5–7 mm × 1–3.5 mm, hairy below, attenuated at the base, serrate at margin, and acute at apex, and presents 3–4 pairs of secondary nerves. The inflorescences are verticils, which are 3–8-flowered. The calyx is minute, hairy, bilabiate, upper lip 3-toothed, and the lower lip 2-toothed. The corolla is tubular, bilabiate, pale pink or mauve, about 5 mm long, the upper lip straight and oblong and the lower lip 3-lobed. The fruits are minute and ovoid nutlets.

Medicinal uses: diarrhea, gonorrhea

Pharmacology: Extracts of the plant exhibited anti-inflammatory, anti-angiogenic, anti-nociceptive[416] sedative, and gastroprotective[417] activities. The plant produces homoplantaginin, which is hepatoprotective.[418]

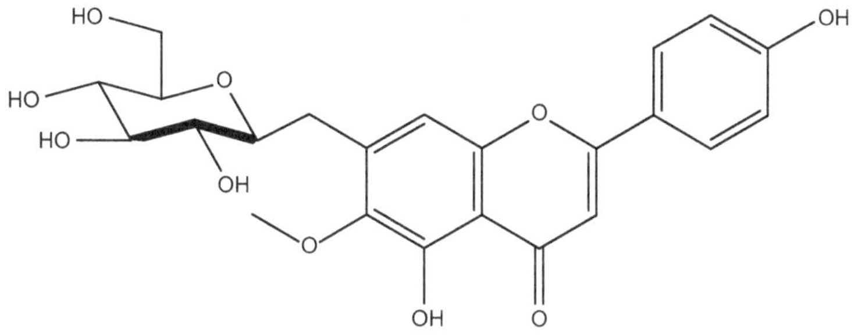

Structure of homoplantaginin.

41 Family Lauraceae A.L. de Jussieu 1789

41.1 *Cinnamomum tamala* (Buch.-Ham.) T. Nees & Nees

Synonyms: *Laurus cassia* Nees & T. Nees; *Laurus tamala* Buch.-Ham.

Local names: Tejpat; tamala

Common name: Indian bay leaf

Botanical description: It is a tree that grows up to 10 m tall. The leaves are simple, spiral, and exstipulate. The petiole is slender and up to 1.5 cm long. The blade is lanceolate, aromatic when crushed, light green, coriaceous, 2.5–8 cm × 7.5–25 cm, acute at base, long acuminate at apex, and presents a pair of secondary nerves originating from the base. The inflorescences are axillary panicles, which are lax and about 10 cm long. The perianth comprises 6 tepals that are oblong, 3–4 mm long, oblong, and yellowish green. The androecium includes 9 stamens. The style is thick, as long as the ovary and develops a stigma, which is small and peltate. The fruits are about 1 cm long, glossy, and green and ellipsoid berries.

Medicinal uses: diarrhea, chicken pox, intestinal worms, tuberculosis

Pharmacology: Extracts of the plant exhibited gastroprotective,[419] anti-diabetic,[420] and antibacterial[421] effects.

41.2 *Litsea glutinosa* (Lour.) C.B. Rob.

Synonym: *Sebifera glutinosa* Lour.

Local names: Kukur chita; moner moton gash; sukujja gash

Common name: Common tallow laurel

Botanical description: It is a tree that grows up to 10 m tall. The stems are hairy. The leaves are simple, spiral, and exstipulate. The petiole is 1–2.5 cm long and hairy. The blade is elliptic-lanceolate, obovate, 3.5–10 cm × 1.5–11 cm, hairy below, with about 10 pairs of secondary nerves, cuneate at base, obtuse at apex, glossy, and coriaceous. The inflorescences are umbels on about 3-cm long peduncles. The perianth includes 4 tepals that are spoon-shaped and about 5 mm long. The androecium comprises about 15 stamens. The ovary is ovoid and develops a slender style. The fruits are globose, about 8 mm in diameter.

Medicinal uses: ringworm, cuts, wounds

Pharmacology: Extracts of the plant exhibited anti-diabetic[422] and antibacterial[423] activites.

42 Family Lecythidaceae A. Richard 1825

42.1 *Barringtonia acutangula* (L.) Gaertn.

Synonym: *Eugenia acutangula* L.

Local name: Hijla daru

Common name: Cut nut

Botanical description: It is a tree that grows up to 8 m tall. The leaves are simple, sessile, spiral, and exstipulate. The blade is obovate, 6–15 cm × 2–6 cm, attenuated at base, acute to acuminate at apex, glossy, and coriaceous. The inflorescences are pendulous racemes that are terminal or axillary and up to 50 cm long. The calyx includes 4 sepals which are, obtuse, and minute. The corolla comprises 4 petals that are elliptic and 1.2 cm long. The androecium includes numerous showy stamens. The ovary produces a 1–2 cm long style. The fruit is ovoid, 2–6 cm × 1–2.5 cm long, woody, quadrangular and contains an ovoid, grooved seed which is up to 4 cm long.

Medicinal use: putrefied wounds

Pharmacology: The plant produces saponins.[424] Extracts of the plant demonstrated antibacterial,[425] anti-nociceptive, anti-diarrheal, and neuropharmacological activities.[426]

42.2 *Barringtonia racemosa* (L.) Spreng

Synonyms: *Barringtonia ceylanica* (Miers) Gardner ex C.B. Clarke; *Barringtonia elongata* Korth.; *Barringtonia timorensis* Blume; *Butonica apiculata* Miers; *Butonica ceylanica* Miers; *Butonica inclyta* Miers; *Butonica terrestris* Miers; *Eugenia racemosa* L.

Local names: Dedaowi; moha shomudro gach

Common name: Fish-killer tree

Botanical description: It is a tree that grows up to 6 m tall. The leaves are simple, spiral, and exstipulate. The petiole is winged and up to 1.5 cm long. The blade is obovate, membranous, glossy, 20–35 cm × 6–14 cm, cuneate at base, serrate at margin, acute at apex, and presents about 15 pairs of secondary nerves. The inflorescences are pendulous, terminal or axillary racemes. The calyx is 2–4-lobed and about 1.2 cm in diameter. The corolla includes 4 green or tinged red or yellow petals, which are oblong, and about 1–1.3 cm × 0.5–0.8 cm. The androecium includes numerous stamens, which are showy, white or pink, and 3–3.5 cm long. The ovary is 2–4-celled, and develops a slender style that is 4–6 cm long. The fruits are 4-angled, 5–9 cm × 3–4 cm, woody, and contain a single seed, which is ovoid and 2–4 cm long.

Medicinal uses: bone fracture, pain, snake bites

Pharmacology: The plant produces bartogenic acid which is anti-inflammatory.[427] Extracts of the plant exhibited anti-nociceptive,[428] antibacterial,[429] and anti-tumor[430] effects.

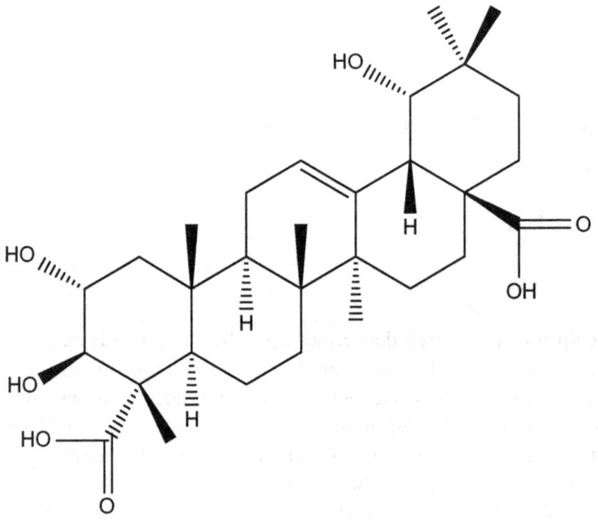

Strucure of bartogenic acid

43 Family Leeaceae Dumortier 1829

43.1 *Leea indica* (Burm. f.) Merr.

Synonyms: *Aquilicia otillis* Gaertn.; *Aquilicia sambucina* L.; *Leea biserrata* Miq.; *Leea celebica* C.B. Clarke; *Leea divaricata* T. & B.; *Leea expansa* Craib; *Leea fuliginosa* Miq.; *Leea gigantea* Griff.; *Leea gracilis* Lauterb.; *Leea longifolia* Merr.; *Leea naumannii* Engl.; *Leea novoguineensis* Val.; *Leea ottilis* (Gaertn.) DC.; *Leea palambanica* Miq.; *Leea pubescens* Zipp. ex Miq.; *Leea ramosii* Merr.; *Leea robusta* Blume; *Leea roehrsiana* Sanders ex Masters; *Leea sambucifolia* Salisb.; *Leea sambucina* (L.) Willd.; *Leea staphylea* Roxb.; *Leea sumatrana* Miq.; *Leea sundaica* Miq.; *Leea umbraculifera* C.B. Clarke; *Leea viridiflora* Planch.; *Staphylea indica* Burm. f.

Local names: Kukur-jhiwa; hashkurobak; hatubhanga; murka; hoti gach; sara gash

Common name: Bandicoot berry

Botanical description: It is a shrub that grows up to 3 m tall. The stems are terete, glossy, and glabrous. The leaves are compound, alternate, and stipulate. The stipules are broadly obovate, 2.5–4.5 cm × 2–3.5 cm, apex rounded, and glabrous. The petiole is 13–23 cm long. The blade is 2- or 3-pinnate and the rachis is 14–30 cm long. The folioles are elongated and elliptical, 6–32 cm × 2.5–8 cm, base rounded, margin serrate, apex acuminate, and with 6–11 pairs of secondary nerves. The inflorescences are opposite to the leaves and corymbose. The calyx is tubular and develops 5 lobes, which are minute and triangular. The corolla includes 5 petals, which are basally united, elliptic, minute, and greenish white. The androecium includes 5 stamens. The ovary is globose and develops a short style. The fruit is a 1 cm in diameter berry containing 4–6 seeds.

Medicinal uses: bone fracture, painful joint, putrefied abscess

Pharmacology: Extracts of the plant exhibited analgesic effects.[431]

43.2 *Leea macrophylla* Roxb.

Synonyms: *Leea angustifolia* P. Lawson; *Leea aspera* Wall. ex G. Don; *Leea cinarea* P. Lawson; *Leea coriacea* P. Lawson; *Leea diffusa* P. Lawson; *Leea integrifolia* Roxb.; *Leea latifolia* Wall. ex Kurz; *Leea pallida* Craib *Leea parallela* Wallich ex Lawson; Leea robusta Roxb.; *Leea simplicifolia* Griff.; *Leea talbotii* King ex Talbot; *Leea venkobarowii* Gamble

Local names: Ash gach; toolsoo moodryia; harmadare

Common name: Large-leaved Leea

Botanical description: It is a shrub that grows 3 m tall. The stems are terete, with longitudinal ridges, and hairy. The leaves are simple or compound, alternate, and stipulate. The stipules are obovate, 4–6 cm × 2–6 cm, and caducous. The petiole is 15–20 cm long and hairy. The blade is broadly ovate, 3-foliolate, or 1–3 pinnate, 40–65 cm × 35–60 cm, base rounded, margin serrate, apex acuminate, hairy, and with 12–15 pairs of secondary nerves. The inflorescences are opposite to leaves and corymbose on a 20–25 cm hairy peduncle. The calyx is tubular, hairy, and develops 5 minute lobes. The corolla includes 5 petals that are basally united, elliptic, greenish white, and about 4 mm long. The androecium includes 5 minute stamens. The ovary is globose and develops a short style and a capitate stigma. The fruit is a berry, which is ovoid, 1.3 cm long, and contains 6 seeds.

Medicinal uses: tonsillitis, tetanus, ringworm, chest pain, wounds

Pharmacology: Extracts of the plant demonstrated anti-inflammatory,[432] renoprotective,[433] and antibacterial[434] properties.

44 Family Liliaceae A.L. de Jussieu 1789

44.1 *Allium cepa* L.

Local names: Piyanj; piaj; palandu

Common name: Onion

Botanical description: It is an herb that grows to about 80 cm tall from a pungent bulb, which is ovoid, solitary, papery and brownish, and glossy. The leaves are simple, entire, linear, terete, and fistulose. The scapes grow to about 1 m tall, terete, and is fistulose. The inflorescences are globose umbels. The perianth includes 6 tepals, which are oblong-ovate, white, and about 4 mm long. The androecium includes 6 stamens. The ovary is sub-globose and minute.

Medicinal uses: antiseptic, cold, diuretic, malaria, rheumatism, headache, blood pressure, vomiting, constipation

Pharmacology: The plant has been subjected to numerous pharmacological investigations.[435] Extracts of the plant demonstrated anti-inflammatory[436] and anti-diabetic[437] effects.

44.2 *Allium sativum* L.

Synonym: *Allium pekinense* Prokhanov

Local names: Rashun; rasun; krachaaipru

Common name: Garlic

Botanical description: It is an herb that grows up to 50 cm tall from a bulb with 6–10 scaly pungent bulblets and covered with a white papery tunic. The leaves are simple, entire, linear, flattened, and form sheaths. The scapes are 25–50 cm, longer than the leaves, and terete. The inflorescences are globose umbels. The perianth includes 6 tepals, which are lanceolate, acuminate, whitish, and about 4 mm long. The androecium includes 6 stamens. The ovary is globose and minute.

Medicinal uses: gastritis, blood pressure, dysentery, rheumatism, tuberculosis, ulcers, leucorrhea, faintness, earache, dysuria, leprosy, whooping cough, tuberculosis, high blood pressure

Pharmacology: The pharmacological properties of this plant have been well studied.[438,439] Of note, the plant has cardioprotective and anti-diabetic effects[440,441] on account of flavonoids and sulfur compounds.[442,443]

44.3 *Asparagus racemosus* Willd.

Synonym: *Protasparagus racemosus* Oberm.

Local name: Sattis chara gach

Common name: Indian asparagus

Botanical description: It is a shrub that grows to about 1.5 m in height from an elongated rhizomes. The stem is erect and develops spiny, woody, and sharp leaf spurs that are straight, 1.5–2 cm on main stems, and 5–10 mm on branches. The cladodes are arranged in fascicles of 3–6, are linear, 1–2.5 cm long, and flat. The inflorescences are up to 4 cm long clusters. The perianth includes 6 tepals, which are minute and white. The androecium includes 6 stamens, which are minute and white. The ovary is obovoid, minute, and somewhat yellowish and develops a trilobed stigma. The fruits are berries, which are red, glossy, about 8 mm in diameter, and trilobed.

Medicinal uses: stomach disorder, diabetes, leucorrhea, vaginitis, dysentery, hematemesis

Pharmacology: The plant has been the subject of numerous pharmacological studies.[444] Extracts of the plant demonstrated anti-depressant,[445] anti-inflammatory,[446] anti-diabetic,[447] and antibacterial properties.[448]

45 Family Lythraceae Jaume Saint-Hilaire 1805

45.1 *Lawsonia inermis* L.

Synonyms: *Alkanna spinosa* Gaertn.; *Lawsonia alba* Lam.; *Lawsonia speciosa* L.; *Lawsonia spinosa* L.; *Rotantha combretoides* Baker

Local names: Mehedi; methi; mendi

Common name: Henna

Botanical description: It is a slender shrub that grows up to 2 m tall. The stems are smooth and brownish red. The leaves are simple, subsessile, opposite, and exstipulate. The blades are elliptic, 0.8–4.5 cm × 0.4–2.2 cm, dull green, tapering at base, and round at the apex. The inflorescences are up to 25 cm long terminal panicles. The calyx includes 4 sepals that are ovate. The corolla includes 4 petals, which are about 5 mm long and white. The androecium includes 8 stamens, which are about 4 mm long. The ovary is obovate and minute. The fruits are capsules, which are about 7 mm long, globose dehiscent, and contain minute seeds.

Medicinal uses: dandruff, skin diseases, jaundice, enlarged spleen, wound, burning sensation, dandruff and footsore, leucorrhea

Pharmacology: The plant has been the subject of numerous pharmacological investigations.[449] It contains the dye lawsone which is antimicrobial[450] and hepatoprotective.[451] Extracts of the plant exhibited anti-inflammatory, anti-pyretic, and analgesic effects.[452]

45.2 *Sonneratia apetala* Buch.-Ham.

Local name: Kerpa

Common name: Mangrove apple

Botanical description: It is a mangrove tree that grows about 8 m tall and developing pneumatophores, which reach about 1 m tall. The leaves are simple, opposite, and exstipulate. The petiole is 5–10 mm long. The blade is elliptic to spathulate, woody, dull green, 5–13 cm × 1.5–4 cm, attenuate at base, and obtuse or rounded at apex. The inflorescences are terminal cymes. The calyx comprises 4 woody, greening, bullet-shaped sepals, which are about 1.5 cm long. There are no petals. The androecium includes numerous stamens that are white and showy. The ovary develops a peltate stigma. The fruits are globose, woody, dull light green, and flattened, and contain numerous U-shaped seeds in a pulp.

Medicinal use: leprosy

Pharmacology: Extract of the plant demonstrated anti-diabetic and antibacterial activities.[453]

46 Family Malpighiaceae A.L. de Jussieu 1789

46.1 *Hiptage benghalensis* (L.) Kurz

Synonyms: *Banisteria benghalensis* L.; *Gaertnera racemosa* (Cav.) Roxb.; *Hiptage madablota Gaertn.*

Local name: Madhobi lota

Common name: Hiptage

Botanical description: It is a stout climber that grows to about 7 m long in length. The stems are hairy. The leaves are simple and opposite. The petiole is 5–10 mm and channeled above. The blade is coriaceous, elliptic-oblong, 9–18 cm × 3–7 cm, broadly cuneate at base, acuminate at apex and with 6 or 7 pairs of secondary nerves. The inflorescences are axillary or terminal, and about 5–10 cm long racemes of fragrant flowers. The calyx comprises 5 sepals which are ovate, 5–6 mm long, and hairy. The corolla includes 5 petals that are white, yellow or pink at base, orbicular, 8–15 mm × 5–10 mm, and membranous. The androecium includes 10 stamens, which are up to 1.3 cm long. The ovary develops a style, which is about 1 cm long and circinate. The fruits are samaras which are 3.5–5 cm × 1–1.6 cm.

Medicinal uses: gastrointestinal tract, skin infection, leprosy

Pharmacology: Extracts of the plant demonstrated anti-inflammatory[454] and insecticidal[455] properties.

47 Family Malvaceae A.L. de Jussieu 1789

47.1 *Abelmoschus moschatus* Medik.

Synonyms: *Abelmoschus abelmoschus* (L.) H. Karst.; *Abelmoschus officinalis* Endl.; *Hibiscus abelmoschus* (L.) H. Karst.; *Hibiscus abelmoschus* L.; *Hibiscus chinensis* Roxb.; *Hibiscus collinsianus* Nutt. ex Torr. & A. Gray; *Hibiscus moschatus* (Medik.) Salisb.

Local name: Mushak-dana

Common name: Musk mallow

Botanical description: It is an erect herb that grows up to 2 m tall. The stems are terete and hairy. The leaves are simple, spiral, and stipulate. The stipules are linear and 6–12 mm long. The petiole is straight, 2–30 cm long and hairy. The blade is up to about 20 cm long, producing 3–7 lobes, which are deltoid and serrate. The flowers are axillary, solitary, and on 2–8 cm long peduncles. The calyx is tubular and 2–3.5 cm long. The corolla comprises 5 membranous petals, which are yellow, 7–9 cm long, and purple at base. The androecium includes numerous stamens joined in a staminal column, which is 1.5–2 cm long. The fruit is a bullet-shaped, coriaceous, 5–8 cm long, dehiscent capsule, containing numerous globose seeds.

Medicinal uses: indigestion, snake bites

Pharmacology: Extracts of the plant exhibited anti-oxidant and cytotoxic properties.[456] Anti-diabetic activity was reported to be on account of myricetin.[457]

47.2 *Abutilon indicum* (L.) Sweet

Synonyms: *Abutilon asiaticum* (L.) Sweet; *Abutilon badium* S.A. Husain & Baquar; *Abutilon cavaleriei* H. Lév.; *Abutilon cysticarpum* Hance ex Walp.; *Abutilon populifolium* (Lam.) G. Don; *Abutilon populifolium* (Lam.) Sweet; *Sida asiatica* L.; *Sida indica* L.; *Sida populifolia* Lam.

Local names: Jhumka; bon kapas; petari

Common name: Indian Abutilon

Botanical description: It is an erect herb that grows to about 1 m tall. The leaves are simple, spiral, and stipulate. The stipules are minute. The petiole is 2–4 cm long and hairy. The blade is ovate-orbicular, 3–9 cm × 2.5–7 cm, hairy below, cordate at base, margin serrate, and acute or acuminate at apex. The flowers are solitary, axillary on 4 cm long and hairy peduncles. The calyx is 5-lobed, about 1 cm in diameter, and hairy. The corolla comprises 5 yellow petals that are 7–8 mm long. The androecium includes numerous stamens joined in a staminal column.

The ovary includes numerous carpels. The fruit is cylindrical, includes 15–20 mericarps and numerous minute seeds.

Medicinal uses: abscess, leucorrhea, cuts, diabetes, toothache, boils, fever, chest pain, urethritis

Pharmacology: Extracts of the plant elicited hepatoprotective,[458] hypoglycemic,[459] and larvicidal properties:[460] The plant contains sesquiterpene lactones.[461]

47.3 *Gossypium arboreum* L.

Local name: Nurma

Common name: Bengal cotton; tree cotton

Botanical description: It is a shrub that grows up to 2 m. The stems are hairy. The leaves are simple, spiral, and stipulate. The stipules are linear. The petiole is 1.5–10 cm long. The blade produces 3–5 lobes that are oblong-lanceolate, hairy below, and about 4–10 cm long. The flowers are solitary, axillary, and on a 1.5–2.5 cm long and hairy peduncle. The calyx is tubular, about 5 mm long, and 5-lobed. The corolla comprises 5 yellowish petals, which are 3–5 cm long, purplish at the base, and membranous. The androecium includes numerous stamens joined in a staminal column, which is 1.5–2 cm long. The ovary develops a slender style. The fruits are ovoid capsules, which are about 3 cm long, dehiscent, and contain numerous seeds in embedded in white wool.

Medicinal use: leucorrhea

Pharmacology: The plant contains antibacterial flavonoids.[462,463]

47.4 *Sida acuta* Burm.f.

Synonyms: *Malvastrum carpinifolium* (Medik.) A. Gray; *Malvinda carpinifolia* Medik.; *Sida balbisiana* DC.; *Sida berlandieri* Turcz.; *Sida bodinieri* Gand.; *Sida carpinifolia* Bourg. ex Griseb.; *Sida carpinifolia* L. f.; *Sida chanetii* Gand.; *Sida disticha* Sessé & Moc.; *Sida frutescens* Cav.; *Sida garckeana* Pol.; *Sida lancea* Gand.; *Sida lanceolata* Retz.; *Sida planicaulis* Cav.; *Sida scoparia* Lour.; *Sida spiraeifolia* Link; *Sida stauntoniana* DC.; *Sida stipulata* Cav.; *Sida trivialis* Macfad.; *Sida ulmifolia* Willd.; *Sida vogelii* Hook. f.

Local names: Kureta; sipsedip

Common name: Common wireweed

Botanical description: It is an erect herb that grows up to 1 m tall. The leaves are simple, spiral, and stipulate. The stipules are linear, 4–6 mm long, and persistent. The petiole is 4–6 mm long. The blade is oblong, lanceolate, 2–5 cm × 0.4–1 cm, obtuse at base, margin serrate, and acute at apex. The flowers are axillary on a peduncle which is 4–12 mm long and hairy. The calyx is tubular, 6 mm long and produces 5 lobes. The corolla consists of 5 petals, which are yellow, 6–7 mm long, and membranous. The androecium includes numerous stamens joined in a column, which is 4 mm long. The fruit is a globose schizocarp, which is about 8 mm in diameter and contains minute seeds.

Medicinal uses: headache, fever, dysentery, abscess, rheumatism

Pharmacology: Extracts of the plant exhibited anti-malarial,[464] antimicrobial,[465] and hepatoprotective[466] effects.

47.5 *Urena lobata* L.

Synonyms: *Urena americana* L. f.; *Urena grandiflora* DC.; *Urena reticulata* Cav.; *Urena trilobata* Vell.

Local names: Jangligagra; ban-bhenda; bedijone; bunkra

Common names: Aramina; caesar weed; burr mallow

Botanical description: It is an herb that grows up to 1 m tall. The stems are terete and hairy. The leaves are simple, spiral, and stipulate. The stipules are filiform, minute, and caducous. The petiole is 1–4 cm and hairy. The blades are orbicular to ovate, 4–7 cm × 3–6.5 cm, rounded or nearly cordate at the base, margin serrate, 3-lobed at apex, 5–7 cm × 3–6.5 cm, and hairy. The flowers are solitary and axillary. The calyx is tubular, 5-lobed, and hairy. The corolla comprises 5 membranous petals, which are about 1.5 cm long, pink, and obovate. The androecium includes numerous stamens joined into a column, which is 1.5 cm long. The ovary develops 10 styles, which are hairy or hirsute. The fruit is a flattened globose, about 1 cm in diameter schizocarp, which is spiny.

Medicinal use: wounds

Pharmacology: Extracts of the demonstrated antibacterial[467] and anti-diabetic[468] properties.

48 Family Meliaceae A.L. de Jussieu 1789

48.1 *Azadirachta indica* A. Juss.

Synonyms: *Melia azadirachta* L.; *Melia indica* (A. Juss.) Brandis

Local names: Neem; inkbow; tamakha

Common names: Margosa tree; neem tree

Botanical description: It is a graceful tree that grows to about 10 m tall. The stems are smooth and reddish-brown. The leaves of the tree are imparipinnate, spiral, and exstipulate. The blade comprises about 5–7 pairs of folioles, which are 2.5–7 cm × 1.5–4 cm, falcate and serrate. The inflorescences are axillary panicles, which are lax, light yellowish green, pendulous, and bear few white, sweet-scented little flowers. The calyx comprises 5 sepals, which are obovate and minute. The corolla consists of 5 petals, which are about 6 mm long, spathulate, and white. The androecium includes 10 stamens joined in a staminal tube, which is about 5 mm long. The ovary is sub-globose, and develops a linear style and a trifid stigma. The fruits are olive-shaped, light yellowish green drupe, which are up to about 1.5 cm long and smooth.

Medicinal uses: skin diseases, toothbrush to develop strong teeth, fever, pimples, gastric acidity, chicken pox, smallpox, measles, intestinal worms, eczema, wounds, blood purifier, syphilis, ringworms, lice, scabies, tumors, rheumatism, leprosy, insecticide

Pharmacology: The plant has been the subject of numerous pharmacological studies.[469,470] The plant produces limonoids with anti-diabetic,[471] antibacterial,[472] insecticidal,[473] and anti-inflammatory[474] effects.

48.2 *Melia azedarach* L.

Synonyms: *Melia orientalis* M. Roem.; *Melia toosendan* Siebold & Zucc.

Local names: Sadi rayssia; bokain

Common names: Common bead-tree; Persian lilac

Botanical description: It is a magnificent tree that grows up to 10 m tall. The stems are hairy when young. The leaves are bipinnate, spiral, and exstipulate. The blade grows up to about 50 cm long. The folioles are opposite, elliptic, 2.5–5 cm × 5–19 mm, serrate, acuminate, somewhat dark green, and glossy. The inflorescences are axillary panicles of beautiful, small lilac, sweet-scented, flowers. The calyx is 5–6-lobed and minute. The corolla comprises 5 petals, which are 7–9 mm long, spathulate to oblong and purplish-white. The androecium includes 10 stamens joined in a purplish staminal tube, which is 6–7 mm long. An annular disc is present.

The ovary produces a style, which is 4–5 mm long, and a capitate stigma. The fruit is a berry, which is green to orange, glossy, ovoid, 1.5–2 cm long, and contains 3–6 seeds.

Medicinal uses: intestinal worms, jaundice, smallpox, indigestion, skin diseases

Pharmacology: The plant has been the subject of numerous pharmacological investigations.[475] Extracts of the plant demonstrated hepatoprotective,[476] insecticidal,[477] and antimicrobial[478] effects.

49 Family Menispermaceae A.L. de Jussieu 1789

49.1 *Cissampelos hirsuta* DC.

Synonyms: *Cissampelos pareira* L.; *Cissampelos pareira* sensu Hook. f. & Thoms.

Local names: Akanbindi; niltat

Common name: False pareira root

Botanical description: It is a woody climber that grows to about 5 m in length. The leaves are simple, spiral, and exstipulate. The petiole is hairy. The blade is peltate, rotund, 2.5–12 cm × 2.5–11.5 cm, with 5–7 pairs of secondary nerves, cordate at base, hairy below, papery, and emarginate at apex. The inflorescences are axillary or corymbose cymes of minute flowers. The calyx includes 4 obovate-oblong, hairy sepals. The corolla is tubular, 4-toothed and hairy. The androecium includes 4 stamens, united into a short column. The gynoecium is made of a single carpel, which is hairy and develops a trifid style. The fruit is a drupe, which is about 5 mm long, red, and shelters a horseshoe-shaped seed.

Medicinal uses: malaria, general weakness, diabetes, diuretic

Pharmacology: Extracts of the plant demonstrated gastroprotective, analgesic, anti-arthritic,[479] and anti-inflammatory[480,481] effects.

49.2 *Cocculus hirsutus* (L.) Diels

Synonyms: *Cocculus villosus* DC.; *Menispermum hirsutum* L.

Local names: Huyer; daikhai; jalajmani

Common name: Broom creeper

Botanical description: It is a climber with hairy stems. The leaves are simple, spiral, and exstipulate. The petiole is 0.5–2.5 cm long. The blade is 4–8 cm × 2.5–6 cm, somewhat deltoid, cordate at the base or truncate, obtuse at apex, hairy below, and presents 2–5 pairs of secondary nerves. The inflorescence is axillary and cymose. The calyx comprises 6 sepals, which are minute and hairy. The corolla includes 6 petals, which are triangular, emarginate and minute. The androecium includes 6 stamens. The gynoecium includes 3 carpels. The fruits are drupes that are dark purple, 4–8 mm long, and contain a horseshoe-shaped seed.

Medicinal use: leucorrhea

Pharmacology: Extracts of the plant exhibited diuretic, laxative,[482,483] and antimicrobial[484] properties.

49.3 *Cyclea barbata* Miers

Synonyms: *Cyclea ciliata* Craib; *Cyclea wallichii* Diels

Local names: Patalpur; wambokhor

Botanical description: It is a climber that grows up to about 5 m in length. The young stems are hairy. The leaves are simple, spiral, and exstipulate. The petiole is 1–5 cm long and hairy. The blade is peltate, 4–10 cm × 2.5–8 cm, hairy below, round or emarginated at base, obtuse at apex, and with about 5 pairs of secondary nerves. The inflorescences are axillary or cauliflorous clusters of minute flowers. The calyx is cupular, minute, and 4–5-lobed. The corolla is cup-shaped, turbinate, and includes 4–5 petals. The ovary is hairy and develops stigmas, which are 3-laciniate. The fruits are red drupes containing horseshoe-shaped seeds.

Medicinal use: allergies

Pharmacology: The plant produces bis-benzylisoquinoline alkaloids,[485] which have cytotoxic and anti-malarial[486] properties. Tetrandrine has anti-inflammatory,[487] antibacterial,[488,489] anti-fungal,[490,491] and antiviral[492] properties.

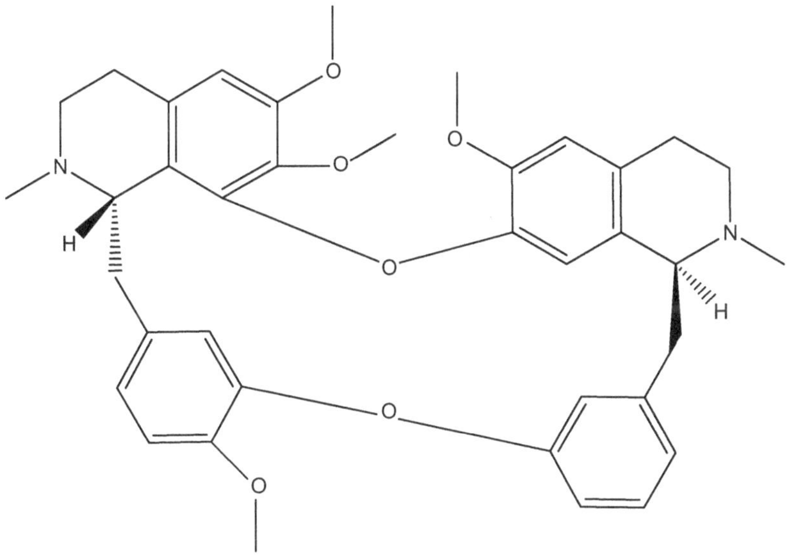

Structure of tetrandrine

49.4 *Stephania japonica* (Thunb.) Miers

Synonym: *Menispermum japonicum* Thunb.

Local names: Akanadi; nimuka; maknadi; thaya nuya

Common name: Snake vine

Botanical description: It is a climber that grows up to about 5 m in length. The leaves are simple, spiral, and exstipulate. The petiole is 3–12 cm long. The blade is peltate, triangular-rotund, 5–12 cm in diameter, glaucous below, glossy above, base rounded, apex acute, and with 8–11 pairs of secondary nerves. The inflorescences are axillary cymes. The calyx includes 3–8 sepals, which are yellowish-green and minute. The corolla includes 3 or 4 petals, which are yellow. The androecium is ovoid, glossy, and develops a stigma that is lacerate. The fruits are drupes, which are red, obovate, 6–8 mm long, and contain horseshoe-shaped seeds.

Medicinal uses: diuretic, piles

Pharmacology: Extracts of the plant exhibited anti-nociceptive[493] and anti-inflammatory[494] activities. The plant contains bis-benzylisquinoline alkaloids increasing the efficacy of cytotoxic agents on cancer cells.[495]

49.5 *Tinospora cordifolia* (Willd.) Miers ex Hook. f. & Thomson

Synonyms: *Menispermum cordifolium* Willd.; *Tinospora malabarica* (Lam.) Hook. f. & Thomson

Local names: Gulancha; gurjalong; vanrui; gulancha-lata; goloncha; dusa sandari; gurach nati

Common name: Heart-leaved moonseed

Botanical description: It is a climber that grows to about 8 m in length. The stems are terete, smooth, and glossy. The leaves are simple, spiral, and exstipulate. The petiole is 5–10 cm long. The blade is cordate, 7.5–13.8 cm × 9–17 cm, with about 3–4 pairs of secondary nerves, and hairy below. The inflorescences are solitary or in clusters and are 7–14 cm long. The calyx includes 6 sepals, which are membranous and about 5 mm long. The corolla includes 6 petals, which are smaller than the sepals. The androecium includes 6 stamens. The gynoecium includes 6 carpels. The fruits are glossy and red drupes, which are ovoid, 6–9 mm long, and contain a horseshoe-shaped seed.

Medicinal uses: gastric acidity, intestinal worms, urinary tract diseases, fatigue, weakness, malaria, galactagogue, diabetes, fever, leucorrhea, jaundice, rheumatism

Pharmacology: Extracts of the plant displayed anti-diabetic,[496] immunomodulatory,[497] and hepatoprotective[498] effects.

50 Family Molluginaceae Bartling 1825

50.1 *Glinus oppositifolius* (L.) Aug. DC.

Synonyms: *Mollugo oppositifolia* L.; *Mollugo spergula* L.

Local names: Gima sask; duserasag

Common name: Bitter leaf

Botanical description: It is an herb that grows up to 40cm long. The stems are hairy when young. The leaves are simple, sessile, opposite or whorled, and exstipulate. The blade is elliptic, 1–2.5cm × 3–6mm, attenuated at the base, laxly serrate, and the apex is obtuse. The inflorescence is a cyme on a peduncle, which is up to about 1.5cm long. The perianth includes 5 tepals, which are greenish-white, oblong, about 5mm long, somewhat membranous and with 3 nerves. The androecium presents 3–5 stamens. The gynoecium develops 3 styles. The fruits are capsular ellipsoid, 3- or 4-valved and contain minute seeds that are reniform.

Medicinal uses: leucorrhea, dysentery, skin diseases, wounds

Pharmacology: The plant produces anti-protozoal saponins.[499] Extracts of the plant exhibited antibacterial effects.[500]

51 Family Moraceae Link 1831

51.1 *Ficus benghalensis* L.

Synonyms: *Ficus indica* L.; *Urostigma benghalense* (L.) Gasp.

Local names: Bar; bot

Common name: Banyan tree

Botanical description: It is a gloomy tree that grows up to 15 m tall, with an anastomosis of branches giving the plant a sinister aspect. The bark is grey and smooth. The leaves are simple, alternate, and stipulate. The stipules are coriaceous, stout, and about 2 cm long. The petiole is 2–6 cm long and hairy. The blade is coriaceous, ovate, 10–20 cm × 8–15 cm, base rounded, apex obtuse, with 4–7 pairs of secondary nerves, dark green, and glossy. The inflorescence is a hypanthodium which is sessile, in axillary pairs, depressed-globose, and about 2.5 cm in diameter. The calyx comprises 2–3 or 3–4 sepals. The androecium presents a single stamen. The ovary develops an elongated style. The fruits are globose, dull red figs, which are 1.5–2.5 cm in diameter.

Medicinal uses: rheumatism, diabetes, diarrhea, dysentery, to promote secretion of milk, toothache, leucorrhea

Pharmacology: Extracts of the plant demonstrated anti-diabetic,[501] antimicrobial,[502] and wound healing[503] activities.

51.2 *Ficus hispida* L.f.

Synonyms: *Covellia hispida* (L. f.) Miq.; *Ficus compressa* S.S. Chang; *Ficus daemonum* K.D. Koenig ex Vahl; *Ficus heterostyla* Merr.; *Ficus letaquii* H. Lév. & Vaniot; *Ficus oppositifolia* Willd.; *Ficus sambucixylon* H. Lév.

Local names: Kuchuli; thupak phang; dumur; kakdumur

Common name: Devil fig

Botanical description: It is a tree that grows up to 8 m tall. The stems are hairy, pithed, and articulate. The leaves are simple, opposite, and stipulate. The stipules are lanceolate, up to 2 cm long, hairy, and caducous. The petiole is up to 10 cm long. The blade is ovate-oblong 10–30 cm × 2.5–20 cm, base obtuse or round, margin serrate, apex acute to acuminate, and with 5–9 pairs of secondary nerves. The inflorescence is a hypanthodium that is obovoid and about 1.5 cm in diameter, green, and lenticellate. The calyx comprises 3 sepals. The androecium includes 1 stamen. The ovary is globose with a hairy style. The fruits are depressed-globose figs, which are 2–3 cm in diameter and dull pale-green.

Medicinal uses: diabetes, leucorrhea, fever, jaundice

Pharmacology: Extracts of the plant demonstrated anti-diarrheal[504] and hepatoprotectory[505] properties. The plant produces hispidacine with vasorelaxant property and hispiloscine with anti-proliferative property.[506]

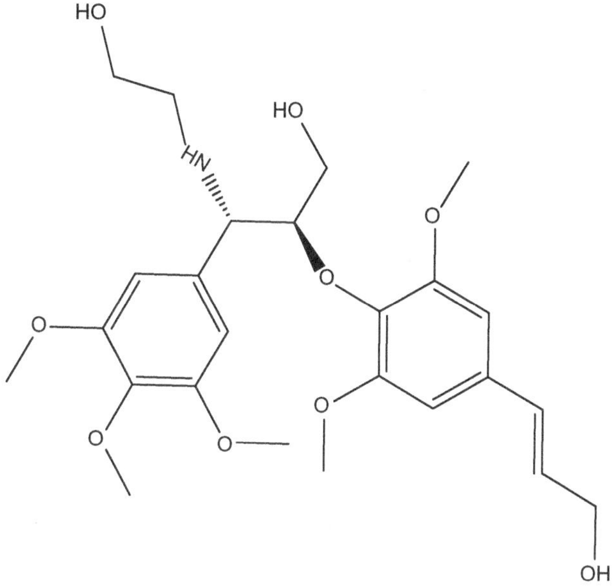

Structure of hispidacine

51.3 *Ficus racemosa* L.

Synonyms: *Covellia glomerata* (Roxb.) Miq.; *Ficus glomerata* Roxb.

Local names: Norpudi tida; jagna dumur; jagnya dumur; dumur; jaggodumur

Common names: Indian fig tree; cluster fig tree

Botanical description: It is a tree that grows up to 10 m tall. The bark is greyish and smooth. The stems are articulate and hairy. The leaves are simple, spiral, and stipulate. The stipules are lanceolate, 1.5–2 cm long, and hairy. The petiole is 2–3 cm long. The blade is elliptic, 10–14 cm × 3–4.5 cm, coriaceous, hairy below, base cuneate to obtuse, apex acuminate to obtuse, and with 4–9 pairs of secondary nerves. The inflorescence is a hypanthodium, which is cauliflorous, pear-shaped, and about 2–2.5 cm in diameter. The calyx includes 4 sepals. The androecium comprises 2 stamens. The ovary develops a slender style and clavate stigma. The fruits are figs, which are red to green, glossy, depressed-globose, and up to 3 cm across.

Medicinal uses: tonsillitis, intestinal worms, leucorrhea

Pharmacology: Extracts of the plant demonstrated anti-diabetic[507,508] and cardioprotective[509] activities.

51.4 *Ficus religiosa* L.

Synonym: *Urostigma religiosum* (L.) Gasp.

Local names: Aswat; Aswathha; pakur; pipul; asath; jil

Common name: Sacred fig tree

Botanical description: It is a graceful tree that grows up to 7 m tall. The bark is grey and smooth. The leaves are simple, spiral, and stipulate. The stipules are ovate. The petiole is 5–10 cm long. The blade is deltoid-ovate, 9–17 cm × 8–12 cm, glossy, cordate at base, membranous, margin entire or wavy, apex caudate, and with 5–10 pairs of secondary nerves. The inflorescence is a hypanthodium , paired or solitary depressed-globose, 1–1.5 cm in diameter, and smooth. The calyx comprises 2–4 sepals. The androecium includes 1 stamen. The ovary is globose and develops a short bilobed style. The fruits are figs, which are depressed-globose, about 1.5 cm in diameter, and dark purple.

Medicinal use: leucorrhea

Pharmacology: Extracts of the plant exhibited anti-diabetic,[510] antiviral,[511,512] and cytotoxic,[513] effects.

51.5 *Ficus scandens* Lam

Synonyms: *Ficus pumila* L.; *Ficus scandens* Roxb.; *Urostigma scandens* (Lam.) Liebm.

Local names: Bera-guarder; lata dumur; shefung

Common name: Creeping fig

Botanical description: It is a climber that grows to about 10 m in length and often covers entire walls or rocks. The stems are woody, very hard, and develop roots. The leaves are simple, alternate, and stipulate. The stipules are hairy and minute. The petiole is up to 2.5 cm long and hairy. The blade is coriaceous, ovate-oblong, 4–10 cm × 2.5–6 cm, cordate at base, entire, obtuse at apex, marginally wavy, hairy below and with 3–6 pairs of secondary nerves. The inflorescence is a hypanthodium, which is solitary and axillary, somewhat obconic, glaucous, and about 2.5–5 cm × 2–3 cm. The calyx includes 4–6 sepals. The androecium presents 2–3 stamens. The ovary develops a slender lateral style. The figs are obconic to tear-shaped, 3.5–7 cm long, and purplish-black.

Medicinal uses: bone fractures, snake bites

Pharmacology: The plant contains flavonoids[514] and an antibacterial and anti-fungal triterpene.[515] Extracts of the plant exhibited anti-inflammatory effects.[516]

51.6 *Streblus asper* Lour.

Synonym: *Diplothorax tonkinensis* Gagnep.

Local names: Shaora; wainghini; sharbo gash; rupashi; sheora; saora; horma

Common name: Siamese rough bush

Botanical description: It is a tree that grows up to about 5 m tall. The stems are hairy. The leaves are simple, alternate, and stipulate. The stipules are minute, linear, and caducous. The petiole is minute. The blade is elliptic, 2.5–6 cm × 1.5–3.5 cm, coriaceous, dark green and glossy, cuneate at base, margin laxly serrate, apex shortly acuminate, and with 4–7 pairs of secondary nerves. The inflorescences are capitate. The calyx comprises 4 sepals. The androecium includes 4 stamens. The ovary is globose and develops a filiform style. The fruits are glossy drupes, which are yellow, globose, and about 1.2 cm long.

Medicinal uses: earache, dysuria, quenches thirst

Pharmacology: Extracts of the plant exhibited filaricidal,[517] antimicrobial,[518] anti-inflammatory,[519] and anti-diabetic[520] activities.

52 Family Moringaceae Martynov 1820

52.1 *Moringa oleifera* Lam.

Synonyms: *Guilandina moringa* L.; *Hyperanthera moringa* (L.) Vahl; *Moringa erecta* Salisb.; *Moringa moringa* (L.) Millsp.; *Moringa pterygosperma* Gaertn.; *Moringa zeylanica* Burmann

Local names: Sajina; mung daru; sazina; dendalum; munga; mungdodare

Common name: Moringa

Botanical description: It is a resinous shrub that grows to about 5 m tall. The bark is light greyish-brown, and smooth. The stems are hairy when young. The leaves are tripinnately imparipinnate, at apex of stems, spiral, and exstipulate. The petiole is 4–15 cm long. The blade is up to about 45 cm long, with 5–11 pinnae and 3–11 dull green folioles, which are 1–1.5 cm × 0.5–1.8 cm, elliptic, edible, and membranous. The inflorescences are axillary panicles, which are 8–30 cm long, and bear white flowers that are 2.5 cm in diameter. The calyx tube is hairy and develops 5 lobes, which are lanceolate and about 1.3–1.5 cm long. The corolla includes 5 petals, which are white, membranous, spathulate with prominent nervation, about 1.2–1.8 cm long, and ephemeral. The androecium includes 5 stamens that are 1 cm long. The ovary is oblong, about 5 mm long and develops a cylindrical style. The fruits are capsular, 9-ribbed, 30–45 cm long, fusiform and contain dark brown seeds, which are globose, about 5 mm in diameter, 3-angled, winged, and which have the property to clean dirty water.

Medicinal uses: sex stimulant, headache, coughs, mucus, indigestion, weakness, blindness, headache, snake-bite, high blood pressure, smallpox

Pharmacology: Extracts of the plant exhibited hypocholesterolemic,[521] hepatoprotective,[522] hypotensive,[523] and antimicrobial[524] properties.

53 Family Musaceae A.L. de Jussieu 1789

53.1 *Musa paradisiaca* L.

Synonym: *Musa sapientum* L.

Local names: Kala; kola; kela; lai fung

Common names: Banana; plantain

Botanical description: It is a fleshy tree which looks like a giant grass that grows up to 8 m tall. The bole is fibrous, fleshy, green, juicy and consists of persistent leaf sheaths. Fibers from the trunk are used in Cebu (Philippines) to make strings by skilled underpaid Filipinos used by Japanese industries. The leaves are simple and spiral. The blade can grow up to about 2 m long, it is oblong, somewhat rectangular, light green, glossy, fleshy, and with a deeply sunken midrib. The inflorescences are spikes, which are stout, about 1 m long, human spine-like, and present burgundy red bracts that are 15–20 cm long, ovate, concave, and fleshy. The inflorescence attracts numerous insects. The perianth is about 2 cm long consists of 5 fused tepals forming a golden yellow lamina. The androecium comprises 5 stamens. The ovary develops a style with a trilobed stigma. The fruits are berries, which are yellow, slightly falcate, oblong, fleshy, 5–15 cm long containing a few seeds in a whitish and delicious pulp which are black, glossy, and about 4 mm across.

Medicinal uses: diarrhea, constipation, sunstroke, hematuria, dysentery

Pharmacology: The plant has been the subject of numerous pharmacological studies.[525] Of note, the plant produces triterpenes with anti-leishmanial properties,[526] and extracts elicited antimicrobial and anti-hyperglycemic activities.[527]

54 Family Myrtaceae A.L. de Jussieu 1789

54.1 *Psidium guajava* L.

Synonyms: *Guajava pyrifera* (L.) Kuntze; *Myrtus guajava* (L.) Kuntze; *Psidium guava* Griseb.; *Psidium guayava* Raddi; *Psidium igatemyense* Barb. Rodr.; *Psidium igatemyensis* Barb. Rodr.; *Psidium pomiferum* L.; *Psidium pumiferum* L.; *Psidium pumilum* Vahl; *Psidium pyriferum* L.

Local names: Peyara; piyara; ombok

Common name: Guava

Botanical description: It is a tree that grows up to 6 m tall. The bark is greyish and peeling. The bole is not straight. The stems are angular and pubescent. The leaves are simple decussate, and exstipulate. The petiole is about 5 mm long. The blade is elliptic, 6–12 cm × 3.5–6 cm, coriaceous, hairy below, with 12–15 pairs of secondary nerves joined into an intramarginal nerve, rounded at base, and acute at apex. The inflorescence is axillary and solitary or in 2–3 flowered cymes. The calyx is tubular and develops 5 lobes. The corolla includes 5 petals, which are white and up to 1.5 cm long. The androecium is showy and includes numerous stamens, which are 6–9 mm long. The ovary develops a slender style and a linear stigma that protrudes out of the corolla. The fruit is a hard, ovoid, 3–8 cm in diameter berry, which is light green, and contain numerous minute seeds in a white, astringent, and delicious pulp.

Medicinal uses: diabetes, contraceptive, stomach ache, dysentery, toothache, vomiting and diarrhea, cold, cough, indigestion

Pharmacology: The plant has been the subject of numerous pharmacological studies.[528] Extracts of the plant exhibited anti-inflammatory and analgesic[529] as well as anti-diabetic and hypotensive activities.[530]

54.2 *Syzygium cumini* (L.) Skeels

Synonyms: *Calyptranthes oneillii Lundell*; *Eugenia cumini* (L.) Druce; *Eugenia jambolana* Lam.; *Myrtus cumini* L.; *Syzygium jambolanum* (Lam.) DC.

Local names: Jam; kalojam; butigajam; guim

Common name: Black plum

Botanical description: It is a tree that grows about 8 m in height. The leaves are simple, opposite, and exstipulate. The petiole is 1–2 cm long. The blade is elliptic to oblong, 6–12 cm × 3.5–7 cm, coriaceous, dark green and glossy, with numerous inconspicuous secondary nerves merged into an intramarginal nerve, base cuneate, and apex rounded

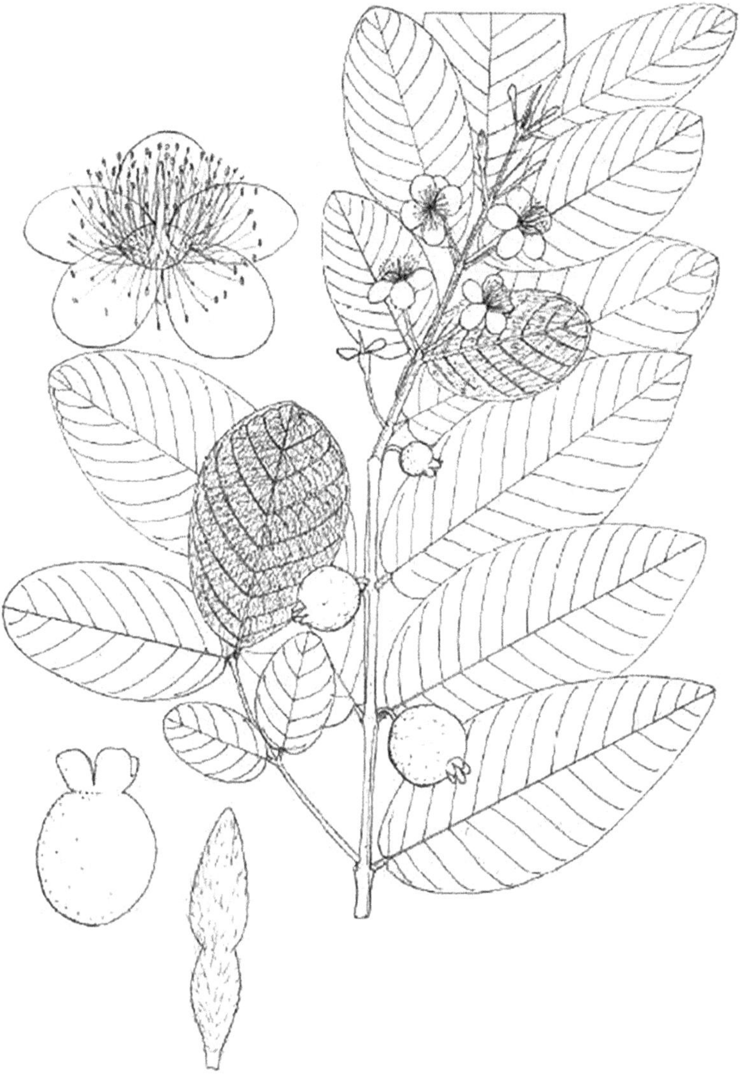

Psidium guajava

to obtuse. The inflorescences are paniculate cymes that are about 10 cm. The calyx is tubular and develops 5 minute lobes. The perianth comprises 5 white petals. The androecium is showy and includes numerous stamens, which are about 5 mm long. The ovary develops a slender style. The fruits are olive-shaped to oblong, globose, glossy, dark blackish-purple, fleshy drupes which are red to black, ellipsoid, and about 1–2 cm long.

Medicinal uses: diabetes, indigestion, dysentery, wound, bronchitis, diabetes

Pharmacology: Extracts of the plant exhibited anti-diabetic,[531] antimicrobial,[532] and gastroprotective[533] effects. The plant produces betulinic acid that is renoprotective.[534]

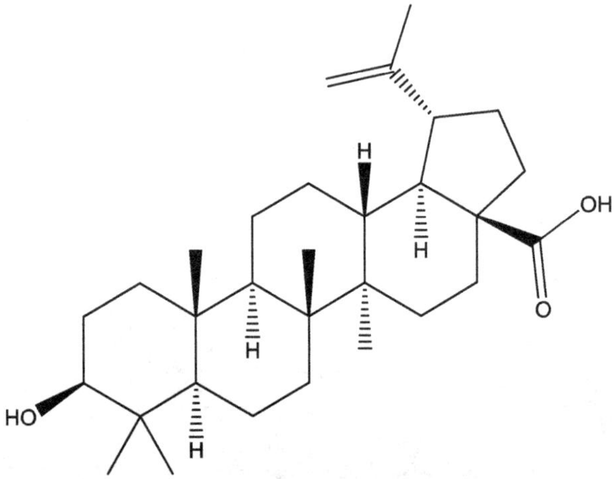

Structure of betulinic acid

54.3 *Syzygium nervosum* DC.

Synonyms: *Calyptranthes mangiferifolia* Hance ex Walp.; *Cleistocalyx cerasoides* (Roxb.) I.M.Turner; (Roxb.) Merr. & L.M. Perry; *Eugenia cerasoides* Roxb.; *Eugenia clausa* C.B. Rob.; *Eugenia divaricatocymosa* Hayata; *Eugenia holtzei* F. Muell.; *Eugenia operculata* Roxb.; *Syzygium angkolanum* Miq.; *Syzygium cerasoides* (Roxb.) Raizada; *Syzygium nodosum* Miq.; *Syzygium operculatum* (Roxb.) Nied.

Local names: Totnopak; rai jam

Common name: Daily river satin-ash

Botanical description: It is a tree that grows about 10 m tall. The stems are flattened and furrowed. The leaves are simple, opposite, and exstipulate. The petiole grows up to about 2 cm long. The blade is oblong spathulate, 10–18 cm × 6.5–7 cm, with 9–13 pairs of secondary nerves, cuneate at base, and acute at apex. The inflorescences are cauliflorous panicles, which grow 10 cm long. The calyx includes 4 sepals that are dome-shaped. The corolla comprises 4 petals, which are minute or obsolete. The androecium includes numerous stamens that are about 6 mm long. The ovary is minute and develops a style, which is about 5 mm long. The fruits are globose, reddish brown, and about 1 cm in diameter.

Medicinal use: dysentery

Pharmacology: Essential oil from this plant exhibited anti-inflammatory activity.[535]

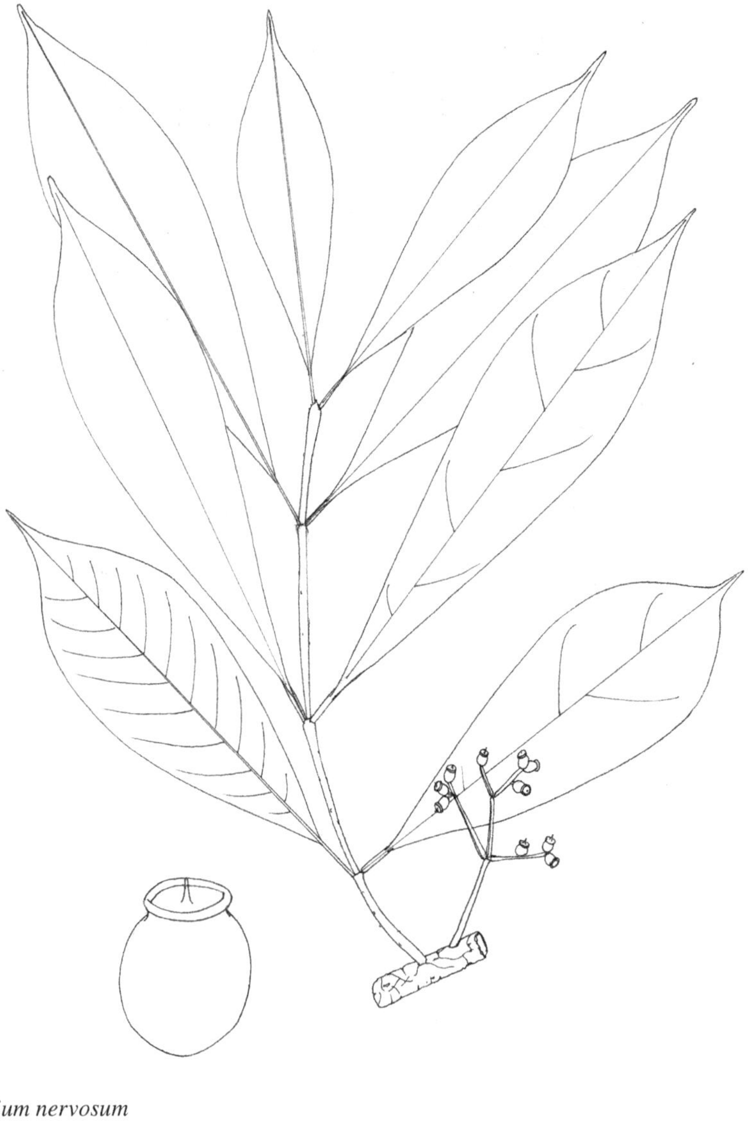

Syzygium nervosum

55 Family Nymphaeaceae R.A. Salisbury 1805

55.1 *Nymphaea nouchali* Burm.f.

Synonym: *Nymphaea stellata* Willd.

Local names: Lal saluki; nil hapla; nil sapla; nil padma

Common name: Blue lotus

Botanical description: It is a magnificent aquatic herb growing from a rhizome. The leaves are simple, alternate, and exstipulate. The petiole is long, terete, fleshy, and slender. The blade is floating, orbicular, sagittate to cordate, 5–20 cm × 10–15 cm, and fleshy. The flowers are graceful, solitary, above the water surface, and 5–15 cm across. The calyx comprises 4 sepals that are up to 3 cm long. The perianth comprises 10–30 petals, which are light blue or yellowish-purple, linear-oblong, and about 10 cm long. The androecium comprises 10–30 stamens, which are about 5 mm long and with blue appendages. The gynoecium includes a syncarpous and many-locular ovary developing 10–20 thorned-shaped stigmas. The fruit is a globose berry, which is about 5 cm in diameter, hard, enclosed by persistent sepals and containing numerous minute seeds.

Medicinal uses: skin infection, dysentery, cardiac problems, blood pressure, diabetes, dyspepsia, diarrhea, piles, diuretic, emollient

Pharmacology: Extracts of the plant demonstrated antibacterial,[536] anti-fungal,[537] anti-diabetic,[538] gastroprotective, anti-inflammatory, and anti-apoptotic effects.[539]

55.2 *Nymphaea pubescens* Willd.

Synonyms: *Nymphaea lotus* var. *pubescens* (Willd.) Hook. f. & Thomson; *Nymphaea rubra* Roxb. ex Salisb.

Local names: Lal saluki; nal; raktahola; lal hapla; lal sapla

Common name: Hairy water lily

Botanical description: It is a beautiful aquatic herb growing from a rhizome. The leaves are simple, alternate, and exstipulate. The petiole is long, terete, fleshy, and slender. The blade is floating, 14–28 cm × 8–26 cm, orbicular, sagittate to cordate, hairy below, serrate at margin, and fleshy. The flowers are graceful, solitary, above the water surface, and 8–15 cm across. The calyx comprises 4 sepals, which are up to 7 cm long and oblong. The perianth comprises numerous petals, which are pure white or light pink, and up to about 7 cm long. The androecium comprises numerous stamens, which are up to 3.5 cm long. The gynoecium includes a syncarpous and many-locular ovary developing 15–30 club-shaped stigmas. The fruit is a globose berry, which is about 5 cm in diameter, hard, enclosed by persistent sepals and containing numerous minute seeds.

Medicinal uses: skin diseases, diarrhea, gynecological diseases, bleedings, skin diseases, diarrhea

Pharmacology: Extracts of the plant exhibited hepatoprotective[540] and anti-diabetic[541] effects.

56 Family Oleaceae Hoffmannsegg et Link 1809

56.1 *Jasminum sambac* (L.) Aiton

Synonym: *Nyctanthes sambac* L.

Local names: Beli; bela; belly; chioy; but moogra

Common names: Arabian jasmine; Hawaiian pikake

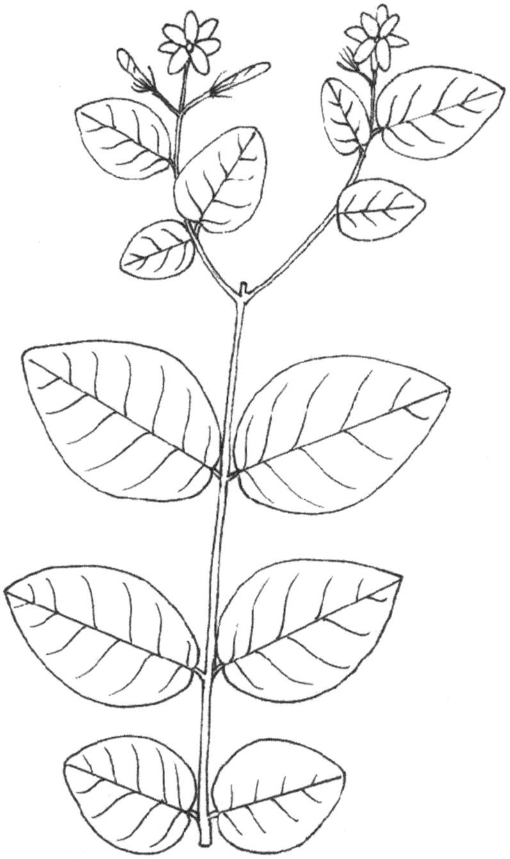

Jasminum sambac

Botanical description: It is a shrub which grows to 3 m tall. The bark is light grey and smooth. The stems are terete. The leaves are simple, opposite, and exstipulate. The petiole is 2 mm long. The blade is entire, elliptic to suborbicular, somewhat coriaceous, obtuse or acute at apex, wavy, up to 9 cm × 6 cm glabrous, and glossy above. The inforescences are few-flowered terminal cymes of fragrant and ephemeral flowers. The calyx comprises 5–9 lobes which are linear about 1 cm long. Corolla is pure white, tubular, with 5–9 lobes which are oblong, and about 1.5 cm long. The fruits are globose, 6 mm in diameter, black berries which are seldom seen.

Medicinal use: asthma

Pharmacology: Extracts of the plant exhibited anti-inflammatory, analgesic, anti-pyretic,[542] and gastroprotective[543] effects.

56.2 *Nyctanthes arbor-tristis* L.

Synonym: *Parilium arbor-tristis* (L.) Gaertn.

Local names: Shefali; shiguri phul gash; shamsihar, sansiari, shinguri, shiwl, seuli; sheuli

Common names: Night jasmine; coral jasmine

Botanical description: It is a tree that grows up to 8 m tall. The stems are angled at the apex. The leaves are simple, opposite, and exstipulate. The petiole is about 1 cm long. The blade is 10 cm × 5 cm, coriaceous, and hairy below. The inflorescences are axillary and terminal cymes, which are few-flowered. The calyx is minute. The corolla is white, tubular, 1 cm long, with about 8 lobes, which are about 7 mm long and with an orange throat. The ovary develops into a cylindrical style and bears a bifid stigma. The fruits are capsules.

Medicinal uses: dysentery, diarrhea, inflammation, fever, cough, cold

Pharmacology: Extracts of the plant exhibited antibacterial,[544] anti-fungal,[545] analgesic, anti-pyretic, ulcerogenic,[546] and anti-inflammatory[547] effects.

57 Family Onagraceae A.L. de Jussieu 1789

57.1 *Ludwigia octovalvis* (N. Jacq.) P.H. Raven

Synonyms: *Jussiaea angustifolia* Lam.; *Jussiaea calycina* C. Presl; *Jussiaea clavata* M.E. Jones; *Jussiaea didymosperma* H. Perrier; *Jussiaea exaltata* Roxb.; *Jussiaea frutescens* J. Jacq. ex DC.; *Jussiaea haenkeana* Steud.; *Jussiaea hirsuta* Mill.; *Jussiaea ligustrifolia* Kunth; *Jussiaea macropoda* C. Presl; *Jussiaea occidentalis* Nutt. ex Torr. & A. Gray; *Jussiaea octofila* DC.; *Jussiaea octonervia* Lam.; *Jussiaea octovalvis* (Jacq.) Sw.; *Jussiaea pubescens* L.; *Jussiaea sagrana* A. Rich.; *Jussiaea suffruticosa* L.; *Jussiaea venosa* C. Presl; *Jussiaea villosa* Lam.; *Ludwigia angustifolia* (Lam.) M. Gómez; *Ludwigia pubescens* (L.) H. Hara; *Ludwigia sagrana* (A. Rich.) M. Gómez; *Ludwigia suffruticosa* (L.) M. Gómez; *Oenothera octovalvis* Jacq.

Local name: Polte pata

Common name: Primrose-willow

Botanical description: It is a water-loving erect herb that grows up to about 50 cm tall. The stems are hairy. The leaves are simple, spiral, and exstipulate. The petiole is about 1 cm long. The blade is narrowly elliptic, 1–14 cm × 0.3–4 cm, with 11–20 pairs of secondary nerves, acute at base, and attenuated at apex. The inflorescence is simple and axillary. The calyx includes 4 sepals that are ovate and up to 1.5 cm long. The corolla includes 4 yellow, obcordate petals, which are up to about 1.5 cm long. The androecium includes 8 stamens. The ovary develops a style, which is minute, and with a 4-lobed stigma. The fruit is a capsule, which is 8-ribbed, terete, up to about 5 cm long and containing numerous minute seeds.

Medicinal use: fever

Pharmacology: The plant produces cytotoxic triterpenes.[548] An extract displayed antibacterial effects.[549]

Ludwigia octovalvis

58 Family Orchidaceae A.L. de Jussieu 1789

58.1 *Aerides odorata* Lour.

Synonym: *Epidendrum odoratum* (Lour.) Poir.

Local name: Pargasa

Common name: Fragrant air plant

Botanical description: It is an epiphytic orchid. The leaves are simple, spiral, and exstipulate. The blade is broadly lorate, 15–20 cm × 2.5–4.6 cm, coriaceous, obtuse, and unequally bilobed. The inflorescence is racemose, 15–30 cm long, pendulous, and bearing 20–30 fragrant, whitish, and tinged pink fragrant flowers that are 1.5–3 cm in diameter. The perianth includes 3 sepals that are elliptic, about 1 cm long, and 2 sepals that are broadly ovate, about 1.2 cm long, and with a labellum which is hooked.

Medicinal uses: arthritis, swelling, wounds, tuberculosis, boils

Pharmacology: Extracts of the plant exhibited antibacterial effects.[550,551]

58.2 *Cymbidium aloifolium* (L.) Sw.

Synonyms: *Cymbidium pendulum* (Roxb.) Sw.; *Cymbidium simulans* Rolfe; *Epidendrum aloifolium* L.; *Epidendrum pendulum* Roxb.

Local names: Surimas; kuntus pata; rashna; auriket; pargasa

Common name: Aloe leaf cymbidium

Botanical description: It is a large epiphytic orchid growing from pseudobulbs that are ovoid and about 10 cm long. The leaves are simple and exstipulate. The blade is linear-oblong, 40–90 cm × 1.5–6 cm, coriaceous, obtuse at apex and unequally 2-lobed. The inflorescence is pendulous, 20–60 cm long, bears numerous brown flowers that are about 2 cm long, and develops from the base of pseudobulb. The perianth comprises 3 tepals, which are narrowly oblong up to 2 cm long, 3 other shorter tepals. The labellum is curved inward and 3-lobed. The fruits are oblong-ellipsoid capsules, which are up to 3 cm long and contain minute seeds.

Medicinal uses: fever, tetanus, asthma, paralysis, cuts, wounds

Pharmacology: Extracts of the plant demonstrated anti-inflammatory effects.[552] The plant produces phenanthrene alkaloids.[553]

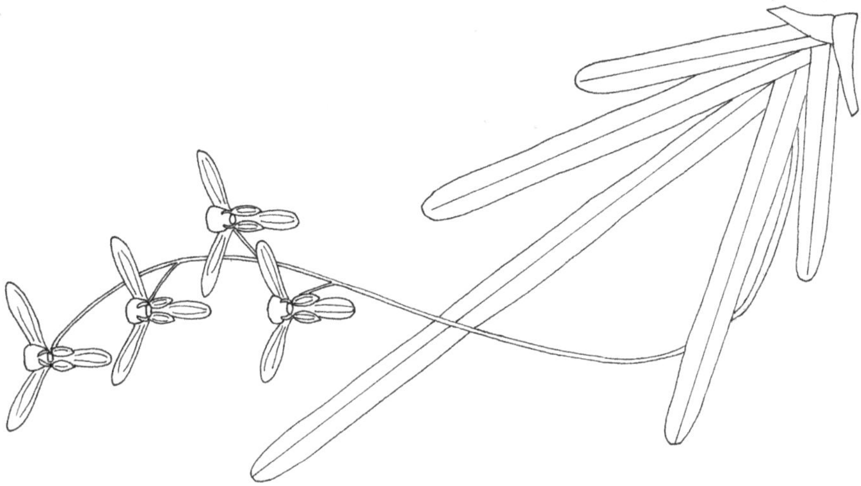

Cymbidium aloifolium

58.3 *Peristylus constrictus* (Lindl.) Lindl.

Synonyms: *Habenaria constricta* (Lindl.) Hook. f.; *Herminium constrictum* Lindl.

Local name: Bhuinora

Common name: Constricted peristylus

Botanical description: It is a terrestrial, erect orchid that grows from tubers to a height of about 80 cm tall. The stem is terete and stout. The leaves are simple, spiral, exstipulate, and form sheaths. The blade is broadly elliptic, 5–13 cm × 3.5–6.5 cm, dull light green with numerous parallel nervations, coriaceous, tapering at base, and acute at apex. The inflorescence is erect, up to about 50 cm long, and many-flowered. The perianth is about 1 cm long, includes 3 tepals that are greenish, and 3 tepals that are pure white. The labellum of is 3-lobed.

Medicinal uses: jaundice, boils

Pharmacology: Unknown.

59 Family Oxalidaceae R. Brown 1818

59.1 *Averrhoa carambola* L.

Synonyms: *Averrhoa acutangula* Stokes; *Averrhoa pentandra* Blanco; *Connaropsis philippica* Fern.-Vill.; *Sarcotheca philippica* (Fern.-Vill.) Hallier f.

Local names: Kordoi; kam-ranga; kamranga

Common names: Carambola; star apple; star fruit

Botanical description: It is a tree that grows up to 5 m tall. The bole is not straight. The leaves are imparipinnate and spiral. The petiole is 2–8 cm long. The blade is 7–25 cm and supports 4–7 sub-opposite pairs of folioles, which are elliptic, glossy, about 3–8 cm × 1.5–4.5 cm, obliquely rounded at base, acute to acuminate at apex, and increasing in size from petiole to the apex of the rachis. The inflorescences are axillary, cymose, reddish, and support numerous, small flowers. The calyx includes 5 sepals that are about 5 mm long. The perianth includes 5 petals, which are light purplish, oblong, recurved, and up to about 1 cm long. The androecium includes 10 stamens. The ovary is hairy and develops 5 styles. The fruit is a yellow, palatable, oblong, deeply 5-ribbed, glossy, and fleshy berry of about 7–13 cm × 5–8 cm, containing numerous minute seeds.

Medicinal uses: jaundice, boils, influenza, fever

Pharmacology: The plant produces hepatoprotective phenolic compounds.[554]

59.2 *Oxalis corniculata* L.

Synonyms: *Oxalis micrantha* Bojko; *Oxalis repens* Thunberg; *Xanthoxalis corniculata* Small

Local names: Amruli; amboti; amilani; amrul

Common name: Creeping wood-sorrel

Botanical description: It is a creeping herb that grows about 50 cm in length. The leaves are compound, alternate, and stipulate. The stipules are minute. The petiole is slender and up to 10 cm long. The blade comprises 3 folioles, which are obcordate, 0.3–1.8 cm × 0.4–2.3 cm, and dull green. The inflorescences are umbellate. The calyx comprises 5 sepals, which are up to about 5 mm long and oblong-lanceolate. The corolla includes 5 petals, which are golden yellow, oblong, and up to about 1 cm long. The fruits are bullet-shaped, vertical capsules, which are up to 2 cm long, pentagonal, dehiscent, hairy, and containing numerous minute seeds.

Medicinal uses: piles, antidote, enteritis, headache, indigestion, diabetes, jaundice, tympanites, infant sickness, dysentery

Pharmacology: Extracts of the plant demonstrated antibacterial,[555] hepatoprotective,[556] and gastroprotective[557] effects.

60 Family Pandanaceae R. Brown 1810

60.1 *Pandanus tectorius* Parkinson

Synonyms: *Pandanus fascicularis* Lam.; *Pandanus odoratissimus* L. f.; *Pandanus tectorius* Sol. ex Balf. f.

Local name: Keya

Common name: Screw pine; beach pandan

Botanical description: It is a tree that grows up to about 4 m tall and develops numerous cylindrical aerial roots. The leaves are simple, spiral, sessile, exstipulate, and at the apex of stems. The blade is linear-ensiform, up to 180 cm × 10 cm, serrate, and coriaceous. The inflorescences are up to 60 cm long and bracteate spikes of minute flowers. The androecium includes 10 stamens. The fruit are ovoid, about 17 cm × 15 cm, yellowish orange, woody, and consists of up to about 50 phalanges which are 5–7 angled.

Medicinal uses: constipation, cold, chicken pox

Pharmacology: The plant produces anti-mycobacterial triterpenes.[558] Extracts of the plant exhibited anti-hyperlipidemic activity.[559]

61 Family Passifloraceae A.L. de Jussieu ex Roussel 1806

61.1 *Passiflora foetida* L.

Synonyms: *Dysosmia foetida* (L.) M. Roem.; *Granadilla foetida* (L.) Gaertn.; *Passiflora hispida* DC. ex Triana & Planch.; *Tripsilina foetida* (L.) Raf.

Local names: Jhumkolata; hurhuna; powmachi

Common name: Fetid passion flower

Botanical description: It is a slender, herbaceous, and slightly malodorant climber. The stems are terete and hairy. The leaves are simple, alternate, and stipulate. The stipules are falcate and laciniated. The petiole is slender, hairy, and 2–4 cm long. The blade is 3-lobed, membranous, cordate at the base, about 4–7 cm × 3–5 cm, and ciliate at margin. The inflorescence is axillary and solitary on up to 4 cm long peduncles. The calyx comprises 5 sepals that are oblong, about 2–2.5 cm long, and white. The corolla comprises 5 petals that are oblong, as long as sepals, obtuse, and mucronate, and white. A multiseriate corona, which is purple to bluish-purple is present. The androecium comprises 5 stamens. The gynoecium includes a globose ovary that is hairy and develops clavate styles. The fruits are ovoid, orange-yellow smooth berries, enclosed by enlarged dissected bracts and about 2 cm in diameter.

Medicinal use: Boils

Pharmacology: The plant accumulates cyanogenic glycosides.[560] Extracts of the plant exhibited analgesic and anti-inflammatory properties.[561] The plant produces the flavonoid ermanin, which is an insect deterrent.[562]

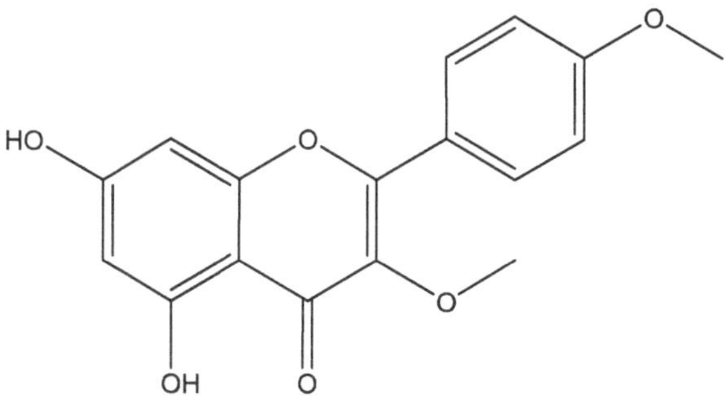

Structure of ermanin

62 Family Phyllanthaceae Martynov 1820

62.1 *Antidesma acidum* Retz.

Synonyms: *Antidesma diandrum* (Roxb.) Roth; *Antidesma diandrum* (Roxb.) Spreng.; *Antidesma wallichianum* C. Presl; *Stilago diandra* Roxb.; *Stilago lanceolaria* Roxb.

Local names: Muta; Elena

Botanical description: It is a shrub that grows up to about 3 m tall. The stems are hairy at the apex. The leaves are simple, alternate, and stipulate. The stipules are linear and up to 1 cm long. The petiole is up to 5 mm long. The blade is obovate to elliptic-oblong, 4–10 cm × 2.5–5 cm, glossy, cuneate at base, rounded at apex, and with 4–9 pairs of secondary nerves. The inflorescences are terminal or axillary, and up to 15 cm long spikes. The flowers are minute. The calyx is 4-lobed. A disc is present. The androecium includes 2 stamens. The ovary is ellipsoid and produces 3 stigmas. The fruits are acidulate drupes, which are ellipsoid, about 5 mm long, reddish-purple, and glossy. The fruits are edible and pleasantly acidic.

Medicinal uses: edema, dysentery, pneumonia, rabies, jaundice

Pharmacology: The plant produces phenolic compounds of which taxifolin is cytotoxic.[563]

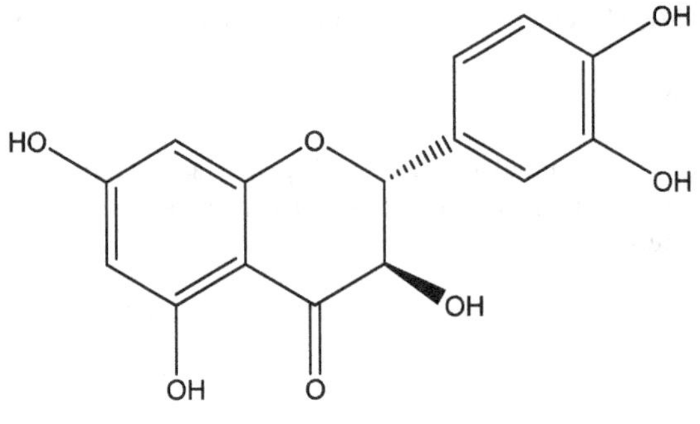

Stucture of taxifoline

62.2 *Antidesma roxburghii* Wall. ex Tul.

Local names: Sadiraissya; Sui Mong

Common name: Roxburgh's Antidesma

Botanical description: It is a shrub, the stems of which are hairy at the apex. The leaves are simple, alternate, and stipulate. The petiole is about 1 cm long. The blade is 5–7 cm × 12–25 cm, obovate, acute at base, acuminate at apex, coriaceous, hairy below, and with 8–10 inconspicuous pairs of secondary nerves. The flowers are axillary and terminal spikes, which are about 15 cm long. The calyx presents 4 lobes. A disc is present. The ovary is minute and hairy. The fruit is ellipsoid and 1 cm long.

Medicinal use: dyspepsia

Pharmacology: Unknown.

62.3 *Bridelia retusa* (L.) A. Juss.

Synonyms: *Bridelia cambodiana* Gagnep.; *Bridelia fordii* Hemsl.; *Bridelia pierrei* Gagnep.; *Bridelia spinosa* (Roxb.) Willd.; *Clutia retusa* L.; *Clutia spinosa* Roxb.

Local names: Shukuja; geio

Common name: Spinous kino tree

Botanical description: It is a tree that grows about 8 m tall. The stems are hairy and present spines. The leaves are simple, alternate, and stipulate. The stipules are minute and triangular. The petiole is up to 1.2 cm long. The blade is obovate, 3–10 cm × 6–25 cm, round at the base, acuminate at apex, and with 16–23 pairs of secondary nerves. The inflorescences are axillary spikes bearing of up to 15 minute flowers. The calyx comprises 5 sepals, which are triangular, velvety, and minute. The corolla includes 5 petals, which are shorter than the petal. A disc is present. The androecium comprises 5 stamens. The ovary is globose and minute and develops 2 styles. The fruits are globose drupes, up to about 1 cm in diameter, dark bluish-black and containing few reddish-brown seeds.

Medicinal uses: skin infection, wounds

Pharmacology: The plant produces anti-fungal phenolics.[564] Extracts of the plant demonstrated anti-inflammatory,[565] analgesic,[566] and hypoglycemic[567] effects.

62.4 *Bridelia stipularis* (L.) Blume

Synonyms: *Bridelia scandens* (Roxb.) Willd.; *Clutia scandens* Roxb.; *Clutia stipularis* L.

Local name: Bangari gach

Common name: Climbing bridelia

Botanical description: It is a shrub that grows about 10 m tall. The leaves are simple, alternate, and stipulate. The stipules are triangular and up to about 1 cm long. The petiole is up to 1.3 cm long. The blade is broadly elliptic, 4–17 cm × 2–11 cm, base obtuse to rounded, apex acute or obtuse, and with 10–14 pairs of secondary nerves. The inflorescences are axillary spikes of minute flowers. The calyx comprises 5 triangular sepals. The corolla includes 5 spathulate petals, which are 3–5 dentate and shorter than calyx lobes. A disc is present. The androecium includes 5 stamens. The ovary is ovoid, with 2 styles and bifid stigmas. The fruits are ovoid drupes, which are 0.7–1.3 cm × 0.6–1.1 cm, green, glossy, and bilocular. The seeds are about 5 mm long and yellowish-brown.

Medicinal use: allergies

Pharmacology: Extracts of the plant displayed antibacterial activity.[568]

62.5 *Glochidion multiloculare* (Rottler ex Willd.) Voigt

Synonyms: *Agyneia multilocularis* Rottler ex Willd.; *Bradleia multilocularis* (Rottler. ex Willd.) Spreng.; *Phyllanthus multilocularis* (Rottler ex Willd.) Mull. Arg.

Local name: Kudurpala

Botanical description: It is a shrub, the stems of which are angular and glossy. The leaves are simple, alternate, and stipulate. The stipules are triangular and minute. The blade is 14–6 cm × 4.5–3 cm, oblong lanceolate, dark green above and glaucous below, cuneate at base and retuse at apex. The inflorescences are slightly pendulous. The calyx comprises 6 sepals in 2 whorls, the outer larger. The androecium comprises 6 stamens united into a column. The ovary is lobed. The fruits are depressed-globose, green, 8–12-lobed, capsules which are 3 cm long and contain numerous seeds.

Medicinal use: diarrhea

Pharmacology: The plant produces triterpenes.[569] Extracts of the plant exhibited anti-inflammatory, analgesic, anti-tumor,[570] and antimicrobial[571] effects.

62.6 *Phyllanthus emblica* L.

Synonyms: *Diasperus emblica* (L.) Kuntze; *Dichelactina nodicaulis* Hance; *Emblica officinalis* Gaertn.; *Phyllanthus mairei* H. Lév.

Local names: Amloki; amlokhi; kada mola; amla; anra; khulu; pyandhum

Common name: Indian gooseberry

Botanical description: It is a tree that grows about 10 m tall. The leaves are simple, subsessile, tightly alternate, and stipulate. The stipules are minute. The blade is linear-oblong, 0.5–1.6 cm × 0.1–0.3 cm. The inflorescences are axillary or cauliflorous. The calyx comprises 6 spathulate sepals, which are minute and whitish. The androecium includes 3 stamens connate into column. A disc is present. The ovary is 6-ribbed, ovoid, and develops 3 styles that are bifid. The fruits are globose, somewhat 3-lobed, 2.5 cm in diameter, fleshy, light green, palatable, glossy, and sheltering a woody and 6-ridged endocarp.

Medicinal uses: insomnia, skin infections, gall pain, leucorrhea, tympanites, urinary tract infection, gastric ulcer, indigestion, intestinal pain, lack of appetite, loss of hair, diseases of mouth, nasal hemorrhage, laxative, jaundice, inflamed eyes, malaria, fever, indigestion

Pharmacology: The plant has been the subject of numerous pharmacological studies.[572] Extracts of the plant demonstrated anti-pyretic, analgesic,[573] cardioprotective,[574] and anti-cancer[575] effects.

62.7 *Phyllanthus niruri* L.

Synonyms: *Diasperus nirur* (L.) Kuntze; *Phyllanthus asperulatus* Hutch.; *Phyllanthus filiformis* Pavon ex Baillon; *Phyllanthus fraternus* G.L. Webster; *Phyllanthus lathyroides* Kunth; *Phyllanthus microphyllus* Mart.

Local name: Bhui amla

Common name: Stonebreaker

Botanical description: It is an invasive herb that grows up to 50 cm tall. The stem is terete, slender, and somewhat lignose. The leaves are simple, alternate, and stipulate. The blade is elliptic, 1.1–2 cm × 4.5–9 mm, rounded at the base, obtuse at apex, dull green and membranous. The inflorescence is axillary. The flowers are minute. The calyx comprises 5 sepals that are green. The androecium includes 3 stamens. The ovary is minute. The fruits are minute, dull light green capsules.

Medicinal uses: fever, infections, dropsy, dysuria, dysentery, gonorrhea, asthma, jaundice, bronchitis

Pharmacology: The plant has been the subject of numerous pharmacological studies.[576] Extracts of the plant exhibited hepatoprotective[577] and hypolipidemic[578] effects. The plant produces the glycoside of a di-cinnamic acid compound, namely niruriside, which evoked anti-HIV properties.[579]

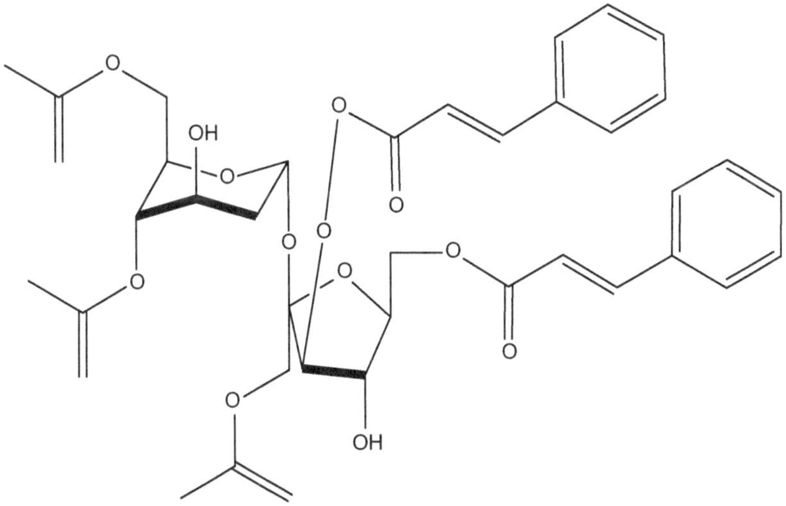

Structure of niruriside

62.8 *Phyllanthus reticulatus* Poir.

Synonyms: *Cicca microcarpa* Benth.; *Cicca reticulata* (Poiret) Kurz; *Glochidion microphyllum* Ridley; *Kirganelia multiflora* Baillon; *Kirganelia reticulata* (Poiret) Baillon; *Kirganelia sinensis* Baillon; *Phyllanthus dalbergioides* Wallich ex J. J. Smith; *Phyllanthus erythrocarpus*

Ridley; *Phyllanthus microcarpus* (Bentham) Müller Argoviensis; *Phyllanthus d multiflorus* Poiret; *Phyllanthus multiflorus* Willd.; *Phyllanthus takaoensis* Hayata.

Local names: Panjuli; panisitki; cirkuti; simikdare

Common name: Black-honey shrub

Botanical description: It is a shrub that grows up to about 3 m tall. The stems are hairy when young. The leaves are simple, alternate, and stipulate. The stipules are lanceolate and minute. The petiole is up to about 5 mm long. The blade is elliptic, 1–5 cm × 0.7–3 cm, membranous, base obtuse, apex acute, and with 5–7 pairs of secondary nerves. The inflorescences are axillary fascicles. The calyx includes 5 or 6 minute sepals. The androecium includes 5 stamens. The ovary develops 3 styles, which are bifid. The fruits are berries, which are globose and about 5 mm in diameter, dark purplish and contain numerous minute seeds.

Medicinal uses: to clean the teeth, malaria

Pharmacology: Extracts of the plant exhibited anti-diabetic,[580] anti-inflammatory, analgesic,[581] and anti-plasmodial[582] activities.

63 Family Plumbaginaceae A.L. de Jussieu 1789

63.1 *Aegialitis rotundifolia* Roxb.

Local name: Banrua

Common name: Club mangrove

Botanical description: It is a seashore shrub which grows up to 3 m tall. The leaves are simple, spiral, and exstipulate. The petiole form sheaths at the stem and is about as long as the blade. The blade is broadly lanceolate-elliptic, fleshy, glossy, 3–9 cm × 2–7 cm with a conspicuous midrib. The flowers are arranged in terminal panicles. The calyx is tubular, about 1 cm long, 5-ribbed, and develops 5 tiny lobes. The corolla is pure white, tubular, 5-lobed, and the lobes are about 1 cm long. The androecium includes 5 stamens. The gynoecium is 0.5 cm long with a capitate stigma. The fruit is a capsule, which is falcate, pentagonal, and up to 5 cm long.

Medicinal use: insect bites

Pharmacology: Extracts exhibited thrombolytic and antibacterial activities.[583,584]

63.2 *Plumbago zeylanica* L.

Synonym: *Plumbago scandens* L.

Local names: Aguni tita; aunitida; chitrak; lalchita; sadachita; chitamul

Common name: Ceylon leadwort

Botanical description: It is an herb that grows up to about 1.5 m tall. The stems are woody at base and striated. The leaves are simple, opposite, and exstipulate. The blade is ovate, 5–10 cm × 2–5 cm, cuneate to obtuse at base, and acuminate and mucronate at apex. The inflorescences are terminal capitate spikes. The calyx is tubular, up to 1.3 cm long, 5-ribbed, covered with glands, and 5-lobed. The corolla is tubular, membranous, with a slender tube that grows up to 2.2 cm long, white to light blue, and developing 5 obovate lobes. The androecium includes 5 stamens with blue anthers. The ovary is 5-angular with filiform stigmas. The fruits are capsular, yellowish-brown, and oblong.

Medicinal uses: abortion, diuretic, fever, scabies, cold, swellings, piles, tumors, dysentery, cuts

Pharmacology: Extracts of the plant demonstrated abortive[585] and wound healing[586] effects. The plant produces the anti-inflammatory naphthoquinone plumbagin.[587]

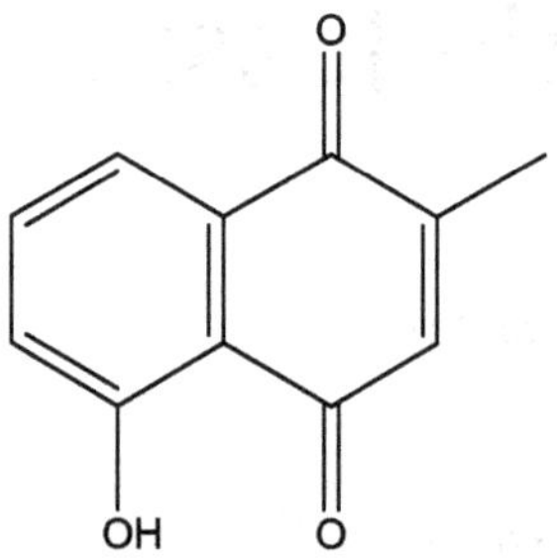

Structure of plumbagin

64 Family Poaceae Barnhart 1895

64.1 *Chrysopogon aciculatus* (Retz.) Trin.

Synonyms: *Andropogon aciculatus* Retz.; *Centrophorum chinense* Trin.

Local name: Chorkanta

Common name: Golden false beardgrass

Botanical description: It is an erect grass that grows up to about 30 cm tall from a rhizome. The leaves form sheaths. A minute ligule is present. The blade is linear, 3–5 cm × 0.4–0.6 cm, and dull green. The inflorescences are lax terminal panicles, which are purplish-brown and grow to about 9 cm long.

Medicinal uses: stomach ache, tonic, chronic fever

Pharmacology: Extracts of the plant demonstrated analgesic properties.[588] The plant produces the C-glycosyl flavonoid aciculatin, which induced apoptosis in cancer cells.[589]

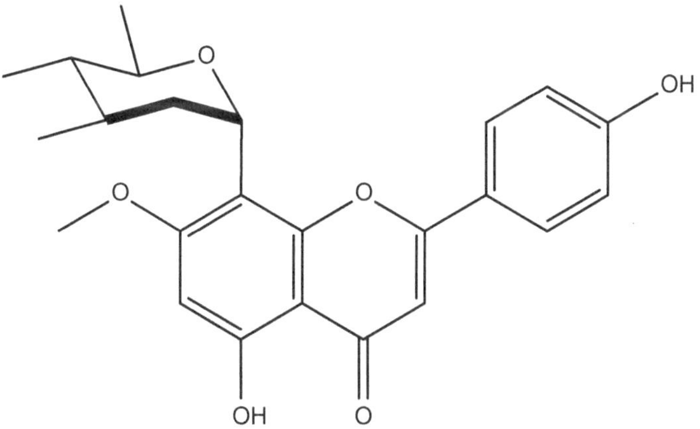

Structure of aciculatin

64.2 *Cynodon dactylon* (L.) Pers.

Synonyms: *Agrostis bermudiana* Tussac ex Kunth; *Agrostis filiformis* J. Koenig ex Kunth; *Capriola dactylon* (L.) Hitchc.; *Capriola dactylon* (L.) Kuntze; *Chloris cynodon* Trin.; *Cynodon aristiglumis* Caro & E.A. Sánchez; *Cynodon aristulatus* Caro & E.A.

Sánchez; *Cynodon erectus* J. Presl; *Cynodon glabratus* Steud.; *Cynodon maritimus* Kunth; *Cynodon occidentalis* Willd. ex Steud.; *Cynodon pascuus* Nees; *Cynodon polevansii* Stent; *Cynodon portoricensis* Willd. ex Steud.; *Cynodon tenuis* Trin. ex Spreng.; *Cynodon umbellatus* (Lam.) Caro; *Cynosurus dactylon* (L.) Pers.; *Cynosurus uniflorus* Walter; *Dactilon officinale* Vill.; *Digitaria dactylon* (L.) Scop.; *Digitaria glumaepatula* (Steud.) Miq.; *Digitaria littoralis* Salisb.; *Digitaria littoralis* Stent; *Digitaria maritima* (Kunth) Spreng.; *Digitaria stolonifera* Schrad.; *Fibichia dactylon* (L.) Beck; *Fibichia umbellata*; *Fibichia umbellata* Koeler; *Milium dactylon* (L.) Moench; *Panicum dactylon* L.; *Panicum glumaepatulum* Steud.; *Paspalum dactylon* (L.) Lam.; *Paspalum umbellatum* Lam.; *Phleum dactylon* Pall. ex Georgi

Local names: Durba ghas; dobigas; doorva; dublo; duglo gach; dupa

Common names: Bermuda grass; Bahama grass

Botanical description: It is an erect grass, which grows up to about 30 cm in height from a rhizome. The stem is somewhat purplish and stoloniferous. The leaves form sheaths. The ligules are hirsute. The blade is linear and about 1–12 cm × 1–4 mm. The inflorescences are terminal clusters of spikes that grow up to 6 cm long.

Cynodon dactylon

Medicinal uses: cough, cold, carminative, pains, inflammation, toothache, cuts, wounds, bleedings, bloody enteritis, dysentery, piles, haematuria, skin infection, to prevent abortion, dysmenorrhea, frequent urination, syphilis

Pharmacology: Extracts of the plant demonstrated antinephrolithoasis,[590] wound healing,[591] styptic,[592] antibacterial,[593] and antiviral[594] effects.

64.3 *Eleusine indica* (L.) Gaertn.

Synonyms: *Cynodon indicus* (L.) Raspail; *Cynosurus indicus* L.

Local names: Sursuri ghas; malkantari

Common name: Indian goose grass

Botanical description: It is a grass that grows up to about 50 cm tall. The leaves form sheaths, which are pilose. The blades are dull green, 10–15 cm × 0.3–0.5 cm, and linear. The inflorescences are 2–7 terminal spikes that are 3–10 cm × 0.3–0.5 cm. The spikelets are elliptic and about 5 mm long.

Medicinal use: snake bites

Pharmacology: The plant produces anti-inflammatory C-glycosyl flavonoids.[595]

64.4 *Vetiveria zizanioides* (L.) Nash

Synonyms: *Agrostis verticillata* Lam.; *Anatherum muricatum* (Retz.) P. Beauv.; *Andropogon muricatus* Retz.; *Andropogon squarrosus* Hook. f.; *Phalaris zizanioides* Linn.; *Sorghum zizanioides* (Linn.) O. Ktze.; *Vetiveria muricata* (Retz.) Griseb.; *Vetiveria odoratissima* Lem.-Lisanc.

Local names: Sirmou; khas khas

Common name: Vetiver

Botanical description: It is a massive tufted grass that grows about 2 m tall from fibrous roots. The leaves form sheaths, which are laxly and up to about 20 cm long. A minute and ciliate ligule is present. The blade is linear and about 30–60 cm × 4–10 mm. The inflorescences are terminal and oblong dark purplish panicles, which are about 20 cm long. Sessile spikelets are about 5 mm long.

Medicinal uses: gastric acidity, dyspepsia, headache, dysuria, headache, vomiting

Pharmacology: The essential oil produced by this plant demonstrated antibacterial[596] and anti-inflammatory activity.[597] The essential oil of the plant is used in perfumery.

65 Family Polygonaceae A.L. de Jussieu 1789

65.1 *Persicaria hydropiper* (L.) Delarbre

Synonym: *Polygonum hydropiper* L.

Local names: Bishkatali; panimarich; chotopanimorich; lalbiskatali

Common names: Annual smartweed; marsh pepper knotweed; mild water pepper

Botanical description: It is an erect herb that grows to about 40 cm in height. The stems are smooth, glossy, fleshy, and somewhat purplish. The leaves are simple, spiral, and stipulate. The stipules are united to form an ochrea, which grows to about 1.5 cm long and develops cilia at apex. The petiole is about 5 mm long. The blade is 1.5–8 cm × 0.4–2 cm, wavy, linear lanceolate, acuminate, and dotted with tiny glands. The inflorescences are 3–7 cm long spikes of minute flowers. The perianth is tubular, white-pinkish, and develops 5 ovate lobes. The androecium includes 8 stamens inserted to the perianth tube. The ovary is trigonous and develops 3 styles with capitate stigmas. The fruits are minute and dark brownish nuts.

Medicinal uses: gout, painful knees, fever, dysentery, tonic, diuretic, bleeding, cough, cold, ulcers, menstrual pain

Pharmacology: Extracts of the plant exhibited antibacterial,[598] anti-nociceptive,[599] and anti-inflammatory[600] activities.

65.2 *Polygonum plebeium* R. Br.

Synonyms: *Polygonum changii* Kitag.; *Polygonum dryandri* Spreng.; *Polygonum parviflorum* Y.L. Chang & S.H. Li; *Polygonum roxburghii* Meisn.

Local names: Chikni-sag; Chemti sak; Raniful

Common name: Small knotweed

Botanical description: It is an herb that grows about 30 cm long. The leaves are simple, spiral, subsessile or sessile, and stipulate. The stipules are united to form an ochrea, which is white, minute, membranous, and lacerated at apex. The blade is narrowly elliptic, somewhat rosemary leaf-like, 5–15 mm × 2–4 mm, base narrowly cuneate, margin entire, apex obtuse or acute. The inflorescences are axillary fascicles of few minute flowers. The perianth is tubular, white or pinkish, and 5-lobed. The androecium comprises 5 stamens attached to the corolla tube. The ovary is minute and develops 3 styles with capitate stigmas. The fruits are minute achenes, which are dark brown, glossy, broadly ovoid, and trigonous.

Medicinal use: pneumonia

Pharmacology: Unknown.

66 Family Ranunculaceae A.L. de Jussieu 1789

66.1 *Naravelia zeylanica* (L.) DC.

Synonym: *Atragene zeylanica* L.

Local names: Toilakti; kubronten

Common name: Ceylon naravelia

Botanical description: It is a woody climber that grows up to about 4 m long. The stem is terete, striated, and subglabrous. The leaves comprise basally 2 folioles with 1.5–2.5 cm long petiolules, and apically 3 folioles developed into tendrils with 3–7 cm long petiolules. The folioles are ovate, 6–11 cm × 6–10 cm, rounded to cordate at base, and present 2 pairs of secondary nerves originating from the base. The inflorescences are terminal or axillary and up to 40 cm long and bear a few flowers that are about 1 cm in diameter. The flower peduncles are 1–1.5 cm long. The calyx comprises 4 sepals, which are light yellowish, elliptic, about 1 cm long, and hairy. The corolla includes 8–10 petals, which are about 1 cm long and spathulate. The androecium comprises numerous linear and minute stamens. The gynoecium consists of hairy and showy carpels. The fruits are 5 mm long, pilose, and fusiform achenes.

Medicinal uses: vertigo and weakness, bone fracture, pains

Pharmacology: Extracts of the plant exhibited anti-inflammatory,[601] anti-ulcer,[602] and antimicrobial[603] properties.

67 Family Rhamnaceae A.L. de Jussieu 1789

67.1 *Ziziphus mauritiana* Lam.

Synonyms: *Paliurus mairei* H. Lév.; *Rhamnus jujuba* L.; *Ziziphus abyssinica* Hochst.; *Ziziphus jujuba* (L.) Gaertn.; *Ziziphus jujuba* (L.) Lam.; *Ziziphus maire* (H. Lév.) Browicz & Lauener

Local names: Boroi; kul; jon jamun

Common name: Indian jujube

Botanical description: It is a tree that grows up to about 8 m tall. The stems are hairy and with 5 mm long woody thorns. The leaves are simple and alternate. The petiole is up to 1.5 cm long and hairy. The blade is elliptic-ovate, 2–9 cm × 1.5–5 cm, asymmetrical at the base, obtuse, rounded shortly and acuminated at the apex, serrate, dark green and glossy above, greyish pale and hairy below. The inflorescences are axillary and hairy cymes bearing minute greenish-yellow flowers. The calyx includes 5 sepals that are minute. The corolla includes 5 petals, which are clawed, reflexed, and minute. The androecium comprises 5 stamens. The ovary is merged into a 10-lobed disc and develops a bifid style. The fruit is globose, light green turning orange, glossy, smooth 1.5–3.5 cm × 1.5–2.5 cm, and contain a rugose endocarp.

Medicinal uses: blood dysentery, headache, diarrhea, fever, wounds, ulcers, vomiting, leucorrhea, intestinal worms

Pharmacology: Extracts of the plant demonstrated anti-allergic and anti-inflammatory,[604] anti-diabetic[605] properties. The plant produces anti-plasmodial and anti-mycobacterial cyclopeptide alkaloids.[606]

68 Family Rosaceae A.L. de Jussieu 1789

68.1 *Cydonia oblonga* Mill.

Synonyms: *Cydonia vulgaris* Pers.; *Pyrus cydonia* L.

Local name: Bihi dana

Common name: Quince

Botanical description: It is a tree that grows up to 5 m tall. The stems are terete, purplish, and hairy. The leaves are simple, spiral, and stipulate. The stipules are ovate and caducous. The petiole is about 1 cm long and hairy. The blade is oblong, wavy, 5–10 cm × 3–5 cm, hairy below, glossy above, base rounded, apex acute and with about 6 pairs of secondary nerves that are inconspicuous. The inflorescences are axillary or terminal with solitary showy flowers, which are 4–5 cm in diameter. The calyx includes 5 sepals of about 5 mm long. The corolla includes 5 white or pinkish, membranous petals which are 1.5 cm long. The androecium includes 20 stamens of about 7 mm long. The ovary develops 5 free styles. The fruits are yellow, dull yellow, pear-shaped, bumpy, edible, and 3–5 cm in diameter.

Medicinal use: Cough

Pharmacology: The plant has been the subject to numerous pharmacological studies.[607] Extracts of the plant exhibited cardioprotective,[608] anti-cancer,[609] hypolipidemic, hepatoprotective, and renoprotective[610] properties.

68.2 *Rosa centifolia* L.

Local name: Golap

Common name: Cabbage rose

Botanical description: It is a spiny rose shrub that grows up to 2 m tall. The leaves are imparipinnate, alternate, and stipulate. The inflorescence is terminal and solitary. The flowers are graceful, fragrant, pinkish-yellow, and showy. A cup-shaped hypanthium is present. The calyx includes 5 sepals, which are up to about 2.5 cm long. The corolla includes 5 petals of 5–4 cm × 1.3–2.5 cm and membranous. The androecium includes numerous stamens of about 5 mm long. A disc is present. The gynoecium presents numerous free carpels. The fruits are minute achenes packed in the hypanthium.

Medicinal uses: Leucorrhea, inflammation, diabetes, AIDS, dementia, cardiovascular problems

Pharmacology: Extracts of the plant exhibited anti-inflammatory,[611] anti-tussive,[612] and antimicrobial[613] effects.

69 Family Rubiaceae A.L. de Jussieu 1789

69.1 *Adina cordifolia* (Roxb.) Hook. f. ex Brandis

Synonym: *Haldina cordifolia* (Roxb.) Ridsdale

Local names: Karam; haldu

Common names: Kelikadum; yellow teak

Botanical description: It is a magnificent buttressed timber tree that grows up to 40 m tall. The outer bark is reddish-brown while the inner bark is reddish. The leaves are simple, opposite, and stipulate. The stipules are spathulate, ovate, or oblong-oblanceolate, 10–12 mm × 5–12 mm, usually strongly keeled, hairy, and broadly rounded. The petiole is up to 12 cm long and hairy. The blade is broadly cordate, about 5–25 cm in diameter, hairy below, with 6–10 pairs of conspicuous secondary nerves, and the margin is wavy. The flowers are arranged in globose heads of about 1 cm diameter on 10 cm long peduncles. The calyx comprises 5 linear and minute lobes. The corolla is tubular, 5 mm long. The androecium comprises 5 stamens attached to the corolla. The fruiting heads are about 1 cm across. The fruits are capsules, which are 5 mm long and contain minute seeds.

Medicinal uses: infection, fever, headache

Pharmacology: The plant contains anti-amoebic coumarins.[614]

69.2 *Anthocephalus cadamba* (Roxb.) Miq.

Synonyms: *Anthocephalus indicus* A. Rich.; *Neolamarckia cadamba* (Roxb.) Bosser; *Nauclea cadamba Roxb.*; *Sarcocephalus cadamba* (Roxb.) Kurz

Local names: Kadam; karam; sanko

Common name: Burflower tree

Botanical description: It is a timber tree that grows to about 30 m tall. The leaves are simple, opposite, and stipulate. The stipules are large and triangular. The petiole is 1.5–2 cm long. The blade is elliptic-oblong, 12–25 cm × 6–10 cm, coriaceous, with 8–12 pairs of secondary nerves. The inflorescence is a terminal, globose, head that is orangish, beautiful, and up to 4.5 cm in diameter. The calyx develops 5 minute teeth. The corolla is tubular, about 8 mm long, and 5-lobed. The androecium includes 5 stamens, which are inserted on the corolla tube. The ovary develops a slender style that protrudes out of the corolla. Fruits are aggregate of capsules packed into a fleshy globose head, which reaches about 6.5 cm in diameter, yellow when ripe and palatable.

Medicinal uses: toothache, laxative, diuretic, diarrhea

Pharmacology: Extracts displayed anti-diabetic[615] and antibacterial,[616] effects. It contains indole alkaloids.[617]

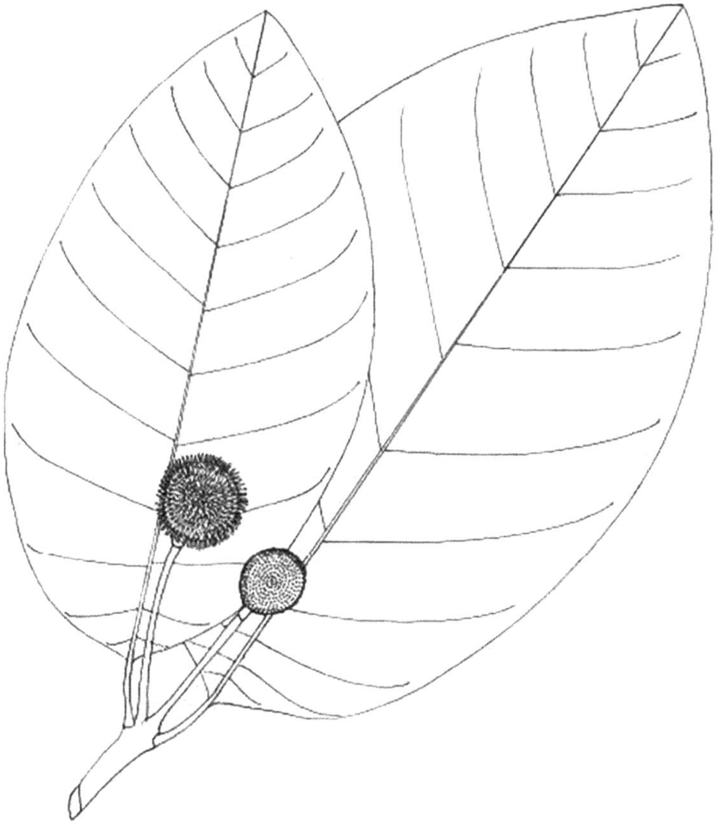

Anthocephalus cadamba

69.3 *Borreria articularis* (L. f.) F.N. Williams

Synonym: *Spermacoce articularis* L. f.

Local name: Todarjil sak

Common name: Jointed buttonweed

Botanical description: It is an erect herb that grows up to 30 cm tall. The stem is quadrangular and hairy. The leaves are simple, decussate, and stipulate. The stipules are minute and hairy. The blade is somewhat elliptic to spatulate, fleshy, 2–3.5 cm × 1–1.5 cm, hairy, with 3–5 pairs of secondary nerves. The inflorescences are axillary verticils of sessile flowers. The calyx develops 4 linear lobes that are minute. The corolla is tubular, light pink, about 1 cm long, and develops 4 lobes. The androecium includes 4 stamens attached to the corolla tube. The ovary produces a slender style with a bifid stigma. The fruit is a capsule, which is minute, globose, and contains minute rough seeds.

Medicinal uses: fever, pain, headache

Pharmacology: Extracts of the plant displayed antibacterial, anti-fungal,[618] and anti-inflammatory[619] properties.

69.4 *Hedyotis scandens* Roxb.

Synonym: *Oldenlandia scandens* (Roxb.) Kuntze

Local names: Bishma; bashpuisak

Common name: climbing hedyotis

Botanical description: It is an herb that grows to 2 m long. The stem is terete or angular and somewhat lignose. The leaves are simple, decussate, and stipulate. The stipules are up to 5 mm long. The petiole is minute. The blade is narrowly elliptic, dark green, glossy, somewhat coriaceous, 5–9 cm × 1.5–3 cm, base acute, apex acuminate, and with 3–5 pairs of secondary nerves. The inflorescences are terminal cymes that grow up to about 10 cm long and bear small flowers. The calyx is cupular and develops 4 triangular lobes. The corolla is tubular, hairy at the throat, whitish, with 4 oblong lobes which are about 5 mm long. The androecium includes 4 stamens, which are inserted on the corolla tube. The ovary develops a slender style and a bifid stigma. The fruits are ovoid capsules, which are about 5 mm long, and contain numerous minute black seeds.

Medicinal uses: stomach pain, aptha, malaria, abscesses

Pharmacology: The plant produces phenolic glycosides including grevilloside G, which exhibited antiviral activity.[620]

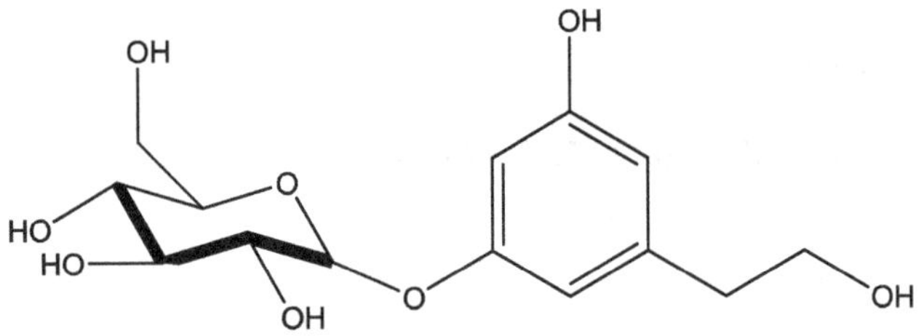

Structure of grevilloside G

69.5 *Hymenodictyon orixense* (Roxb.) Mabb.

Synonyms: *Cinchona excelsa* Roxb.; *Cinchona orixensis* Roxb.; *Hymenodictyon excelsum* (Roxb.) Wall.

Local name: Latikarum

Common name: Bridal couch tree

Botanical description: It is a massive timber tree that grows up to 30 m tall. The bark is fissured and thick. The leaves are simple, opposite, and stipulate. The stipules are triangular, about 1 cm long and serrate. The petiole is 5–8 cm long. The blade is 15–30 cm × 10–17 cm, with 6–7 pairs of secondary nerves, elliptic, tapering at the base and acute to acuminate at apex. The inflorescences are large terminal and axillary panicles of small flowers. The calyx is 5-lobed and minute. The corolla is tubular, about 5 mm long, and 5-lobed. The androecium includes 5 stamens inserted on the corolla tube. The ovary develops a slender style that extrudes out of the corolla. The fruits are elongated capsules, which are about 1.5 cm long and contain numerous minute winged seeds.

Medicinal use: goiter

Pharmacology: Extracts of the plant exhibited anti-inflammatory effects.[621]

69.6 *Ixora coccinea* L.

Synonyms: *Ixora bandhuca* Roxb.; *Ixora grandiflora* Ker Gawl.; *Ixora incarnata* Roxb. ex Sm.; *Pavetta bandhuca* Miq.

Local name: Rongon

Common name: Flame of the woods

Botanical description: It is a shrub that grows up to 2 m tall. The leaves are simple, decussate, sessile, and stipulate. The stipules are interpetiolar and oblong. The blade is 4–10 cm × 1.5–5 cm, glossy, coriaceous, with about 6–10 pairs of secondary nerves, acute at the base and apex. The inflorescences are showy, terminal cymes, which bear numerous flowers. The calyx presents 4 lobes. The corolla is tubular, dull red, which is slender, about 2.5–3 cm long tube developing 4 elliptic lobes. The androecium comprises 4 stamens inserted at the corolla throat. The ovary develops a slender style, which protrudes out of the corolla. The fruits are fleshy, globose, red, glossy, about 8 mm in diameter and contain 2 seeds.

Medicinal uses: dysentery, leucorrhea, dysmenorrhea, hemoptysis, fever, gonorrhea, diarrhea, bronchitis

Pharmacology: Extracts of the plant demonstrated anti-inflammatory,[622] antimicrobial,[623] chemoprotective,[624] and cytotoxic[625] activities.

69.7 *Morinda angustifolia* Roxb.

Synonym: Morinda squarrosa Buch.-Ham.

Local names: Koba bena; rang gach; muli

Common name: Narrow-leaved Indian mulberry

Botanical description: It is a tree that grows up to 5 m tall. The stems are quadrangular and somewhat swollen at the nodes. The leaves are simple, opposite, and stipulate. The stipules are interpetiolar, triangular, and about 5 mm long. The petiole is about 8 mm long. The blade

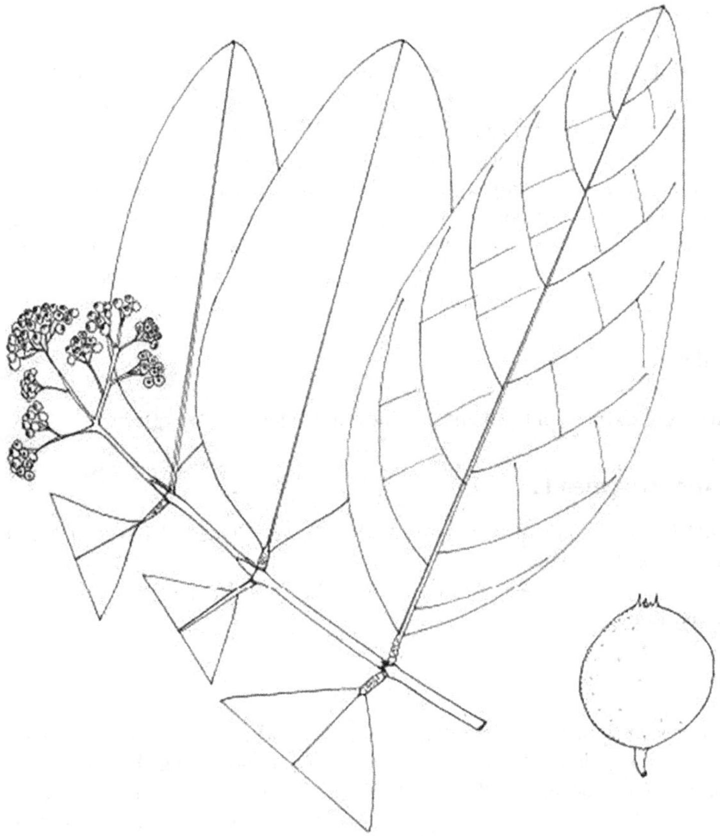

Ixora coccinea L

is elliptic, wavy, 10–20 cm × 4–12.5 cm, base attenuate, apex acuminate with about 10 pairs of secondary nerves. The inflorescences consist of incompletely fused globose, glossy, and green receptacles forming a head which is about 1.5–3 cm in diameter. The calyx is obscure. The corolla is tubular, pure white, about 2 cm long and develops 5 lobes, which are lanceolate. The ovary develops a slender style. The fruits are aggregates of partially fused to nearly separate green and glossy drupes, which are about 1 cm long.

Medicinal uses: dysuria, indigestion, bone fracture

Pharmacology: The plant produces antimicrobial anthraquinones.[626]

69.8 *Mussaenda glabrata* (Hook. f.) Hutch. ex Gamble

Local name: Gach ranitak

Common name: White lady

Botanical description: It is a shrub that grows up to about 2 m tall. The leaves are simple, opposite, and stipulate. The petiole is about 1 cm. The stipule is about 1 cm long and bifid. The blade is elliptic, green and glossy above, light yellow below, soft, 8–12 cm × 3–6.5 cm, with about 11 pairs of conspicuous secondary nerves, tapering at the base and acuminate at apex. The inflorescences are cymose and terminal. The calyx includes 5 linear sepals, which are about 8 mm long. The corolla is tubular, about 2–2.5 cm long, and develops 5 triangular lobes, which are bright orange. The fruits are oblong, green, fleshy, and about 1 cm long.

Medicinal use: jaundice, leprosy

Pharmacology: Extracts of the plant exhibited cytotoxic effects.[627]

Mussaenda glabrata

69.9 *Neonauclea sessilifolia* (Roxb.) Merr.

Synonyms: *Adina sessilifolia* (Roxb.) Hook. f. ex Brandis; *Nauclea sericea* Wall. ex G. Don; *Nauclea sessilifolia* Roxb.

Local names: Kam gash; kum

Botanical description: It is a tree that grows up to about 8 m tall. The stems are angled. The leaves are simple, decussate, and stipulate. The stipules are interpetiolar, showy, elliptic, and about 2 cm long. The petiole is minute to none. The blade is broadly elliptic, 10–20 cm × 5–10 cm, base rounded or slightly cordate, apex obtuse, and with 6–10 pairs of secondary nerves. The inflorescences are terminal heads, which are about 2.5 cm in diameter. The calyx is 5-lobed, minute, and hairy. The corolla is tubular, about 5 mm long, brownish, and with 5 triangular lobes. The androecium includes 5 stamens attached to the corolla throat. The ovary develops a slender style, which is whitish and protrudes from the corolla. The fruits are capsules, which are about 1 cm long and packed in a globose fruiting head, which is about 3 cm in diameter.

Medicinal uses: ringworm, skin infection

Pharmacology: Unknown. The plant produces chromone-secoiridoid glycosides[628] and triterpenoid saponins.[629]

69.10 *Paederia foetida* L.

Synonym: *Paederia scandans* (Lour.) Merr.

Local names: Padbailudi; gandal; gadalpata

Common name: Stinkvine

Botanical description: It is a malodorous, slender climber that grows up to about 3 m long. The stem is terete and hairy. The leaves are simple, opposite, and stipulate. The stipule is triangular and about 5 mm. The petiole is about 2.5–5.5 cm and hairy. The blade is lanceolate, 5–10 cm × 1–5 cm, base slightly cordate or rounded, apex acuminate, and with 4–6 pairs of secondary nerves. The inflorescences are axillary or terminal, pendulous clusters with a few flowers. The calyx is cupular and minute. The corolla is tubular, membranous, whitish with a purple and hairy throat, and developing 5 square-like, incised, and wavy lobes. The androecium comprises 5 stamens inserted at the throat of the corolla. The ovary elongates into a deeply bifid style. The fruits are globose, glossy, orangish-brown, and about 8 mm in diameter.

Medicinal uses: dysentery, enteritis, diarrhea-infected sores, rheumatisms, leucorrhea

Pharmacology: Extracts of the plant demonstrated anti-inflammatory[630,631] and anti-diarrheal[632] effects.

70 Family Rutaceae A.L. de Jussieu 1789

70.1 *Aegle marmelos* (L.) Corrêa

Synonyms: *Crataeva marmelos* L.; *Crateva marmelos* L.

Local names: Bel; shepalpubaong; singadare; srifal; waresi apang

Common names: Stone apple; Bengal quince

Botanical description: It is a tree that grows up to 10 m tall. The stems present straight thorns that are up to 3 cm long. The leaves are spiral, exstipulate, and 3-foliolate. The folioles

Aegle marmelos (L.) Corrêa

are broadly lanceolate, 4–12 cm × 2–6 cm, the margin is crenulate, apex acuminate. The calyx comprises 5 lobes that are minute. The corolla consists of 5 petals that are white and up to 1.3 cm long. The androecium includes 50 stamens nearly as long as petals. The gynoecium is 6 mm long and comprises around 20 locules. The fruit is greenish yellow, about 14 cm in diameter, globose, woody, and contains a sticky and delicious pulp in which numerous flat seeds are immersed. These seeds are up to 0.8 cm long.

Medicinal uses: constipation, gastralgia, dysentery, fever, abscesses, urinary problems, bradycardia, depression, leucorrhea, insomnia

Pharmacology: The plant has been the subject of numerous pharmacological studies.[633] Extracts displayed gastroprotective,[634,635] and anti-diabetic[636] effects.

70.2 *Clausena heptaphylla* (Roxb.) Wight et Arn. ex Steudel.

Synonyms: *Amyris heptaphylla* Roxb. ex DC.; *Amyris anisata* Willd.; *Clausena anisata* (Willd.) Hook. f. ex Benth.

Local names: Alkatra; arhit thi; kebutaye; karanfal

Botanical description: It is a treelet that grows up to about 4 m. The leaves are imparipinnate, spiral, and exstipulate. The blade grows up to about 30 cm and comprises 3–5 subopposite folioles plus a terminal one. The folioles are 4–7 cm × 2–15 cm, elliptic, with about 5 discrete pairs of secondary nerves, asymmetrical, tapering at the base and acuminate at apex. The inflorescences are terminal panicles of minute flowers. The calyx comprises 4 sepals. The corolla includes 4 petals, which are about 5 mm long, pure white, and elliptic. The androecium includes 8 stamens. The ovary is minute and develops a short style. The fruit is a berry, which is glossy, ovoid, and about 1 cm long.

Medicinal use: headache

Pharmacology: The plant produces antimicrobial carbazole alkaloids.[637,638]

70.3 *Feronia elephantum* Corrêa

Synonyms: *Feronia limonia* (L.) Swingle; *Limonia acidissima* L.; *Schinus limonia* L.

Local name: Kothbel

Common name: Wood apple

Botanical description: It is a tree that grows up to about 7 m tall. The stems are spiny at node, the spines are woody and about 2 cm long. The leaves are imparipinnate, spiral, and exstipulate. The petiole and rachis are winged and fleshy. The blade presents 2–4 opposite folioles plus a terminal one, which are 2.5–3.5 cm × 1.2–2.5 cm, glossy, fleshy, with 3–5 pairs of secondary nerves, spathulate and about 5 mm long emarginated at the apex. The inflorescences are terminal or axillary panicles of minute reddish flowers. The calyx is minute and includes 6 lobes that are triangular. The corolla presents 6 petals, which are elliptic and about 5 mm long. The androecium includes 12 stamens. The ovary is minute. The fruit is globose, light brown, rough, about 10 cm in diameter, woody, and contains seeds in an edible pulp.

Medicinal uses: diarrhea, dysentery, diuretic, vomiting, leucorrhea

Pharmacology: The plant produces an essential oil with antibacterial and anti-fungal activities.[639] Extracts of this plant demonstrated hepatoprotective[640] and insecticidal[641] effects.

70.4 *Glycosmis pentaphylla* (Retz.) DC.

Synonyms: *Glycosmis arborea* (Roxb.) DC.; *Glycosmis chylocarpa* Wight & Arn.; *Glycosmis quinquefolia* Griff.; *Limonia arborea* Roxb.; *Limonia pentaphylla* Retz.; *Myxospermum chylocarpum* (Wight & Arn.) M. Roem.

Local names: Ashaora; atissorah; athishadla; bujuraful; datmajan; tatiang

Common name: Orange berry

Botanical description: It is a tree that grows up to about 4 m tall. The leaves are imparipinnate, spiral, and exstipulate. The blade includes 2–4 folioles plus a terminal one that are 10–20 cm × 3–6 cm, glossy, somewhat asymmetrical, cuneate and asymmetrical at base, acute at apex, and with about 8–12 pairs of secondary nerves. The inflorescences are axillary or terminal panicles. The calyx comprises 5 sepals that are minute and ovate. The corolla includes 5 petals, which are pure white, about 5 mm long and elliptic. The androecium includes 10 stamens. The ovary is minute, globose, and develops a short style. The fruits are globose, light pink, somewhat translucent, and about 8 mm in diameter.

Medicinal uses: fever, jaundice, dysentery, cough, fever, skin infections, cleaning teeth, painful menses, anemia

Pharmacology: Extracts of the plant demonstrated hepatoprotective[642] and wound healing[643] effects.

70.5 *Murraya koenigii* (L.) Spreng.

Synonyms: *Bergera koenigii* L.; *Chalcas koenigii* (L.) Kurz

Local name: Karipatta

Common name: Curry leaf tree

Botanical description: It is a treelet that grows up to about 3 m tall. The stems are hairy at apex. The leaves are imparipinnate, spiral, and exstipulate. The blade presents 4–13 pairs of folioles, which are subopposite, 2–3.5 cm × 1–1.8 cm, asymmetrical at base, serrate, fleshy, slightly coriaceous, aromatic, and acute to acuminate at the apex. The inflorescences are terminal cymes of minute flowers. The calyx presents 5 sepals. The corolla includes 5 white, recurved, and oblong petals that are about 5 mm long. The androecium includes 10 stamens. The ovary develops a short style and a lobed stigma. A disc is present. The fruit is a berry, which is oblong-ovoid, about 5 mm long, glossy, containing few seeds, and dotted with minute oil glands.

Medicinal uses: fever, diarrhea, diabetes

Pharmacology: This plant has been the subject of numerous pharmacological studies.[644] Extracts of the plant exhibited anti-inflammatory and analgesic[645] activities.

71 Family Santalaceae R. Brown 1810

71.1 *Santalum album* L.

Local names: Chandan; sada chandan; shet chandan

Common name: Indian sandalwood

Botanical description: It is a tree that grows up to about 6 m tall. The leaves are simple, decussate, and exstipulate. The petiole is slender and grows up to 2.5 cm long. The blade is lanceolate, obovate, or elliptic, 3.5–6 cm × 1.5–2.5 cm, coriaceous, fleshy, light green, obtuse at base and apex, and with inconspicuous secondary nerves. The inflorescences are axillary or terminal panicles of minute flowers. The perianth is tubular, develops 4 triangular and recurved lobes, which about 5 mm long, and burgundy to red. The androecium comprises 4 stamens. The ovary is minute and develops a slender style, which slightly protrudes out of the corolla, and a 4-lobed stigma. The fruits are ovoid, about 1 cm long, and smooth.

Medicinal uses: gout, dyspepsia, dysentery

Pharmacology: Essential oil extracted from the plant is used in perfumery on account of α- santalol, which also has neuroleptic[646] and cytotoxic[647] activities. Extracts of the plant evoked anti-cancer[648] and anti-diarrheal[649] activities. The plant produces cytotoxic lignans.[650]

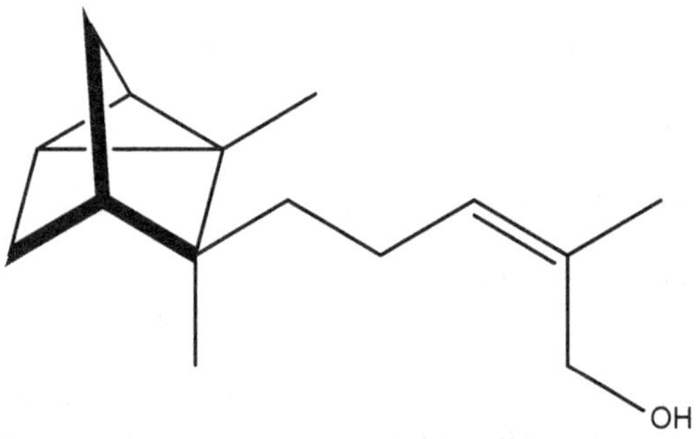

Structure of α- santalol

72 Family Sapindaceae A.L. de Jussieu 1789

72.1 *Cardiospermum halicacabum* L.

Synonyms: *Cardiospermum corindum* L.; *Cardiospermum corycodes* Kuntze; *Cardiospermum glabrum* Schumach. & Thonn.; *Cardiospermum luridum* Blume; *Cardiospermum microcarpum* Kunth; *Cardiospermum molle* Kunth

Local names: Bhado; ketafoksa

Common name: Balloon vine

Botanical description: It is a slender climber with tendrils that grows up to about 3 m long. The stems are terete, striated, and hairy at the apex. The leaves are bipinnate, alternate, and exstipulate. The petiole is about 3.5 cm long. The blade is pinnatisect and about 6 cm long, and membranous. The inflorescences are pendulous axillary cymes, which are about 10 cm long. The calyx includes 4 sepals of inequal length, which grow up to 4 mm long. The corolla presents 4 petals, which are spathulate, whitish, and about 4 mm long. The androecium includes 8 stamens. A disc is present. The ovary is ovoid and minute. The fruits are capsules that are sharply 3-lobed, obcordate, membranous, light green, lantern-like, up to about 5 cm long, and containing a few globose, dull black seeds, which are about 5 mm in diameter.

Medicinal uses: snake bites, chicken pox, diabetes

Pharmacology: Extracts of the plant displayed anti-malarial,[651] anti-inflammatory,[652] and anti-fungal[653] effects.

72.2 *Schleichera oleosa* (Lour.) Oken

Synonym: Schleichera oleosa (Lour.) Merr.

Local name: Kusum

Common name: Macassar oil tree

Botanical description: It is a massive tree that grows to about 30 m tall. The bark is ash-colored. The stems are terete and hairy at the apex. The leaves are pinnate, spiral, and exstipulate. The blade presents 2–3 pairs of oblong-obovate folioles which are 5–20 cm × 2.5–10 cm, coriaceous, reddish when young, acute at base, round or emarginate at apex, and with about 15 pairs of secondary nerves. The inflorescences are axillary, spike-like, and up to 15 cm long. The perianth comprises 5 minute and yellowish-green lobes. The androecium includes

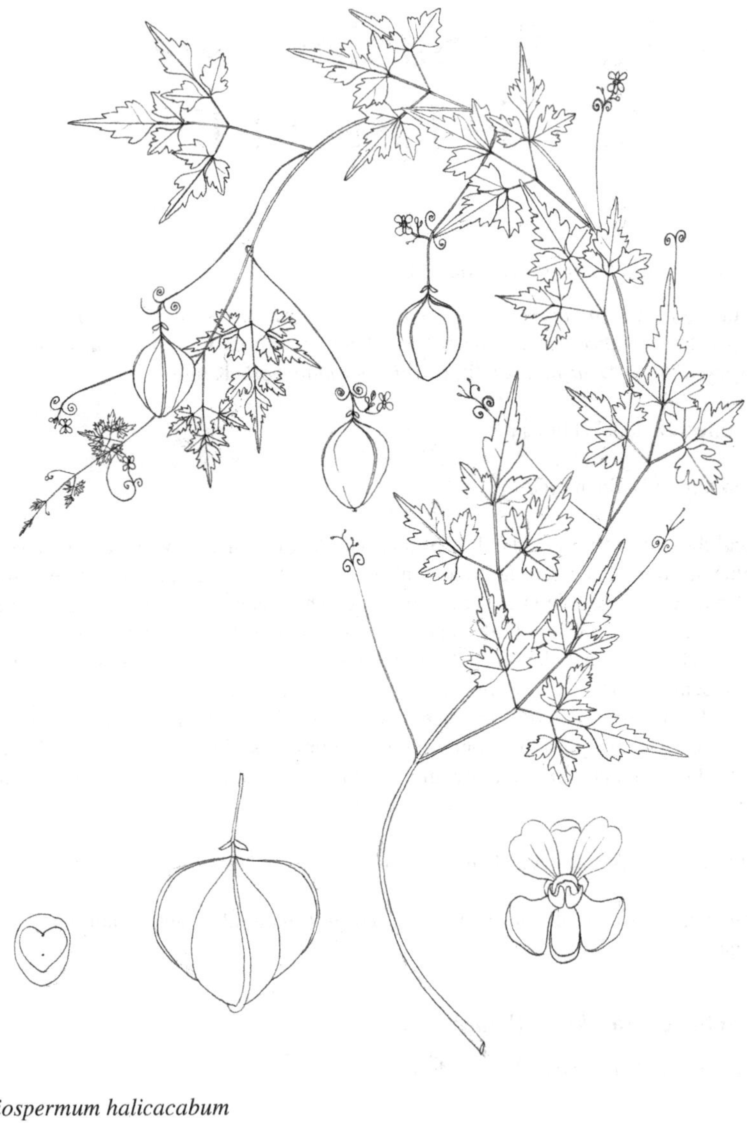

Cardiospermum halicacabum

8–9 slender stamens. A disc is present. The ovary is ovoid and minute. The fruit is globose, about 1.5 cm long, beaked, and contains up to two seeds.

Medicinal uses: hair care, skin diseases, rheumatism

Pharmacology: Extracts of the plant exhibited anti-inflammatory effects.[654] The plant produces triterpenes with cytotoxic[655] and antibacterial[656] activities.

73 Family Sapotaceae A.L. de Jussieu 1789

73.1 *Madhuca indica* J.F. Gmel.

Synonyms: *Bassia longifolia* Koenig; *Madhuca longifolia* (J. Koenig ex L.) J.F. Macbr.

Local names: Mahua; mahul; mohua; mool

Common names: Moa tree; Indian butter tree

Botanical description: It is a laticiferous tree that grows up to about 10 m tall. The stems are hairy at apex. The leaves are simple, spiral, and stipulate. The stipules are small and caducous. The petiole is slender, straight, and up to about 3 cm long. The blade is coriaceous, 7–20 cm × 5–9 cm, elliptic-lanceolate, and with about 15 pairs of secondary nerves. The inflorescences are pendulous, hairy, fascicles of flowers on slender and about 5 cm long peduncles. The calyx comprises 4 elliptic and hairy sepals. The corolla is tubular, white, and develops 8 lobes which are ovate and acuminate. The androecium comprises about 20 stamens or more inserted on the corolla tube. The ovary develops a slender and sharp style that protrudes out of the corolla. The fruits are ovoid, globose, or pear-shaped, about 1.5 cm long, dull green to brownish, beaked berries containing a few oblong and glossy seeds.

Medicinal uses: rheumatism, skin diseases, boils, fever, cough

Pharmacology: Extracts of the plant demonstrated gastroprotective[657] and antibacterial[658] properties.

73.2 *Mimusops elengi* L.

Synonyms: *Kaukenia elengi* (L.) Kuntze; *Mimusops elengi* Bojer

Local names: Bakul; Bohur

Common name: Spanish cherry

Botanical description: It is a laticiferous tree that grows up to about 10 m tall. The leaves are simple, spiral, and stipulate. The stipules are minute and caducous. The petiole is about 3 cm long. The blade is coriaceous, glossy, 3–5 cm × 7–12.5 cm, acute at base, shortly acuminate at apex, and with numerous inconspicuous secondary nerves. The flowers are slightly fragrant, axillary and solitary on a 2 cm long peduncle or in small fascicles. The calyx present 8 lobes. The corolla is white, 8-lobed and the lobes are deeply incised. The androecium includes 8 stamens joined in a cone. The ovary is minute. The fruits are ovoid, bullet-shaped, dull red-orange, about 3 cm long, and contain a few tear-shaped glossy brown seeds.

Medicinal uses: toothache, gingivitis, laxative, fever, dysentery, diarrhea

Pharmacology: Extracts of the plant demonstrated anti-tumor,[659] anti-inflammatory, analgesic and anti-pyretic,[660] and anthelmintic[661] properties.

74 Family Scrophulariaceae A.L. de Jussieu 1789

74.1 *Limnophila indica* (L.) Druce

Synonyms: *Hottonia indica* L.; *Limnophila gratioloides* R. Br.

Local name: Karpur

Common name: Indian marshweed

Botanical description: It is an aquatic herb. The submerged stems are articulate, stoloniferous, and develop pinnatisect, 6-whorled, exstipulate, capillary-like leaves that grow up to about 3 cm long. The aerial leaves are simple, 4-whorled, and exstipulate. The blade of aerial leaves is elliptic-lanceolate, serrate, lobed, up to 2 cm long, and light dull green. The inflorescences are axillary and solitary on a 1 cm long peduncle. The calyx is about 5 mm long, tubular, and develops 5 triangular lobes. The corolla is tubular, bilabiate, white to light purple, 1 cm long. The upper lip has 2 lobes while the lower lip is 3-lobed. The lobes are round. The androecium comprises 4 stamens attached to the corolla tube. The ovary is ovoid and develop a slender style. The fruits are minute and ovoid capsules.

Medicinal uses: infections, fever

Pharmacology: The plant produces the antibacterial triterpene 3-oxo-olean-12(13), 18(19)-dien-29α-carboxylic acid[662] and the antibacterial and anti-fungal flavonoid 5,6-dihydroxy-4′,7,8-trimethoxyflavone.[663]

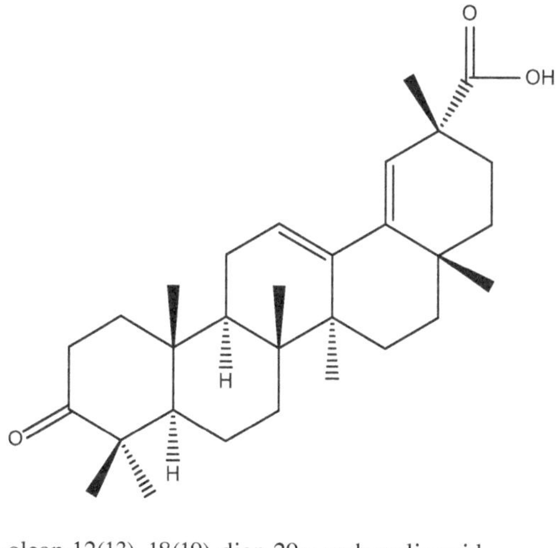

Structure of 3-Oxo-olean-12(13), 18(19)-dien-29α-carboxylic acid

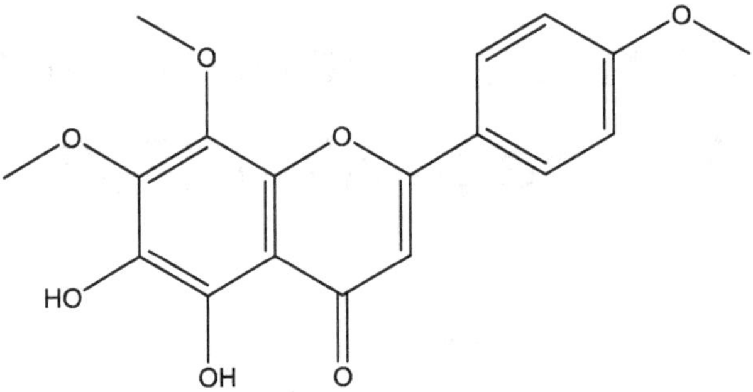

Structure of 5,6-dihydroxy-4′,7,8-trimethoxyflavone

74.2 *Scoparia dulcis* L.

Synonyms: *Scoparia grandiflora* Nash.; *Capraria dulcis* (L.) Kuntze; *Gratiola micrantha* Nutt.

Local names: Bandhonia; bondana; midaraissa; mithapata; oyashipene

Common name: Licorice weed

Botanical description: It is an erect herb that grows up to about 30 cm tall. The stems are angled, somewhat rigid, upright and smooth. The leaves are simple, 3-whorled, subsessile, and exstipulate. The blades are narrowly elliptic, serrate, tapering at base, about 4 cm × 1.5 cm, and acute at apex. The inflorescence is axillary and solitary on a slender, 1 cm long peduncle. The calyx is 4-lobed and minute. The corolla is white, tubular, 4-lobed, and with numerous spreading hairs around the throat. The androecium presents 4 stamens. The ovary develops a slender style. The fruits are capsules, which are about 1.5 cm long.

Medicinal uses: diabetes, indigestion, boils, tumors, lung infection, bladder and kidney stones, toothache, mouth ulcers, diabetes, diarrhea, dysentery, menorrhagia, swellings, "repels evil spirits" (!)

Pharmacology: Extracts of the plant demonstrated insulin secretagogue,[664] analgesic, anti-inflammatory,[665] and anti-ulcer[666] activities.

75 Family Smilacaceae Ventenat 1799

75.1 *Smilax zeylanica* L.

Synonyms: *Smilax elliptica* Desv. ex Ham.; *Smilax macrophylla* Roxb.; *Smilax ovalifolia* Roxb.

Local names: Antikinari; kumarika; kankomaicha; kanku maicha; ramdantan; radamtun; rampaun

Common name: Ceylon smilax

Botanical description: It is a climber that grows up to about 3 m long. The stems are lignose. The leaves are simple, alternate, and exstipulate. The petiole is about 1.5 cm long and curved. The blade is broadly elliptic, coriaceous, glossy, 6.5–15 cm × 4–12.5 cm, rounded at base and apex, and with 3–5 longitudinal nervations. The inflorescences are axillary solitary umbels on 3 cm long peduncles. The perianth consists of 6 whitish and about 5 mm long tepals. The androecium consists of 6 stamens. The ovary is minute and the stigma is sessile. The fruits are globose, green, glossy, and about 1 cm in diameter.

Medicinal uses: boils, leucorrhea, dysentery, abscesses, indigestion, gastralgia, venereal disease, swellings, diarrhea, tooth infection

Pharmacology: Extracts of the plant demonstrated hepatoprotective[667] and cytotoxic[668] effects. The plant produces diosgenin,[669] which improves fertility[670] and has anti-inflammatory[671] activity.

76 Family Solanaceae A.L. de Jussieu 1789

76.1 *Capsicum frutescens* L.

Synonyms: *Capsicum annuum* L.; *Capsicum conicum* G. Mey.; *Capsicum longum* A. DC.

Local name: Marish

Common name: Chili pepper; tabasco pepper

Botanical description: It is an herb that grows up to about 1 m tall. The leaves are simple, spiral, and exstipulate. The petiole is 4–6 cm long, terete, smooth, and fleshy. The blade is lanceolate, wavy, fleshy, pendulous, and about 4–12 cm × 1.5–5 cm. The inflorescences are axillary and solitary, or few-flowered clusters. The flowers are somewhat bending down. The calyx is tubular, minute, and five-lobed. The corolla is tubular, five-lobed, pure white, the lobes broadly lanceolate. The androecium includes five stamens that are purplish and joined in a cone. The ovary is minute and develops a slender and straight style. The fruits are irregularly fusiform, somewhat erect, pungent, red, glossy, and edible berries that grow up to about 10 cm long and contain numerous minute yellowish seeds.

Medicinal uses: headache, sores, bronchitis, indigestion

Pharmacology: The plant has been the subject of numerous pharmacological investigations.[672] It contains the pungent capsaicin, which is anti-inflammatory[673] and antibacterial.[674]

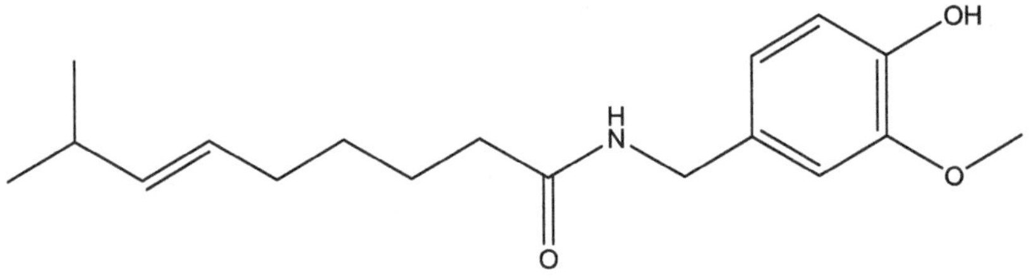

Structure of capsaicin

76.2 *Datura metel* L.

Synonyms: *Datura alba* Rumph. ex Nees; *Datura fastuosa* L.; *Datura hummatu* Bernh.; *Datura nilhummatu* Dunal.

Local names: Dhutra; dothura; dhutura

Common name: Angel's trumpet

Botanical description: It is a dreadfully poisonous herb that grows up to 2 m tall. The stems are terete, green to light purple, hairy, and woody at base. The leaves are simple, spiral, and exstipulate. The petiole is up to about 15 cm long. The blade is wavy, broadly lanceolate, fleshy, dull green, 7–17 cm × 4.5–13 cm, base somewhat asymmetrical, and apex acute to acuminate, and with about 5–11 pairs of secondary nerves. The inflorescence is axillary and solitary and bears magnificent yet eerie showy flowers. The flower peduncle is about 2 cm long. The calyx is tubular, cylindrical, 5-lobed. The lobes are triangular and about 2 cm long. The corolla tube is up to 20 cm long, white, funnel-like, straight, and develops 5–10 broadly triangular and acuminate lobes. The androecium includes 4 slender stamens. The fruits are globose, nodding, irregularly dehiscent, spiny, about 5 cm diameter capsules containing numerous seeds, which are about 5 mm long.

Medicinal uses: fever, lunacy, flue, skin diseases

Pharmacology: The plant produces the anti-cholinergic tropane alkaloids scopolamine and atropine used for therapeutic purposes.[675] Extracts of the plant demonstrated analgesic[676] and anti-fungal[677] properties. The plant produces anti inflammatory withanolides.[678]

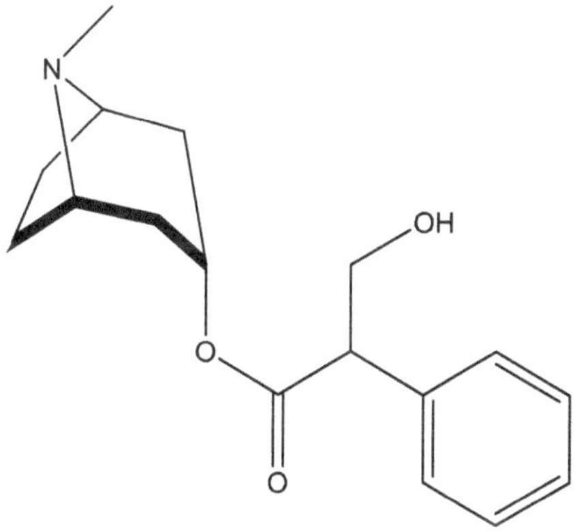

Structure of atropine

76.3 *Physalis minima* L.

Synonyms: *Physalis lagascae* Roem. & Schult.; *Physalis parviflora* R. Br.

Local names: Bantepari; chattojhara; handikundi; kopalfuta, kopalfutki

Common name: Pygmy groundcherry

Botanical description: It is a fleshy herb that grows up to about 30 cm tall. The leaves are simple, spiral, and exstipulate. The petiole grows up to about 1.5 cm long. The blade is elliptic, ovate-lanceolate, 2–3.5 cm × 1–2 cm, asymmetrical, round or tapering at base, laxly incised at margin, and acuminate at apex. The inflorescence is axillary and solitary. The calyx is tubular, 5-lobed, and minute. The corolla is tubular, yellowish, funnel-shaped, and about 5 mm long. The androecium includes 5 stamens. The fruit is a globose, red, glossy, 5 mm in diameter berry enclosed in an accrescent calyx that is ribbed, somewhat lamp-shaped, green, and up to 2 cm long.

Medicinal uses: gastralgia, earache, skin diseases, indigestion

Pharmacology: The plant elaborates cytotoxic[679] and anti-inflammatory,[680] anti-leishmanial[681] sterols.

76.4 *Withania somnifera* (L.) Dunal

Synonyms: *Physalis somnifera* L.; *Withania kansuensis* Kuang & A. M. Lu; *Withania microphysalis* Suess.

Local name: Ashwagandha

Common name: Indian ginseng

Botanical description: It is an herb that grows up to about 1 m tall. The stems are terete and hairy. The leaves are simple, spiral, and exstipulate. The petiole grows up to about 1.5 cm long. The blade is lanceolate, tapering at base, acute at apex, 2.5–12 cm × 1.5–6.5 cm, hairy, and with about 5 to 6 pairs of secondary nerves. The inflorescences are axillary few-flowered fascicles. The calyx is campanulate, hairy, about 5 mm long, and develops 5 lobes. The corolla is tubular, 5-lobed, about 1 cm long and the lobes are triangular. The androecium includes 5 stamens that are attached to the corolla tube and hairy at base. The ovary is minute and develop short style. The fruit is a globose, red, glossy, 8 mm in diameter berry enclosed in an enlarged calyx, which is ovoid, brown, and up to 2.5 cm long.

Medicinal uses: sexual impotence in male, ulcer, cancer, seizures

Pharmacology: The plant has been the subject of numerous pharmacological studies.[682] Extracts of the plant demonstrated anxiolytic-anti-depressant,[683] and spermatogenic[684] activities. The plant produces anti-tumor withanolides.[685]

77 Family Sterculiaceae E.P. Ventenat ex Salisbury 1807

77.1 *Abroma augusta* (L.) L. f.

Synonym: *Theobroma augusta* L.

Local names: Udurful; Ulat-kambal; olot-kambal; ulat-chandal; ulut-kumbul

Common name: Devil's cotton

Botanical description: It is a sinister shrub that grows up to about 3 m tall. The stems are terete and hairy. The plant has somewhat a strange architecture. The leaves are simple, alternate, and stipulate. The stipules are narrowly triangular, about 1 cm long, and caducous. The petiole grows up to about 10 cm long. The blade is lanceolate to palmately lobed, 6.5–20 cm × 4–15 cm, hairy below, with 5–7 pairs of secondary nerves, cordate at base, and acute or acuminate at apex. The inflorescences are axillary and solitary with pendulous showy flowers. The calyx comprises 5 sepals that are lanceolate, up to 2 cm long, and hairy. The corolla presents 5 petals that are dull purplish, about 2.5 cm long, and spathulate-clawed. The androecium includes 5 groups of 3 stamens. The ovary is oblong and develops a 5-lobed stigma. The fruits are upright brownish, papery, and freaky capsules that are 5-partite, obconic, about 5 cm in diameter, dehiscent, hairy within, and containing minute black seeds.

Medicinal uses: dysmenorrhea, mucosal irritations, diabetes, impotency, dysentery, fatigue, leucorrhea, piles, diarrhea, gonorrhea

Pharmacology: The plant exhibited anti-diabetic properties[686,687] on account of taraxerol.[688]

78 Family Tiliaceae A.L. de Jussieu 1789

78.1 *Grewia hirsuta* Vahl

Synonyms: *Grewia hirsuta* Roxb.; Grewia hirsuta Sm.

Local names: Seta kata; seta andir

Botanical description: It is a shrub that grows up to about 3 m tall. The stems are hairy. The leaves are simple, alternate, and stipulate. The stipules are minute. The petiole is minute and hairy. The blade is lanceolate, 5–15 cm × 2.5–3.5 cm, coriaceous, hairy below, with 4–5 pairs of secondary nerves, minutely serrate at margin, round or cordate and asymmetrical at base, and acute at apex. The inflorescences are axillary cymes of few flowers. The calyx includes 5 sepals, which are lanceolate, hairy, and about 1 cm long. The corolla is white and consists of 5 oblong petals, which are shorter than the sepals. The androecium includes 40 stamens that are about 5 mm long. The ovary is hairy and develops a slender style and a 4-lobed stigma. The fruit is a globose drupe.

Medicinal use: fever

Pharmacology: Extracts of the plant exhibited hypoglycemic effects.[689]

79 Family Ulmaceae Mirbel 1815

79.1 *Trema orientalis* (L.) Bl.

Synonyms: *Celtis orientalis* L.; *Celtis discolor* Brongn.; *Celtis rigida* Blume; *Sponia argentea* Planch.; *Sponia orientalis* (L.) Decne.; *Sponia wightii* Planch.; *Trema polygama* Z.M. Wu & J.Y. Lin.

Local names: Khaksi daru; sugarar amila

Common name: Charcoal tree

Botanical description: It is a tree that grows up to about 15 m tall. The stems are hairy. The leaves are simple, alternate, and stipulate. The stipules are narrowly triangular and about 1 cm long. The petiole grows up to 2 cm long. The blade is oblong-lanceolate, 10–20 cm × 5–10 cm, hairy below, cordate and asymmetrical at base, serrate at margin, acuminate at apex, and with 4–6 pairs of secondary nerves. The inflorescences are axillary clusters of minute flowers. The perianth includes 4–5 hairy tepals. The androecium comprises 5 stamens. The ovary is minute and develops a bifid style. The fruits are black, which are about 5 mm long and contain minute seeds.

Medicinal uses: menorrhagia, putrefied abscess

Pharmacology: Extracts of the plant demonstrated analgesic, anti-inflammatory,[690] anti-diabetic,[691] and anti-diarrheal[692] properties.

80 Family Urticaceae A.L. de Jussieu 1789

80.1 *Boehmeria nivea* (L.) Gaudich.

Synonyms: *Ramium niveum* (L.) Kuntze; *Urtica nivea* L.; *Urtica tenacissima* Roxb.

Local name: Hurumbuto pada

Common name: Chinese grass

Botanical description: It is an herb that grows up to about 1.5 m tall. The stem is hairy. The leaves are simple, spiral, and stipulate. The stipules are hairy, lanceolate, bifid, and about 1 cm long. The petiole is straight, slender, and up to 10 cm long. The blade is elliptic, ovate, deltoid, 5–15 cm × 3.5–14 cm, hairy below, papery, dull green, wedge-shaped at the base or cordate, serrate at margin, cuspidate at apex, and with 2–4 pairs of secondary nerves. The inflorescences are axillary spikes that are up to about 10 cm long. The flowers are minute. The perianth includes 4 tepals. The androecium consists of 4 stamens. The ovary is minute. The fruits are minute achenes.

Medicinal use: infected wounds

Pharmacology: Extracts of the plant exhibited hepatoprotective,[693] anti-HBV,[694] and antibacterial[695] effects.

81 Family Verbenaceae J. Saint-Hilaire 1805

81.1 *Callicarpa macrophylla* Vahl

Synonyms: *Callicarpa dunniana* H. Lév.; *Callicarpa incana* Roxb.

Local names: Kojagaicho; dubhosa

Common name: Long-leaved beauty berry

Botanical description: It is a graceful shrub that grows up to about 2 m tall. The stems are hairy. The leaves are simple, opposite, and exstipulate. The petiole is hairy and up to 1.5 cm long. The blade is 10–25 × 2.5–10.5 cm, elliptic to lanceolate, tapering or cuneate at base, serrate, acute at apex, hairy below, and with about 5–8 pairs of secondary nerves. The inflorescences are axillary and hairy cymes that grow up to 8.5 cm long, and bear minute pinkish flowers. The calyx is tubular, hairy, and minute. The corolla is tubular, about 5 mm in diameter, and develops 4 rounded lobes. The androecium includes 4 stamens. The ovary develops a linear style and a bifid stigma. The fruits are globose, whitish, and minute.

Medicinal uses: intestinal worms, fever

Pharmacology: The plant produces diterpenes[696,697] with anti-inflammatory activities.[698]

81.2 *Clerodendrum indicum* (L.) Kuntze

Synonyms: *Clerodendrum siphonanthus* R. Br.; *Clerodendrum verticillatum* D. Don; *Ovieda mitis* L.; *Siphonanthus indicus* L.

Local name: Ekdaira gach

Common names: Turk's turban; tube flower

Botanical description: It is a shrub that grows up to about 2 m tall. The stems are quadrangular at the apex and striated. The leaves are simple, 3–5-whorled, sessile, and exstipulate. The blade is narrowly elliptic, 10–20 cm × 1.5–2.5 cm, membranous, tapering at base, acute at apex, and with about 10 pairs of secondary nerves. The inflorescences are terminal and about 30 cm long thyrses. The calyx is tubular, 5-lobed, and up to 1.5 cm long. The corolla is tubular, bending, whitish, and develops a slender, up to about 10 cm long tube and 5 elliptic lobes. The androecium comprises 5 stamens that protrude out of the corolla. The ovary develops a style longer than stamen and a bilobed stigma. The fruit is a bluish, glossy, and globose drupe, which is about 1.5 cm in diameter and seated on a red and glossy accrescent star-shaped calyx.

Medicinal use: carbuncles

Pharmacology: The plant produces antibacterial flavonoids[699] as well as cytotoxic triterpenes.[700]

81.3 *Gmelina arborea* Roxb. ex Sm.

Synonyms: *Gmelina rheedii* Hook.; *Premna arborea* Farw.

Local names: Gamarigach; gamari

Common name: Gmelina

Botanical description: It is a tree that grows to about 10 m tall. The stems are hairy and quadrangular at apex and lenticelled. The leaves are simple, decussate, and exstipulate. The petiole is hairy, slender, and up to 10 cm long. The blade is broadly ovate, wavy, 8–20 cm × 4.5–15 cm, wedge-shaped at base, acuminate at apex, and with 3-5 pairs of secondary nerves. The inflorescence are hairy, lax, terminal panicles that are about 30 cm long. The calyx is tubular, about 5 mm long and 5 lobed. The corolla is yellowish to brown, 3–4 cm, bilabiate, the lower lip 3-lobed, and the upper lip entire or 2-lobed. The androecium comprises 4 stamens. The ovary is minute and develops a bilobed stigma. The fruits are glossy, poisonous, yellowish-green, ovoid drupes, and are up to about 2.5 cm long.

Medicinal uses: jaundice, stomach disorders, vomiting, diarrhea, diabetes

Pharmacology: Extracts of the plant exhibited anti-inflammatory[701] and anti-diabetic[702] activities.

81.4 *Premna esculenta* Roxb.

Synonym: *Gumira esculenta* (Roxb.) Kuntze.

Local names: Silazra; lelom pada; lamur; tatui

Botanical description: It is a shrub that grows up to about 2 m tall. The stems are quadrangular at apex and smooth. The leaves are simple, opposite, subsessile, edible and exstipulate. The blade is elliptic, 12–25 cm × 5–12.5 cm, base acute, margin serrate, and apex acuminate. The flowers are terminal and dense corymbs of minute yellowish-white flowers. The calyx is minute and presents 5 lobes. The corolla is bilabiate, the upper lip is 3-lobed, and hairy at the throat. The androecium includes 4 stamens. The ovary is minute. The fruits are globose drupes, which are minute, glossy, and purple.

Medicinal uses: bacterial and fungal infection, bleeding at delivery

Pharmacology: Extracts of the plant exhibited analgesic, anti-inflammatory, thrombolytic, and hepatoprotective effects.[703]

81.5 *Vitex negundo* L.

Synonyms: *Vitex arborea* Desf.; *Vitex bicolor* Willd.; *Vitex paniculata* Lam.

Local names: Nishinda; bara nishinda; nisinda

Common names: Chinese chaste tree; five-leaf chaste tree

Botanical description: It is a graceful tree that grows up to about 5 m tall. The stems are hairy and quadrangular at apex. The leaves are imparipinnate, decussate, and exstipulate. The petiole is up to 6 cm long and slender. The blade comprises 3–5 folioles that are lanceolate, thin, 5–10 cm × 1.5–3 cm broad, obtuse at base, acute at apex, and with about 4–9 pairs of secondary nerves. The inflorescences are lax and terminal panicles, which grow up to about 20 cm long. The calyx is tubular, minute, and 5-lobed. The corolla is tubular, bluish, bilabiate, the upper lip is somewhat 4-lobed. The androecium includes 4 stamens. The ovary is minute and develops a slender style and bifid stigma. The fruits are glossy drupes, which are globose and about 5 mm in diameter.

Medicinal uses: rheumatism, body pain, cold, worms, headaches, whitening of hair, memory loss, cancer, gout, cold, cough, asthma, tonic, fever, diuretic, sores, intestinal worms, fever, post-partum, flu

Pharmacology: Extracts of the plant exhibited anti-inflammatory and analgesic[704,705] activities. The plant produces antibacterial,[706] and cytotoxic[707] flavonoids.

Vitex negundo

82 Family Vitaceae A.L. de Jussieu 1789

82.1 *Cissus quadrangularis* L.

Synonyms: *Cissus edulis* Dalziel; *Vitis quadrangularis* (L.) Wall. ex Wight.

Local names: Hadjorha lata; hasjora; harjora; harjoralata; pyandhum

Common names: Veldt grape; devil's backbone

Botanical description: It is a climber that grows up to about 5 m long. The stems are fleshy, smooth, articulated, 4-winged, and developing tendrils opposite to the leaves. The leaves are simple, alternate, and stipulate. The stipules are minute and caducous. The petiole is up to 1.2 cm long. The blade is ovate or kidney-shaped, fleshy, serrate, light green, glossy, and 1.5–2.5 cm × 3–5 cm. The inflorescences are cymes of minute pinkish flowers on an about 2.5-cm long peduncle facing the leaves. The calyx is cupular. The corolla comprises 4 petals. A disc is present. The androecium comprises 4 stamens. The ovary is minute and develops a slender style. The fruits are berries, which are red, and about 1 cm long.

Medicinal uses: bone fractures, intestinal problems, remove thorns

Pharmacology: Extracts of the plant demonstrated anti-inflammatory[708] and bone-healing[709] properties.

83 Family Zingiberaceae Martynov 1820

83.1 *Alpinia conchigera* Griff.

Synonym: *Languas conchigera* (Griff.) Burkill.

Local names: Khetranga; Ketranga

Common name: the lesser Alpinia

Botanical description: It is a plant that grows up to about 2 m tall from a rhizome. The leaves are simple, form sheaths, and with an entire ligule, which is about 5 mm long. The petiole is up to 1 cm long. The blade is elliptic, glossy, 20–30 cm × 10–15 cm, obtuse at base, and acute at apex. The inflorescences are panicles which are about 25 cm long. The calyx is tubular and 3-lobed. The corolla is about 1 cm long, with an obovate and emarginate labellum marked with red lines. The androecium comprises 1 conspicuous stamen. The ovary is minute. The fruits are oblong capsules which are about 1 cm long.

Medicinal uses: dysentery, indigestion

Pharmacology: The plant produces acetoxyeugenol acetate, which is cytotoxic and anti-fungal.[710] Essential oil of the plant demonstrated antimicrobial activity.[711] acetoxychavicol from this plant is cytotoxic.[712]

83.2 *Alpinia galanga* (L.) Willd.

Synonym: *Maranta galanga* L.

Local names: Hoimboti-boch; mohavori-boch

Common name: Galanga

Botanical description: It is an herb that grows up to about 2 m tall from a rhizome. The leaves are simple with a ligule, which is about 5 mm, and a petiole, which is about the same length. The blade is oblong, 20–30 cm × 7–12 cm, attenuate at base, and acute or acuminate at apex. The inflorescences are terminal panicles which are about 25 cm long. The calyx is tubular and about 1 cm long. The corolla is tubular, with a white labellum which is marked with lines, spathulate, about 2 cm long and bifid. The androecium comprises 1 stamen, which is about 1.5 cm long. The ovary is minute and develops a slender style. The fruit is a capsule, which is globose, dull red, and up to 1.5 cm in diameter.

Medicinal uses: goiter, eye infection

Pharmacology: The plant produces phenolic compounds of which acetoxychavicol acetate is cytotoxic and inhibits the growth of *Mycobacterium tuberculosis*.[713,714]

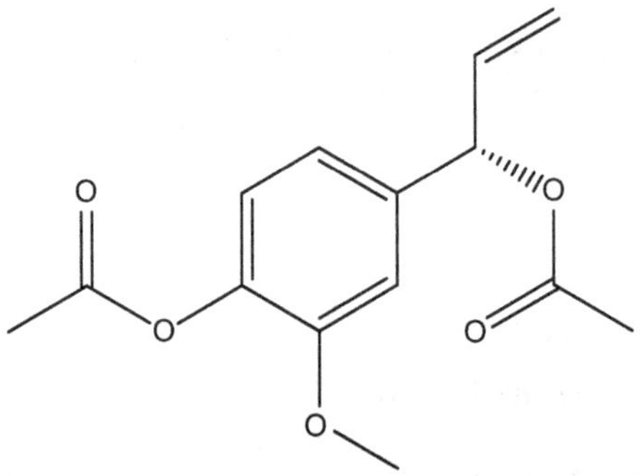

Structure of Acetoxyeugenol acetate

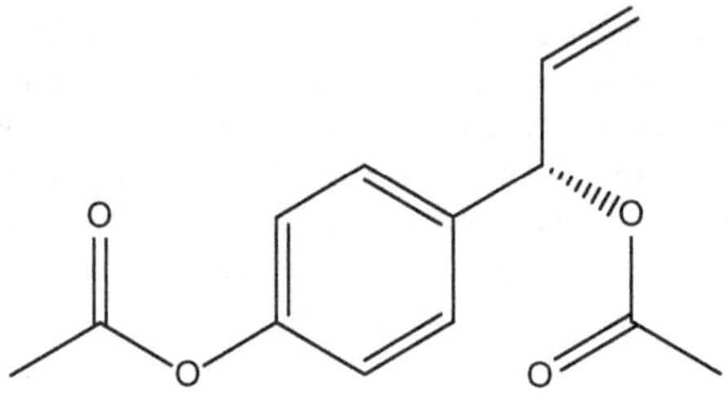

Structure of Acetoxychavicol acetate

83.3 *Alpinia nigra* (Gaertn.) B. L. Burtt.

Synonyms: *Alpinia allughas* (Retz.) Roscoe; *Alpinia aquatica* (J. Koenig) Roscoe; *Heritiera allughas* Retz.; *Languas allughas* (Retz.) Burkill; *Languas aquatica* J. Koenig; *Zingiber nigrum* Gaertn.

Local names: Bhulchengi; khetranga; tara; taruka

Common name: Black-fruited galangal

Botanical description: It is an herb that grows up to about 2.5 m tall from an aromatic rhizome. The leaves are simple, with a ligule, which are about 5 mm and comprise a short petiole. The blade is elliptic, glossy, 25–35 cm × 6–10 cm, and acute at base and apex. The inflorescences are terminal panicles that are about 30 cm long. The calyx is tubular, 3-lobed, and about 1.5 cm long. The corolla is tubular, pinkish, with a white labellum which is about 1.5 cm long and bifid. The androecium comprises 1 stamen, which is about 1.5 cm long. The ovary is minute and develops a slender style. The fruit is a capsule, which is globose, black, and up to 1.5 cm in diameter.

Medicinal uses: jaundice, gastric ulcers

Pharmacology: Extracts of the plant displayed anti-inflammatory and anthelmintic properties.[715,716] The plant produces flavonoids and antibacterial essential oil.[717,718]

83.4 *Amomum dealbatum* Roxb.

Synonym: *Cardamomum dealbatum* (Roxb.) Kuntze.

Local name: Palachengay

Common name: Insipid amomum

Botanical description: It is an herb that grows up to about 2.5 m tall from a rhizome. The leaves are simple, form sheaths, and with an orbicular, bifid, ligule, which is about 1.5 cm long. The petiole is up to 3 cm long. The blade is broadly lanceolate, 50–70 cm × 6–15 cm, hairy below, cuneate at base, and acuminate. The inflorescences are spikes, which are about 5 cm in diameter. The calyx is tubular and 3-lobed. The corolla is tubular, pure white, about 2.5 cm long, and develops a labellum, which is elliptic, marked with a yellow patch, and emarginate. The androecium is a single conspicuous stamen. The fruits are capsules, which are ribbed and about 3 cm long.

Medicinal use: abscess

Pharmacology: Unknown.

83.5 *Curcuma angustifolia* Roxb.

Synonym: *Curcuma angustifolia* Dalzell & A. Gibson.

Local name: Palo

Common name: East Indian arrowroot

Botanical description: It is an herb that grows up to about 1.5 m tall from a rhizome. The leaves are simple and with a slender petiole that grows up to about 30 cm long. The blade is narrow, lanceolate, glossy, and 10–30 cm × 3–9 cm. The inflorescences are spikes, which are about 30 cm long and bear delightful purplish-pink bracts that are oblong and about 3 cm long. The calyx is tubular, 3-lobed, and about 1.5 cm long. The corolla is tubular, yellow, 3-lobed, and develops a broad golden-yellow, about 2.5 cm long, and bifid labellum. The androecium comprises 1 stamen. The ovary is minute and develops a slender style and a globular stigma. The fruit is an oblong and red capsule.

Medicinal use: enteritis and chronic dysentery

Pharmacology: Extracts of the plant exhibited anti-ulcer property.[719] Essential oils from this plant displayed anti-fungal property.[720]

83.6 *Kaempferia rotunda* L.

Synonym: *Kaempferia rotunda* Blanco.

Local names: Chakma-kala halud; adakomol; bhooi-champa; bhoo-champa

Common name: Indian crocus

Botanical description: It is a small herb that grows from an aromatic rhizome. The leaves are simple, few, and comprise a minute ligule and a petiole that is about 2 cm long. The blade is elliptic, 15–25 cm × 7.5–10 cm, dark green, glossy, darker around the midrib, and purplish below, oblong-lanceolate, cuneate at base, and acute at apex. The inflorescences arise from the rhizome and bear beautiful purple bracts which are about 2 cm long and bifid. The calyx is tubular and 3-lobed and about 6 cm long. The corolla is tubular, purplish, about as long as the calyx, slender, with 2 linear lobes and a labellum which is 2.5 cm long, bifid, and purple. The androecium comprises a single stamen. The ovary is about 5 mm long. The fruit is a capsule.

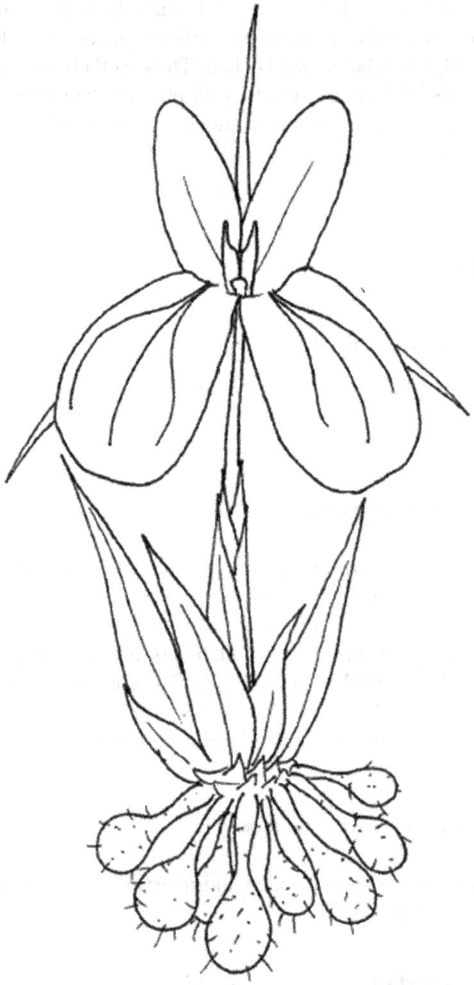

Kaempferia rotunda L

Medicinal uses: insect bite, goiter

Pharmacology: The plant produces insecticidal phenolics[721] as well as cytotoxic polyoxygenated cyclohexanes.[722]

83.7 *Zingiber cassumunar* Roxb.

Synonyms: *Amomum cassumunar* (Roxb.) Donn; *Cassumunar roxburghii* (Roxb.) Colla; *Amomum montanum* J. Koenig ex Retz.; *Zingiber montanum* (J. Koenig ex Retz.) Link ex A. Dietr.

Local names: Banada; bun ada

Common name: Cassumunar ginger

Botanical description: It is an herb that grows to about 2 m tall from a rhizome which is aromatic and yellow inside. The leaves are simple, with a minute ligule, subsessile, 20–35 cm × 7.5–15 cm, oblong, acute or round at base, acuminate at apex, glossy, form sheaths, and hairy below. The inflorescence is a terminal, fusiform, red, somewhat scaly, and about 15 cm longfleshy spike on a 20 cm tall stem emerging from the rhizome. The calyx is tubular and about 2 cm long. The corolla is tubular, white, and develops a yellow and bilobed labellum which is about 5 cm long. The androecium includes a stamen which about 1 cm long and slender. The ovary is about 5 mm long and develops a slender style and an obconic stigma. The fruit is a globose capsule, which is about 1.5 cm in diameter.

Medicinal uses: anorexia, analgesic, diarrhea

Pharmacology: The plant produces the sesquiterpene zerumbone, which is anti-fungal,[723] as well as cytotoxic[724] and anti-inflammatory,[725] phenolics.

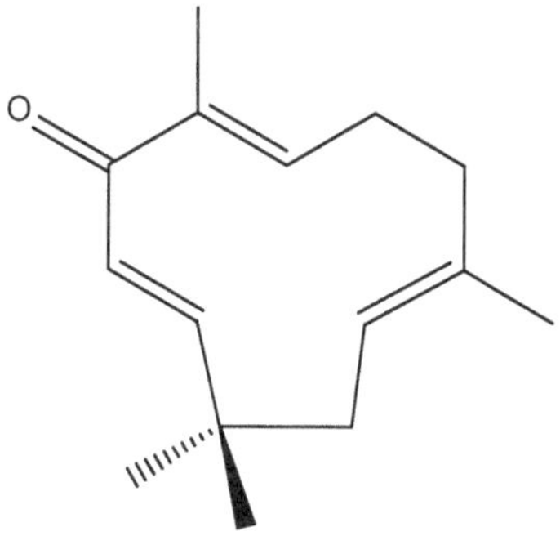

Structure of zerumbone

References

1. Wei, P.H., Wu, S.Z., Mu, X.M., Xu, B., Su, Q.J., Wei, J.L., Yang, Y., Qin, B. and Xie, Z.C., 2015. Effect of alcohol extract of *Acanthus ilicifolius* L. on anti-duck hepatitis B virus and protection of liver. *Journal of Ethnopharmacology*, *160*, pp. 1–5.
2. Wai, K.K., Liang, Y., Zhou, L., Cai, L., Liang, C., Liu, L., Lin, X., Wu, H. and Lin, J., 2015. The protective effects of *Acanthus ilicifolius* alkaloid A and its derivatives on pro-and anti-inflammatory cytokines in rats with hepatic fibrosis. *Biotechnology and Applied Biochemistry*, *62*(4), pp. 537–546.
3. Ravikumar, S., Raja, M. and Gnanadesigan, M., 2012. Antibacterial potential of benzoate and phenylethanoid derivatives isolated from *Acanthus ilicifolius* L. leaf extracts. *Natural Product Research*, *26*(23), pp. 2270–2273.
4. Dai, Y., Chen, S.R., Chai, L., Zhao, J., Wang, Y. and Wang, Y., 2018. Overview of pharmacological activities of *Andrographis paniculata* and its major compound andrographolide. *Critical Reviews in Food Science and Nutrition*, pp. 1–42.
5. Yu, B.C., Chen, W.C. and Cheng, J.T., 2003. Antihyperglycemic effect of andrographolide in streptozotocin-induced diabetic rats. *Planta Medica*, *69*(12), pp. 1075–1079.
6. Sheeja, K. and Kuttan, G., 2007. Activation of cytotoxic T lymphocyte responses and attenuation of tumor growth in vivo by *Andrographis paniculata* extract and andrographolide. *Immunopharmacology and Immunotoxicology*, *29*(1), pp. 81–93.
7. Banerjee, M., Parai, D., Chattopadhyay, S. and Mukherjee, S.K., 2017. Andrographolide: antibacterial activity against common bacteria of human health concern and possible mechanism of action. *Folia Microbiologica*, *62*(3), pp. 237–244.
8. Calabrese, C., Berman, S.H., Babish, J.G., Ma, X., Shinto, L., Dorr, M., Wells, K., Wenner, C.A. and Standish, L.J., 2000. A phase I trial of andrographolide in HIV positive patients and normal volunteers. *Phytotherapy Research*, *14*(5), pp. 333–338.
9. Sule, A., Ahmed, Q.U., Latip, J., Samah, O.A., Omar, M.N., Umar, A. and Dogarai, B.B.S., 2012. Antifungal activity of *Andrographis paniculata* extracts and active principles against skin pathogenic fungal strains in vitro. *Pharmaceutical Biology*, *50*(7), pp. 850–856.
10. Singh, A. and Handa, S.S., 1995. Hepatoprotective activity of Apium graveolens and *Hygrophila auriculata* against paracetamol and thioacetamide intoxication in rats. *Journal of Ethnopharmacology*, *49*(3), pp. 119–126.
11. Vijayakumar, M., Govindarajan, R., Rao, G.M.M., Rao, C.V., Shirwaikar, A., Mehrotra, S. and Pushpangadan, P., 2006. Action of *Hygrophila auriculata* against streptozotocin-induced oxidative stress. *Journal of Ethnopharmacology*, *104*(3), pp. 356–361.
12. Hussain, M.S., Azam, F., Ahamed, K.N., Ravichandiran, V. and Alkskas, I., 2016. Anti-endotoxin effects of terpenoids fraction from *Hygrophila auriculata* in lipopolysaccharide-induced septic shock in rats. *Pharmaceutical Biology*, *54*(4), pp. 628–636.
13. Singh, B., Bani, S., Gupta, D.K., Chandan, B.K. and Kaul, A., 2003. Anti-inflammatory activity of 'TAF'an active fraction from the plant *Barleria prionitis* L. *Journal of Ethnopharmacology*, *85*(2–3), pp. 187–193.
14. Jaiswal, S.K., Dubey, M.K., Das, S. and Rao, C.V., 2014. Gastroprotective effect of the iridoid fraction from *Barleria prionitis* leaves on experimentally-induced gastric ulceration. *Chinese Journal of Natural Medicines*, *12*(10), pp. 738–744.

15. Chen, J.L., Blanc, P., Stoddart, C.A., Bogan, M., Rozhon, E.J., Parkinson, N., Ye, Z., Cooper, R., Balick, M., Nanakorn, W. and Kernan, M.R., 1998. New iridoids from the medicinal plant *Barleria prionitis* with potent activity against respiratory syncytial virus. *Journal of Natural Products*, *61*(10), pp. 1295–1297.
16. Reegan, A.D., Gandhi, M.R., Sivaraman, G., Cecilia, K.F., Ravindhran, R., Balakrishna, K., Paulraj, M.G. and Ignacimuthu, S., 2016. Bioefficacy of ecbolin A and ecbolin B isolated from *Ecbolium viride* (Forsk.) Alston on dengue vector *Aedes aegypti* L.(Diptera: Culicidae). *Parasite Epidemiology and Control*, *1*(2), pp. 78–84.
17. Lalitha, K.G. and Sethuraman, M.G., 2010. Anti-inflammatory activity of roots of *Ecbolium viride* (Forsk) Merrill. *Journal of Ethnopharmacology*, *128*(1), pp. 248–250.
18. Sarkar, C., Bose, S. and Banerjee, S., 2014. Evaluation of hepatoprotective activity of vasicinone in mice. *Journal of Experimental Biology*, *52*(7), pp. 705–711.
19. Jha, D.K., Panda, L., Lavanya, P., Ramaiah, S. and Anbarasu, A., 2012. Detection and confirmation of alkaloids in leaves of *Justicia adhatoda* and bioinformatics approach to elicit its anti-tuberculosis activity. *Applied Biochemistry and Biotechnology*, *168*(5), pp. 980–990.
20. Pa, R. and Mathew, L., 2012. Antimicrobial activity of leaf extracts of *Justicia adhatoda* L. in comparison with vasicine. *Asian Pacific Journal of Tropical Biomedicine*, *2*(3), pp. S1556–S1560.
21. Chakraborty, A. and Brantner, A.H., 2001. Study of alkaloids from *Adhatoda vasica* Nees on their antiinflammatory activity. *Phytotherapy Research*, *15*(6), pp. 532–534.
22. Kumar, K.S., Sabu, V., Sindhu, G., Rauf, A.A. and Helen, A., 2018. Isolation, identification and characterization of apigenin from *Justicia gendarussa* and its anti-inflammatory activity. *International Immunopharmacology*, *59*, pp. 157–167.
23. Kumar, K.S., Vijayan, V., Bhaskar, S., Krishnan, K., Shalini, V. and Helen, A., 2012. Anti-inflammatory potential of an ethyl acetate fraction isolated from *Justicia gendarussa* roots through inhibition of iNOS and COX-2 expression via NF-κB pathway. *Cellular Immunology*, *272*(2), pp. 283–289.
24. Senthilkumar, N., Varma, P. and Gurusubramanian, G., 2009. Larvicidal and adulticidal activities of some medicinal plants against the malarial vector, *Anopheles stephensi* (Liston). *Parasitology Research*, *104*(2), pp. 237–244.
25. Charoenchai, P., Vajrodaya, S., Somprasong, W., Mahidol, C., Ruchirawat, S. and Kittakoop, P., 2010. Part 1: Antiplasmodial, cytotoxic, radical scavenging and antioxidant activities of Thai plants in the family Acanthaceae. *Planta Medica*, *76*(16), pp. 1940–1943.
26. Alam, M.A., Subhan, N., Awal, M.A., Alam, M.S., Sarder, M., Nahar, L. Sarker, S.D., 2009. Antinociceptive and anti-inflammatory properties of *Ruellia tuberosa*. *Pharmaceutical Biology*, *47*(3), pp. 209–214.
27. Wiart, C., Akaho, E., Hannah, M., Yassim, M., Hamirnah, H., Au, T.S. and Sulaiman, M., 2005. Antimicrobial activity of *Ruellia tuberosa* L. *American Journal of Chinese Medicine*, *33*(4), pp. 683–685.
28. Keller, K. and Stahl, E., 1983. Composition of the essential oil from β-asarone free calamus. *Planta Medica*, *47*(2), pp. 71–74.
29. Wang, N., Han, Y., Luo, L., Zhang, Q., Ning, B. and Fang, Y., 2018. β-asarone induces cell apoptosis, inhibits cell proliferation and decreases migration and invasion of glioma cells. *Biomedicine & Pharmacotherapy*, *106*, pp. 655–664.
30. Zhang, Q.S., Wang, Z.H., Zhang, J.L., Duan, Y.L., Li, G.F. and Zheng, D.L., 2016. β-asarone protects against MPTP-induced Parkinson's disease via regulating long non-coding RNA MALAT1 and inhibiting α-synuclein protein expression. *Biomedicine & Pharmacotherapy*, *83*, pp. 153–159.

31. Jacob, J., Aravind, S.R., Nishanth Kumar, S., Sreelekha, T.T. and Kumar, D., 2015. Asarones from *Acorus calamus* in combination with Azoles and Amphotericin B: a novel synergistic combination to compete against human Pathogenic Candida species in vitro. *Applied Biochemistry and Biotechnology*, *175*(8), pp. 3683–3695.
32. Lee, J.Y., Lee, J.Y., Yun, B.S. and Hwang, B.K., 2004. Antifungal activity of β-asarone from rhizomes of Acorus gramineus. *Journal of Agricultural and Food Chemistry*, *52*(4), pp. 776–780.
33. Yokosuka, A., Mimaki, Y., Kuroda, M. and Sashida, Y., 2000. A new steroidal saponin from the leaves of *Agave americana*. *Planta Medica*, *66*(4), pp. 393–396.
34. Wilkomirski, B., Bobeyko, V.A. and Kintia, P.K., 1975. New steroidal saponins of *Agave americana*. *Phytochemistry*, *14*(12), pp. 2657–2659.
35. Peana, A.T., Moretti, M.D., Manconi, V., Desole, G. and Pippia, P., 1997. Anti-inflammatory activity of aqueous extracts and steroidal sapogenins of *Agave americana*. *Planta Medica*, *63*(3), pp. 199–202.
36. Guleria, S. and Kumar, A., 2009. Antifungal activity of *Agave americana* leaf extract against *Alternaria brassicae*, causal agent of *Alternaria* blight of Indian mustard (*Brassica juncea*). *Archives of Phytopathology and Plant Protection*, *42*(4), pp. 370–375.
37. Sharma, A.K., Agarwal, V., Kumar, R., Balasubramaniam, A., Mishra, A. and Gupta, R., 2011. Pharmacological studies on seeds of *Alangium salvifolium* L. *Acta Poloniae Pharmaceutica*, *68*(6), pp. 897–904.
38. Zahan, R., Nahar, L. and Nesa, M.L., 2013. Antinociceptive and anti-inflammatory activities of flower (*Alangium salvifolium*) extract. *Pakistan Journal of Biological Sciences*, *16*(19), pp. 1040–1045.
39. Wang, Y.L., Zhao, J.C., Liang, J.H., Tian, X.G., Huo, X.K., Feng, L., Ning, J., Wang, C., Zhang, B.J., Chen, G. and Li, N., 2017. A bioactive new protostane-type triterpenoid from Alisma plantago-aquatica subsp. orientale (Sam.) Sam. *Natural Product Research*, pp. 1–6.
40. Zhao, X.Y., Wang, G., Wang, Y., Tian, X.G., Zhao, J.C., Huo, X.K., Sun, C.P., Feng, L., Ning, J., Wang, C. and Zhang, B.J., 2017. Chemical constituents from Alisma plantago-aquatica subsp. orientale (Sam.) Sam and their anti-inflammatory and antioxidant activities. *Natural Product Research*, pp. 1–7.
41. Yoshikawa, M., Yamaguchi, S., Murakami, T., Matsuda, H., Yamahara, J. and Murakami, N., 1993. Absolute stereostructures of trifoliones A, B, C, and D, new biologically active diterpenes from the tuber of *Sagittaria trifolia* L. *Chemical and Pharmaceutical Bulletin*, *41*(9), pp. 1677–1679.
42. Liu, X.T., Pan, Q., Shi, Y., Williams, I.D., Sung, H.H.Y., Zhang, Q., Liang, J.Y., Ip, N.Y. and Min, Z.D., 2006. *ent*-Rosane and labdane diterpenoids from *Sagittaria sagittifolia* and their antibacterial activity against three oral pathogens. *Journal of Natural Products*, *69*(2), pp. 255–260.
43. Kothavade, P.S., Bulani, V.D., Nagmoti, D.M., Deshpande, P.S., Gawali, N.B. and Juvekar, A.R., 2015. Therapeutic effect of saponin rich fraction of *Achyranthes aspera* L. on adjuvant-induced arthritis in Sprague-Dawley rats. *Autoimmune Diseases*, pp. 1–8.
44. Narayan, C. and Kumar, A., 2014. Antineoplastic and immunomodulatory effect of polyphenolic components of *Achyranthes aspera* (PCA) extract on urethane induced lung cancer in vivo. *Molecular Biology Reports*, *41*(1), pp. 179–191.
45. Mukherjee, H., Ojha, D., Bag, P., Chandel, H.S., Bhattacharyya, S., Chatterjee, T.K., Mukherjee, P.K., Chakraborti, S. and Chattopadhyay, D., 2013. Anti-herpes virus activities of *Achyranthes aspera*: an Indian ethnomedicine, and its triterpene acid. *Microbiological Research*, *168*(4), pp. 238–244.

46. Siveen, K.S. and Kuttan, G., 2011. Immunomodulatory and antitumor activity of *Aerva lanata* ethanolic extract. *Immunopharmacology and Immunotoxicology*, *33*(3), pp. 423–432.
47. Chowdhury, D., Sayeed, A., Islam, A., Bhuiyan, M.S.A. and Khan, G.A.M., 2002. Antimicrobial activity and cytotoxicity of *Aerva lanata*. *Fitoterapia*, *73*(1), pp. 92–94.
48. Mandal, A., Ojha, D., Lalee, A., Kaity, S., Das, M., Chattopadhyay, D. and Samanta, A., 2015. Bioassay directed isolation of a novel anti-inflammatory cerebroside from the leaves of *Aerva sanguinolenta*. *Medicinal Chemistry Research*, *24*(5), pp. 1952–1963.
49. Rao, G.V., Kavitha, K., Gopalakrishnan, M. and Mukhopadhyay, T., 2012. Isolation and characterization of a potent antimicrobial compound from *Aerva sanguinolenta* Bl.: an alternative source of bakuchiol. *Journal of Pharmaceutical Research*, *5*(1), pp. 174–176.
50. Wu, C.H., Hsieh, H.T., Lin, J.A. and Yen, G.C., 2013. Alternanthera paronychioides protects pancreatic β-cells from glucotoxicity by its antioxidant, antiapoptotic and insulin secretagogue actions. *Food Chemistry*, *139*(1–4), pp. 362–370.
51. Petrus, A.A. and Seetharaman, T.R., 2005. Antioxidant flavone c-biosides from the aerial parts of *Alternanthera pungens*. *Indian Journal of Pharmaceutical Sciences*, *67*(2), p. 187.
52. Rayees, S., Kumar, A., Rasool, S., Kaiser, P., Satti, N.K., Sangwan, P.L., Singh, S., Johri, R.K. and Singh, G., 2013. Ethanolic extract of *Alternanthera sessilis* (AS-1) inhibits IgE-mediated allergic response in RBL-2H3 cells. *Immunological Investigations*, *42*(6), pp. 470–480.
53. Bhuyan, B., Baishya, K. and Rajak, P., 2018. Effects of *Alternanthera sessilis* on liver function in carbon tetra chloride induced hepatotoxicity in Wister rat model. *Indian Journal of Clinical Biochemistry*, *33*(2), pp. 190–195.
54. Sivakumar, R. and Sunnathi, D., 2016. Phytochemical screening and antimicrobial activity of ethanolic leaf extract of *Alternanthera sessilis* (L.) R. Br. ex Dc and *Alternanthera Philoxeroides* (Mart.) Griseb. *European Journal of Pharmaceutical Sciences*, *3*(3), pp. 409–412.
55. Pandey, S., Ganeshpurkar, A., Bansal, D. and Dubey, N., 2016. Hematopoietic effect of amaranthus cruentus extract on phenylhydrazine-induced toxicity in rats. *Journal of Dietary Supplements*, *13*(6), pp. 607–615.
56. Sani, H.A., Rahmat, A., Ismail, M., Rosli, R. and Endrini, S., 2004. Potential anticancer effect of red spinach (*Amaranthus gangeticus*) extract. *Asia Pacific Journal of Clinical Nutrition*, *13*(4), pp. 396–400.
57. Verma, R.K., Sisodia, R. and Bhatia, A.L., 2002. Radioprotective role of *Amaranthus gangeticus* L.: a biochemical study on mouse brain. *Journal of Medicinal Food*, *5*(4), pp. 189–195.
58. Mondal, A. and Maity, T.K., 2016. Antibacterial activity of a novel fatty acid (14E, 18E, 22E, 26E)-methyl nonacosa-14, 18, 22, 26 tetraenoate isolated from *Amaranthus spinosus*. *Pharmaceutical Biology*, *54*(10), pp. 2364–2367.
59. Ajileye, O.O., Obuotor, E.M., Akinkunmi, E.O. and Aderogba, M.A., 2015. Isolation and characterization of antioxidant and antimicrobial compounds from *Anacardium occidentale* L.(Anacardiaceae) leaf extract. *Journal of King Saud University-Science*, *27*(3), pp. 244–252.
60. Hemshekhar, M., Sebastin Santhosh, M., Kemparaju, K. and Girish, K.S., 2012. Emerging roles of anacardic acid and its derivatives: a pharmacological overview. *Basic & Clinical Pharmacology & Toxicology*, *110*(2), pp. 122–132.
61. Hollands, A., Corriden, R., Gysler, G., Dahesh, S., Olson, J., Ali, S.R., Kunkel, M.T., Lin, A.E., Forli, S., Newton, A.C. and Kumar, G.B., 2016. Natural product anacardic acid from cashew nut shells stimulates neutrophil extracellular trap production and bactericidal activity. *Journal of Biological Chemistry*, *291*(27), pp. 13964–13973.

62. Kubo, I., Muroi, H., Himejima, M., Yamagiwa, Y., Mera, H., Tokushima, K., Ohta, S. and Kamikawa, T., 1993. Structure-antibacterial activity relationships of anacardic acids. *Journal of Agricultural and Food Chemistry*, *41*(6), pp. 1016–1019.
63. Muroi, H. and Kubo, I., 1996. Antibacterial activity of anacardic acid and totarol, alone and in combination with methicillin, against methicillinresistant Staphylococcus aureus. *Journal of Applied Bacteriology*, *80*(4), pp. 387–394.
64. Gandhidasan, R., Thamaraichelvan, A. and Baburaj, S., 1991. Anti-inflammatory action of *Lannea coromandelica* by HRBC membrane stabilization. *Fitoterapia*, *62*, pp. 81–83.
65. Singh, S. and Singh, G.B., 1994. Anti-inflammatory activity of *Lannea coromandelica* bark extract in rats. *Phytotherapy Research*, *8*(5), pp. 311–313.
66. Imam, M.Z. and Moniruzzaman, M., 2014. Antinociceptive effect of ethanol extract of leaves of *Lannea coromandelica*. *Journal of Ethnopharmacology*, *154*(1), pp. 109–115.
67. Majumder, R., Jami, M.S.I., Alam, M.E.K. and Alam, M.B., 2013. Antidiarrheal activity of *Lannea coromandelica* L: bark extract. *American–Eurasian Journal of Scientific Research*, *8*, pp. 128–134.
68. Parvez, G.M., 2016. Pharmacological activities of Mango (*Mangifera indica*): A review. *Journal of Pharmacognosy and Phytochemistry*, *5*(3), p. 1.
69. Muruganandan, S., Srinivasan, K., Gupta, S., Gupta, P.K. and Lal, J., 2005. Effect of mangiferin on hyperglycemia and atherogenicity in streptozotocin diabetic rats. *Journal of Ethnopharmacology*, *97*(3), pp. 497–501.
70. Makare, N., Bodhankar, S. and Rangari, V., 2001. Immunomodulatory activity of alcoholic extract of *Mangifera indica* L. in mice. *Journal of Ethnopharmacology*, *78*(2–3), pp. 133–137.
71. Garrido, G., González, D., Delporte, C., Backhouse, N., Quintero, G., Núñez-Sellés, A.J. and Morales, M.A., 2001. Analgesic and anti-inflammatory effects of *Mangifera indica* L. extract (Vimang). *Phytotherapy Research*, *15*(1), pp. 18–21.
72. Sairam, K., Hemalatha, S., Kumar, A., Srinivasan, T., Ganesh, J., Shankar, M. and Venkataraman, S., 2003. Evaluation of anti-diarrhoeal activity in seed extracts of *Mangifera indica*. *Journal of Ethnopharmacology*, *84*(1), pp. 11–15.
73. Khan, H.B.H., Vinayagam, K.S., Renny, C.M., Palanivelu, S. and Panchanadham, S., 2013. Potential antidiabetic effect of the *Semecarpus anacardium* in a type 2 diabetic rat model. *Inflammopharmacology*, *21*(1), pp. 47–53.
74. Lingaraju, G.M., Krishna, V., Joy Hoskeri, H., Pradeepa, K., Venkatesh and Babu, P.S., 2012. Wound healing promoting activity of stem bark extract of *Semecarpus anacardium* using rats. *Natural Product Research*, *26*(24), pp. 2344–2347.
75. Chitnis, M.P., Bhatia, K.G., Phatak, M.K. and Kesava, K.R., 1980. Anti-tumour activity of the extract of *Semecarpus anacardium* L. nuts in experimental tumor models. *Indian Journal of Experimental Biology*, *18*(1), pp. 6–8.
76. Islam, S.M.A., Ahmed, K.T., Manik, M.K., Wahid, M.A. and Kamal, C.S.I., 2013. A comparative study of the antioxidant, antimicrobial, cytotoxic and thrombolytic potential of the fruits and leaves of *Spondias dulcis*. *Asian Pacific Journal of Tropical Biomedicine*, *3*(9), p. 682.
77. Mohamed, S., Saka, S., El-Sharkawy, S.H., Ali, A.M. and Muid, S., 1996. Antimycotic screening of 58 Malaysian plants against plant pathogens. *Pesticide Science*, *47*(3), pp. 259–264.
78. Jamkhande, P.G. and Wattamwar, A.S., 2015. *Annona reticulata* L.(Bullock's heart): plant profile, phytochemistry and pharmacological properties. *Journal of Traditional and Complementary Medicine*, *5*(3), pp. 144–152.
79. Chavan, M.J., Kolhe, D.R., Wakte, P.S. and Shinde, D.B., 2012. Analgesic and anti-inflammatory activity of kaur-16-en-19-oic acid from *Annona reticulata* L: bark. *Phytotherapy Research*, *26*(2), pp. 273–276.

80. Chang, F.R., Chen, J.L., Chiu, H.F., Wu, M.J. and Wu, Y.C., 1998. Acetogenins from seeds of *Annona reticulata*. *Phytochemistry*, *47*(6), pp. 1057–1061.
81. Chang, F.R. and Wu, Y.C., 2001. Novel cytotoxic annonaceous acetogenins from Annona m uricata. *Journal of Natural Products*, *64*(7), pp. 925–931.
82. Ma, C., Chen, Y., Chen, J., Li, X. and Chen, Y., 2017. A review on *Annona squamosa* L.: phytochemicals and biological activities. *The American Journal of Chinese Medicine*, *45*(5), pp. 933–964.
83. Fiaz, M., Martínez, L.C., da Silva Costa, M., Cossolin, J.F.S., Plata-Rueda, A., Gonçalves, W.G., Sant'Ana, A.E.G., Zanuncio, J.C. and Serrão, J.E., 2018. Squamocin induce histological and ultrastructural changes in the midgut cells of *Anticarsia gemmatalis* (Lepidoptera: Noctuidae). *Ecotoxicology and Environmental Safety*, *156*, pp. wl–8.
84. Souza, M., Bevilaqua, C.M., Morais, S.M., Costa, C.T., Silva, A.R. and Braz-Filho, R., 2008. Anthelmintic acetogenin from *Annona squamosa* L.: seeds. *Anais da Academia Brasileira de Ciências*, *80*(2), pp. 271–277.
85. Wu, P., Wu, M., Xu, L., Xie, H. and Wei, X., 2014. Anti-inflammatory cyclopeptides from exocarps of sugar-apples. *Food Chemistry*, *152*, pp. 23–28.
86. Wu, Y.C., Hung, Y.C., Chang, F.R., Cosentino, M., Wang, H.K. and Lee, K.H., 1996. Identification of ent-16β, 17-dihydroxykauran-19-oic acid as an anti-HIV principle and isolation of the new diterpenoids annosquamosins A and B from *Annona squamosa*. *Journal of Natural Products*, *59*(6), pp. 635–637.
87. Saleem, R., Ahmed, M., Ahmed, S.I., Azeem, M., Khan, R.A., Rasool, N., Saleem, H., Noor, F. and Faizi, S., 2005. Hypotensive activity and toxicology of constituents from root bark of *Polyalthia longifolia* var. pendula. *Phytotherapy Research*, *19*(10), pp. 881–884.
88. Chang, F., Hwang, T., Yang, Y., Li, C., Wu, C., Issa, H.H., Hsieh, W. and Wu, Y., 2006. Anti-inflammatory and cytotoxic diterpenes from formosan *Polyalthia longifolia* var. pendula. *Planta Medica*, *72*(14), p. 1344.
89. Murthy, M.M., Subramanyam, M., Bindu, M.H. and Annapurna, J., 2005. Antimicrobial activity of clerodane diterpenoids from *Polyalthia longifolia* seeds. *Fitoterapia*, *76*(3–4), pp. 336–339.
90. Paul, J.H., Seaforth, C.E. and Tikasingh, T., 2011. Eryngium foetidum L.: a review. *Fitoterapia*, *82*(3), pp. 302–308.
91. Promtes, K., Kupradinun, P., Rungsipipat, A., Tuntipopipat, S. and Butryee, C., 2016. Chemopreventive effects of eryngium foetidum L: leaves on COX-2 reduction in mice induced colorectal carcinogenesis. *Nutrition and Cancer*, *68*(1), pp. 144–153.
92. Iranshahy, M. and Iranshahi, M., 2011. Traditional uses, phytochemistry and pharmacology of asafoetida (Ferula assa-foetida oleo-gum-resin)—A review. *Journal of Ethnopharmacology*, *134*(1), pp. 1–10.
93. Kavoosi, G. and Rowshan, V., 2013. Chemical composition, antioxidant and antimicrobial activities of essential oil obtained from Ferula assa-foetida oleo-gum-resin: effect of collection time. *Food Chemistry*, *138*(4), pp. 2180–2187.
94. Lee, C.L., Chiang, L.C., Cheng, L.H., Liaw, C.C., Abd El-Razek, M.H., Chang, F.R. and Wu, Y.C., 2009. Influenza A (H1N1) antiviral and cytotoxic agents from Ferula assa-foetida. *Journal of Natural Products*, *72*(9), pp. 1568–1572
95. Al Faruk, M., Khan, M.F., Mian, M.Y., Rahman, M.S. and Rashid, M.A., 2015. Analgesic and anti-diarrheal activities of *Aganosma dichotoma* (Roth) K. Schum. in Swiss-albino mice model. *Bangladesh Pharmaceutical Journal*, *18*(1), pp. 15–19.
96. Nayak, S., Nalabothu, P., Sandiford, S., Bhogadi, V. and Adogwa, A., 2006. Evaluation of wound healing activity of *Allamanda cathartica* L. *Laurus nobilis* L. extracts on rats. *BMC Complementary and Alternative Medicine*, *6*(1), p. 12.

97. Baliga, M.S., 2012. Review of the phytochemical, pharmacological and toxicological properties of *Alstonia scholaris* L. R. Br (Saptaparna). *Chinese Journal of Integrative Medicine*, pp. 1–14.
98. Yang, J., Fu, J., Liu, X., Jiang, Z.H. and Zhu, G.Y., 2018. Monoterpenoid indole alkaloids from the leaves of *Alstonia scholaris* and their NF-κB inhibitory activity. *Fitoterapia, 124*, pp. 73–79.
99. Zhao, Y.L., Shang, J.H., Pu, S.B., Wang, H.S., Wang, B., Liu, L., Liu, Y.P., Hong-Mei, S. and Luo, X.D., 2016. Effect of total alkaloids from *Alstonia scholaris* on airway inflammation in rats. *Journal of Ethnopharmacology, 178*, pp. 258–265.
100. Zhang, L., Hua, Z., Song, Y. and Feng, C., 2014. Monoterpenoid indole alkaloids from *Alstonia rupestris* with cytotoxic, antibacterial and antifungal activities. *Fitoterapia, 97*, pp. 142–147.
101. Bonvicini, F., Mandrone, M., Antognoni, F., Poli, F. and Angela Gentilomi, G., 2014. Ethanolic extracts of *Tinospora cordifolia* and *Alstonia scholaris* show antimicrobial activity towards clinical isolates of methicillin-resistant and carbapenemase-producing bacteria. *Natural Product Research, 28*(18), pp. 1438–1445.
102. Liu, L., Chen, Y.Y., Qin, X.J., Wang, B., Jin, Q., Liu, Y.P. and Luo, X.D., 2015. Antibacterial monoterpenoid indole alkaloids from *Alstonia scholaris* cultivated in temperate zone. *Fitoterapia, 105*, pp. 160–164.
103. Zhang, L., Zhang, C.J., Zhang, D.B., Wen, J., Zhao, X.W., Li, Y. and Gao, K., 2014. An unusual indole alkaloid with anti-adenovirus and anti-HSV activities from *Alstonia scholaris*. *Tetrahedron Letters, 55*(10), pp. 1815–1817.
104. Wang, C.M., Chen, H.T., Wu, Z.Y., Jhan, Y.L., Shyu, C.L. Chou, C.H., 2016. Antibacterial and synergistic activity of pentacyclic triterpenoids isolated from *Alstonia scholaris*. *Molecules, 21*(2), p. 139.
105. Gandhi, M. and Vinayak, V.K., 1990. Preliminary evaluation of extracts of *Alstonia scholaris* bark for in vivo antimalarial activity in mice. *Journal of Ethnopharmacology, 29*(1), pp. 51–57.
106. Arulmozhi, S., Mazumder, P.M., Lohidasan, S. and Thakurdesai, P., 2010. Antidiabetic and antihyperlipidemic activity of leaves of *Alstonia scholaris* L. R. Br. *European Journal of Integrative Medicine, 2*(1), pp. 23–32.
107. Deshmukh, P.T., Fernandes, J., Atul, A. and Toppo, E., 2009. Wound healing activity of *Calotropis gigantea* root bark in rats. *Journal of Ethnopharmacology, 125*(1), pp. 178–181.
108. Pathak, A.K. and Argal, A., 2007. Analgesic activity of *Calotropis gigantea* flower. *Fitoterapia, 78*(1), pp. 40–42.
109. Alam, M.A., Habib, M.R., Nikkon, R., Rahman, M. and Karim, M.R., 2008. Antimicrobial activity of akanda (*Calotropis gigantea* L.) on some pathogenic bacteria. *Bangladesh Journal of Scientific and Industrial Research, 43*(3), pp. 397–404.
110. El-Desoky, A.H., Abdel-Rahman, R.F., Ahmed, O.K., El-Beltagi, H.S. and Hattori, M., 2018. Anti-inflammatory and antioxidant activities of naringin isolated from *Carissa carandas* L.: in vitro and in vivo evidence. *Phytomedicine, 42*, pp. 126–134.
111. Begum, S., Syed, S.A., Siddiqui, B.S., Sattar, S.A. and Choudhary, M.I., 2013. Carandinol: first isohopane triterpene from the leaves of *Carissa carandas* L. its cytotoxicity against cancer cell lines. *Phytochemistry Letters, 6*(1), p. 95.
112. Itankar, P.R., Lokhande, S.J., Verma, P.R., Arora, S.K., Sahu, R.A. and Patil, A.T., 2011. Antidiabetic potential of unripe *Carissa carandas* L. fruit extract. *Journal of Ethnopharmacology, 135*(2), pp. 430–433.
113. Verma, S. and Chaudhary, H.S., 2011. Effect of *Carissa carandas* against clinically pathogenic bacterial strains. *Journal of Pharmacy Research, 4*(10), p. 3769.

114. Jain, S., Sharma, P., Ghule, S., Jain, A. and Jain, N., 2013. In vivo anti-inflammatory activity of *Tabernaemontana divaricata* leaf extract on male albino mice. *Chinese Journal of Natural Medicines*, *11*(5), pp. 472–476.
115. Khan, M.S.A., Ahmed, N., Arifuddin, M., Rehman, A. and Ling, M.P., 2018. Indole alkaloids and anti-nociceptive mechanisms of *Tabernaemontana divaricata* (L.) R. Br. flower methanolic extract. *Food and Chemical Toxicology*.
116. Bao, M.F., Yan, J.M., Cheng, G.G., Li, X.Y., Liu, Y.P., Li, Y., Cai, X.H. and Luo, X.D., 2013. Cytotoxic indole alkaloids from *Tabernaemontana divaricata*. *Journal of Natural Products*, *76*(8), pp. 1406–1412.
117. Singh, B., A Sharma, R. and K Vyas, G., 2011. Antimicrobial, antineoplastic and cytotoxic activities of indole alkaloids from *Tabernaemontana divaricata* (L.) R. Br. *Current Pharmaceutical Analysis*, *7*(2), pp. 125–132.
118. Chakraborty, A. and Brantner, A.H., 1999. Antibacterial steroid alkaloids from the stem bark of *Holarrhena pubescens*. *Journal of Ethnopharmacology*, *68*(1–3), pp. 339–344.
119. Dutta, N.K. and Iyer, S.N., 1968. Anti-amoebic value of berberine and kurchi alkaloids. *Journal of the Indian Medical Association*, *50*(8), pp. 349–352.
120. Saha, S. and Subrahmanyam, E., 2013. Evaluation of anti-inflammatory activity of ethanolic extract of seeds of *Holarrhena pubescens*(Buch.-Ham.) Wall. *International Journal of Pharmacy & Pharmaceutical Sciences*, *5*, pp. 915–919.
121. Sadhu, S.K., Khatun, A., Ohtsuki, T. and Ishibashi, M., 2008. Constituents from Hoya parasitica and their cell growth inhibitory activity. *Planta Medica*, *74*(7), pp. 760–763.
122. Kumarappan, C. and Mandal, S.C., 2014. Short communication polyphenol extract of ichnocarpus frutescens leaves modifies hyperglycemia in dexamethasone (dex) treated rats. *Indian Journal of Physiology and Pharmacology*, *58*(4), pp. 441–445.
123. Bandara, V., Weinstein, S.A., White, J. and Eddleston, M., 2010. A review of the natural history, toxinology, diagnosis and clinical management of *Nerium oleander* (common oleander) and *Thevetia peruviana* (yellow oleander) poisoning. *Toxicon*, *56*(3), pp. 273–281.
124. Hussain, M.A. and Gorsi, M.S., 2004. Antimicrobial activity of *Nerium oleander* L. *Asian Journal of Plant Science*, *3*(2), pp. 177–180.
125. Kumar, S. and Anand, G.R., 2010. Evaluation of anti-inflammatory activity of *Nerium oleander*. *Pharmacia*, *1*, pp. 33–36.
126. Dey, A. and De, J.N., 2010. *Rauvolfia serpentina* (L). Benth. ex Kurz.-A review. *Asian Journal of Plant Sciences*, *9*(6), p. 285.
127. Plummer, A.J., Earl, A., Schneider, J.A., Trapold, J. and Barrett, W., 1954. Pharmacology of *Rauwolfia* alkaloids, including reserpine. *Annals of the New York Academy of Sciences*, *59*(1), pp. 8–21.
128. Locket, S., 1955. Oral preparations of *Rauwolfia serpentina* in treatment of essential hypertension. *British Medical Journal*, *1*, p. 809.
129. Govindachari, T.R., Viswanathan, N., Radhakrishnan, J., Pai, B.R., Natarajan, S. and Subramaniam, P.S., 1973. Minor alkaloids of *Tylophora asthmatica*: revised structure of tylophorinidine. *Tetrahedron*, *29*(6), pp. 891–897.
130. Datta, G., Gurnani, S., Sen, G. and Mulchandani, N.B., 1981. The interaction of tylophorinidine, a potential leukemic drug, with bovine serum albumin. *Biochemical and Biophysical Research Communications*, *101*(3), pp. 995–1002.
131. Dhiman, M., Parab, R.R., Manju, S.L., Desai, D.C. and Mahajan, G.B., 2012. Antifungal activity of hydrochloride salts of tylophorinidine and tylophorinine. *Natural Product Communications*, *7*(9), pp. 1171–1172.
132. Ganguly, T. and Sainis, K.B., 2001. Inhibition of cellular immune responses by *Tylophora indica* in experimental models. *Phytomedicine*, *8*(5), pp. 348–355.
133. Roy, A., Biswas, S.K., Chowdhury, A., Shill, M.C., Raihan, S.Z. and Muhit, M.A., 2011. Phytochemical screening, cytotoxicity and antibacterial activities of two Bangladeshi medicinal plants. *Pakistan Journal of Biological Sciences*, *14*(19), p. 905.

134. Peng, Q., Cai, H., Sun, X., Li, X., Mo, Z. and Shi, J., 2013. *Alocasia cucullata* exhibits strong antitumor effect in vivo by activating antitumor immunity. *PLoS One*, *8*(9), p. e75328.
135. Fang, M., Zhu, D., Luo, C., Li, C., Zhu, C., Ou, J., Li, H., Zhou, Y., Huo, C., Liu, W. and Peng, J., 2018. In vitro and in vivo anti-malignant melanoma activity of *Alocasia cucullata* via modulation of the phosphatase and tensin homolog/phosphoinositide 3-kinase/AKT pathway. *Journal of Ethnopharmacology*, *213*, pp. 359–365.
136. Huang, W., Li, C., Wang, Y., Yi, X. and He, X., 2017. Anti-inflammatory lignanamides and monoindoles from Alocasia macrorrhiza. *Fitoterapia*, *117*, pp. 126–132.
137. Elsbaey, M., Ahmed, K.F., Elsebai, M.F., Zaghloul, A., Amer, M.M. and Lahloub, M.F.I., 2017. Cytotoxic constituents of Alocasia macrorrhiza. *Zeitschrift für Naturforschung C*, *72*(1–2), pp. 21–25.
138. Rahman, M.M., Hossain, M.A., Siddique, S.A., Biplab, K.P. and Uddin, M.H., 2012. Antihyperglycemic, antioxidant and cytotoxic activities of *Alocasia macrorrhizos* (L.) rhizomes extract. *Turkish Journal of Biology*, *36*(5), pp. 574–579.
139. Reddy, S.K., Kumar, S.A., Kumar, V.D. and Ganapaty, S., 2012. Anti-inflammatory and analgesic activities of *Amorphophallus bulbifer* (Roxb) Kunth whole plant. *Tropical Journal of Pharmaceutical Research*, *11*(6), pp. 971–976.
140. Chua, M., Baldwin, T.C., Hocking, T.J. and Chan, K., 2010. Traditional uses and potential health benefits of *Amorphophallus konjac* K. Koch ex NE Br. *Journal of Ethnopharmacology*, *128*(2), pp. 268–278.
141. Wu, C., Qiu, S., Liu, P., Ge, Y. and Gao, X., 2018. Rhizoma Amorphophalli inhibits TNBC cell proliferation, migration, invasion and metastasis through the PI3K/Akt/mTOR pathway. *Journal of Ethnopharmacology*, *211*, pp. 89–100.
142. Ansil, P.N., Wills, P.J., Varun, R. and Latha, M.S., 2014. Cytotoxic and apoptotic activities of *Amorphophallus campanulatus* (Roxb.) Bl. tuber extracts against human colon carcinoma cell line HCT-15. *Saudi Journal of Biological Sciences*, *21*(6), pp. 524–531.
143. Shilpi, J.A., Ray, P.K., Sarder, M.M. and Uddin, S.J., 2005. Analgesic activity of *Amorphophallus campanulatus* tuber. *Fitoterapia*, *76*(3–4), pp. 367–369.
144. Khan, A., Rahman, M. and Islam, M.S., 2008. Antibacterial, antifungal and cytotoxic activities of amblyone isolated from *Amorphophallus campanulatus*. *Indian Journal of Pharmacology*, *40*(1), p. 41.
145. Uddin, S.J., Rouf, R., Shilpi, J.A., Alamgir, M., Nahar, L. and Sarker, S.D., 2008. Screening of some Bangladeshi medicinal plants for in vitro antibacterial activity. *Oriental Pharmacy and Experimental Medicine*, *8*(3), pp. 316–321.
146. Policegoudra, R.S., Goswami, S., Aradhya, S.M., Chatterjee, S., Datta, S., Sivaswamy, R., Chattopadhyay, P. and Singh, L., 2012. Bioactive constituents of *Homalomena aromatica* essential oil and its antifungal activity against dermatophytes and yeasts. *Journal of Medical Mycology*, *22*(1), pp. 83–87.
147. Zhao, F., Sun, C., Ma, L., Wang, Y.N., Wang, Y.F., Sun, J.F., Hou, G.G., Cong, W., Li, H.J., Zhang, X.H. and Ren, Y., 2016. New sesquiterpenes from the rhizomes of *Homalomena occulta*. *Fitoterapia*, *109*, pp. 113–118.
148. Yadav, A.K., 2012. Efficacy of *Lasia spinosa* leaf extract in treating mice infected with *Trichinella spiralis*. *Parasitology Research*, *110*(1), pp. 493–498.
149. Bhurat, M.R., Sapkale, H.S., Salunkhe, K.G., Sanghvi, R.S. and Kawatikwar, P.S., 2011. Preliminary chemical evaluation and in vitro anti-inflammatory activity of leaves of *Remusatia vivipara*. *Asian Journal of Biochemical and Pharmaceutical Research*, *2*(1), pp. 303–306.
150. Peng, W., Liu, Y.J., Wu, N., Sun, T., He, X.Y., Gao, Y.X. and Wu, C.J., 2015. *Areca catechu* L.(Arecaceae): a review of its traditional uses, botany, phytochemistry, pharmacology and toxicology. *Journal of Ethnopharmacology*, *164*, pp. 340–356.

151. Jeng, J.H., Chen, S.Y., Liao, C.H., Tung, Y.Y., Lin, B.R., Hahn, L.J. and Chang, M.C., 2002. Modulation of platelet aggregation by areca nut and betel leaf ingredients: roles of reactive oxygen species and cyclooxygenase. *Free Radical Biology and Medicine*, *32*(9), pp. 860–871.
152. Khan, S., Mehmood, M.H., Ali, A.N.A., Ahmed, F.S., Dar, A. and Gilani, A.H., 2011. Studies on anti-inflammatory and analgesic activities of betel nut in rodents. *Journal of Ethnopharmacology*, *135*(3), pp. 654–661.
153. Hada, L.S., Kakiuchi, N., Hattori, M. and Namba, T., 1989. Identification of antibacterial principles against Streptococcus mutans and inhibitory principles against glucosyltransferase from the seed of *Areca catechu* L. *Phytotherapy Research*, *3*(4), pp. 140–144.
154. Yenjit, P., Issarakraisila, M., Intana, W. and Chantrapromma, K., 2010. Fungicidal activity of compounds extracted from the pericarp of *Areca catechu* against *Colletotrichum gloeosporioides* in vitro and in mango fruit. *Postharvest Biology and Technology*, *55*(2), pp. 129–132.
155. Yoshikawa, M., Xu, F., Morikawa, T., Pongpiriyadacha, Y., Nakamura, S., Asao, Y., Kumahara, A. and Matsuda, H., 2007. Medicinal flowers. XII. 1) New spirostane-type steroid saponins with antidiabetogenic activity from *Borassus flabellifer*. *Chemical and Pharmaceutical Bulletin*, *55*(2), pp. 308–316.
156. Keerthi, A.A.P., Mendis, W.S.J., Jansz, E.R., Ekanayake, S. and Perera, M.S.A., 2007. A preliminary study on the effects of an antibacterial steroidal saponin from *Borassus flabellifer* L. fruit, on wound healing. *Journal of the National Science Foundation of Sri Lanka*, *35*(4), pp. 263–265.
157. Révész, L., Hiestand, P., La Vecchia, L., Naef, R., Naegeli, H.U., Oberer, L. and Roth, H.J., 1999. Isolation and synthesis of a novel immunosuppressive 17α-substituted dammarane from the flour of the Palmyrah palm (*Borassus flabellifer*). *Bioorganic & Medicinal Chemistry Letters*, *9*(11), pp. 1521–1526.
158. Hong, E.H., Heo, E.Y., Song, J.H., Kwon, B.E., Lee, J.Y., Park, Y., Kim, J., Chang, S.Y., Chin, Y.W., Jeon, S.M. and Ko, H.J., 2017. Trans-scirpusin A showed antitumor effects via autophagy activation and apoptosis induction of colorectal cancer cells. *Oncotarget*, *8*(25), p. 41401.
159. Bhattacharjee, P. and Bhattacharyya, D., 2013. Characterization of the aqueous extract of the root of *Aristolochia indica*: evaluation of its traditional use as an antidote for snake bites. *Journal of Ethnopharmacology*, *145*(1), pp. 220–226.
160. Desai, D.C., Jacob, J., Almeida, A., Kshirsagar, R. and Manju, S.L., 2014. Isolation, structural elucidation and anti-inflammatory activity of astragalin,(–) hinokinin, aristolactam I and aristolochic acids (I & II) from *Aristolochia indica*. *Natural Product Research*, *28*(17), pp. 1413–1417.
161. Shafi, P.M., Rosamma, M.K., Jamil, K. and Reddy, P.S., 2002. Antibacterial activity of the essential oil from *Aristolochia indica*. *Fitoterapia*, *73*(5), pp. 439–441.
162. Venkatadri, B., Arunagirinathan, N., Rameshkumar, M.R., Ramesh, L., Dhanasezhian, A. and Agastian, P., 2015. In vitro antibacterial activity of aqueous and ethanol extracts of *Aristolochia indica* and *Toddalia asiatica* against multidrug-resistant bacteria. *Indian Journal of Pharmaceutical Sciences*, *77*(6), p. 788.
163. Vogler, B.K. and Ernst, E., 1999. *Aloe vera*: a systematic review of its clinical effectiveness. *British Journal of General Practice*, *49*(447), pp. 823–828.
164. Eshun, K. and He, Q., 2004. *Aloe vera*: a valuable ingredient for the food, pharmaceutical and cosmetic industries—a review. *Critical Reviews in Food Science and Nutrition*, *44*(2), pp. 91–96.
165. Surjushe, A., Vasani, R. and Saple, D.G., 2008. *Aloe vera*: a short review. *Indian Journal of Dermatology*, *53*(4), p. 163.

166. Hossain, H., Karmakar, U.K., Biswas, S.K., Shahid-Ud-Daula, A.F.M., Jahan, I.A., Adnan, T. and Chowdhury, A., 2013. Antinociceptive and antioxidant potential of the crude ethanol extract of the leaves of *Ageratum conyzoides* grown in Bangladesh. *Pharmaceutical Biology*, *51*(7), pp. 893–898.
167. Durodola, J.I., 1977. Antibacterial property of crude extracts from a herbal wound healing remedy–*Ageratum conyzoides*, L. *Planta Medica*, *32*(8), pp. 388–390.
168. De Melo, N.I., Magalhaes, L.G., De Carvalho, C.E., Wakabayashi, K.A., De P Aguiar, G., Ramos, R.C., Mantovani, A.L., Turatti, I.C., Rodrigues, V., Groppo, M. and Cunha, W.R., 2011. Schistosomicidal activity of the essential oil of *Ageratum conyzoides* L.(Asteraceae) against adult *Schistosoma mansoni* worms. *Molecules*, *16*(1), pp. 762–773.
169. Nour, A.M., Khalid, S.A., Kaiser, M., Brun, R., Wai'l, E.A. and Schmidt, T.J., 2010. The antiprotozoal activity of methylated flavonoids from *Ageratum conyzoides* L. *Journal of Ethnopharmacology*, *129*(1), pp. 127–130.
170. Taiwo, O.B., Olajide, O.A., Soyannwo, O.O. and Makinde, J.M., 2000. Anti-inflammatory, antipyretic and antispasmodic properties of *Chromolaena odorata*. *Pharmaceutical Biology*, *38*(5), pp. 367–370.
171. Triratana, T., Suwannuraks, R. and Naengchomnong, W., 1991. Effect of *Eupatorium odoratum* on blood coagulation. *Journal of the Medical Association of Thailand= Chotmaihet thangphaet*, *74*(5), pp. 283–287.
172. Wagner, H., Geyer, B., Kiso, Y., Hikino, H. and Rao, G.S., 1986. Coumestans as the main active principles of the liver drugs *Eclipta alba* and *Wedelia calendulacea* 1. *Planta Medica*, *52*(5), pp. 370–374.
173. Manvar, D., Mishra, M., Kumar, S. and Pandey, V.N., 2012. Identification and evaluation of anti-hepatitis C virus phytochemicals from *Eclipta alba*. *Journal of Ethnopharmacology*, *144*(3), pp. 545–554.
174. Sarveswaran, S., Gautam, S.C. and Ghosh, J., 2012. Wedelolactone, a medicinal plant-derived coumestan, induces caspase-dependent apoptosis in prostate cancer cells via downregulation of PKCε without inhibiting Akt. *International Journal of Oncology*, *41*(6), p. 21.
175. Tewtrakul, S., Subhadhirasakul, S., Cheenpracha, S. and Karalai, C., 2007. HIV-1 protease and HIV-1 integrase inhibitory substances from *Eclipta prostrata*. *Phytotherapy Research*, *21*(11), pp. 1092–1095.
176. Dua, T.K., Dewanjee, S. and Khanra, R., 2016. Prophylactic role of *Enhydra fluctuans* against arsenic-induced hepatotoxicity via anti-apoptotic and antioxidant mechanisms. *Redox Report*, *21*(4), pp. 147–154.
177. Sannigrahi, S., Mazumder, U.K., Mondal, A., Pal, D., Mishra, S.L. Roy, S., 2010. Flavonoids of *Enhydra fluctuans* exhibit anticancer activity against Ehrlich's ascites carcinoma in mice. *Natural Product Communications*, *5*(8), pp. 1239–1242.
178. Yadava, R.N. and Singh, S.K., 2007. Novel bioactive constituents from *Enhydra fluctuans* LOUR. *Natural Product Research*, *21*(6), pp. 481–486.
179. Akter, R., Uddin, S.J., Grice, I.D. and Tiralongo, E., 2014. Cytotoxic activity screening of Bangladeshi medicinal plant extracts. *Journal of Natural Medicines*, *68*(1), pp. 246–252.
180. Aderogba, M.A., McGaw, L.J., Bagla, V.P., Eloff, J.N. and Abegaz, B.M., 2014. In vitro antifungal activity of the acetone extract and two isolated compounds from the weed, *Pseudognaphalium luteoalbum*. *South African Journal of Botany*, *94*, pp. 74–78.
181. Huang, W.C., Wu, L.Y., Hu, S. and Wu, S.J., 2018. Spilanthol inhibits COX-2 and ICAM-1 expression via suppression of NF-κB and MAPK signaling in interleukin-1β-stimulated human lung epithelial cells. *Inflammation*, *41*(5), pp. 1934–1944.

182. Sathyaprasad, S., Jose, B.K. and Chandra, H.S., 2015. Antimicrobial and antifungal efficacy of *Spilanthes acmella* as an intracanal medicament in comparison to calcium hydroxide: an in vitro study. *Indian Journal of Dental Research*, *26*(5), p. 528.
183. Tsai, C.H., Tzeng, S.F., Hsieh, S.C., Lin, C.Y., Tsai, C.J., Chen, Y.R., Yang, Y.C., Chou, Y.W., Lee, M.T. and Hsiao, P.W., 2015. Development of a standardized and effect-optimized herbal extract of *Wedelia chinensis* for prostate cancer. *Phytomedicine*, *22*(3), pp. 406–414.
184. Mottakin, A.K.M., Chowdhury, R., Haider, M.S., Rahman, K.M., Hasan, C.M. and Rashid, M.A., 2004. Cytotoxicity and antibacterial activity of extractives from *Wedelia calendulacea*. *Fitoterapia*, *75*(3–4), pp. 355–359.
185. Sharma, A.K., Anand, K.K., Pushpangadan, P., Chandan, B.K., Chopra, C.L., Prabhakar, Y.S. and Damodaran, N.P., 1989. Hepatoprotective effects of *Wedelia calendulacea*. *Journal of Ethnopharmacology*, *25*(1), pp. 93–102.
186. Verma, A., Singh, D., Anwar, F., Bhatt, P.C., Al-Abbasi, F. and Kumar, V., 2018. Triterpenoids principle of *Wedelia calendulacea* attenuated diethynitrosamine-induced hepatocellular carcinoma via down-regulating oxidative stress, inflammation and pathology via NF-kB pathway. *Inflammopharmacology*, *26*(1), pp. 133–146.
187. Haque, M.E., Rahman, S., Rahmatullah, M. and Jahan, R., 2013. Evaluation of anti-hyperglycemic and antinociceptive activity of *Xanthium indicum* stem extract in Swiss albino mice. *BMC Complementary and Alternative Medicine*, *13*(1), p. 296.
188. Imam, M.Z., Nahar, N., Akter, S. and Rana, M.S., 2012. Antinociceptive activity of methanol extract of flowers of *Impatiens balsamina*. *Journal of Ethnopharmacology*, *142*(3), pp. 804–810.
189. Kim, C.S., Bae, M., Oh, J., Subedi, L., Suh, W.S., Choi, S.Z., Son, M.W., Kim, S.Y., Choi, S.U., Oh, D.C. and Lee, K.R., 2017. Anti-neurodegenerative biflavonoid glycosides from *impatiens balsamina*. *Journal of Natural Products*, *80*(2), pp. 471–478.
190. Yang, X., Summerhurst, D.K., Koval, S.F., Ficker, C., Smith, M.L. Bernards, M.A., 2001. Isolation of an antimicrobial compound from *Impatiens balsamina* L. using bioassay-guided fractionation. *Phytotherapy Research*, *15*(8), pp. 676–680.
191. Sithisarn, P., Nantateerapong, P., Rojsanga, P. and Sithisarn, P., 2016. Screening for antibacterial and antioxidant activities and phytochemical analysis of *Oroxylum indicum* fruit extracts. *Molecules*, *21*(4), p. 446.
192. Singh, J. and Kakkar, P., 2013. Modulation of liver function, antioxidant responses, insulin resistance and glucose transport by *Oroxylum indicum* stem bark in STZ induced diabetic rats. *Food and Chemical Toxicology*, *62*, pp. 722–731.
193. Sun, W., Zhang, B., Yu, X., Zhuang, C., Li, X., Sun, J., Xing, Y., Xiu, Z. and Dong, Y., 2018. Oroxin A from *Oroxylum indicum* prevents the progression from prediabetes to diabetes in streptozotocin and high-fat diet induced mice. *Phytomedicine*, *38*, pp. 24–34.
194. Owoyele, B.V. and Bakare, A.O., 2018. Analgesic properties of aqueous bark extract of *Adansonia digitata* in Wistar rats. *Biomedicine & Pharmacotherapy*, *97*, pp. 209–212.
195. Adeoye, A.O. and Bewaji, C.O., 2018. Chemopreventive and remediation effect of *Adansonia digitata* L. Baobab stem bark extracts in mouse model malaria. *Journal of Ethnopharmacology*, *210*, pp. 31–38.
196. Jain, V. and Verma, S.K., 2012. *Pharmacology of Bombax ceiba L.* Springer Science & Business Media: Berlin, Germany.
197. Wang, G.K., Lin, B.B., Rao, R., Zhu, K., Qin, X.Y., Xie, G.Y. and Qin, M.J., 2013. A new lignan with anti-HBV activity from the roots of *Bombax ceiba*. *Natural Product Research*, *27*(15), pp. 1348–1352.
198. Jalalpure, S.S. and Gadge, N.B., 2011. Diuretic effects of young fruit extracts of *Bombax ceiba* L. in rats. *Indian Journal of Pharmaceutical Sciences*, *73*(3), p. 306.

199. Bhargava, C., Thakur, M. and Yadav, S.K., 2012. Effect of *Bombax ceiba* L. on spermatogenesis, sexual behaviour and erectile function in male rats. *Andrologia*, *44*, pp. 474–478.
200. Saleem, R., Ahmad, M., Hussain, S.A., Qazi, A.M., Ahmad, S.I., Qazi, M.H., Ali, M., Faizi, S., Akhtar, S. and Husnain, S.N., 1999. Hypotensive, hypoglycaemic and toxicological studies on the flavonol C-glycoside shamimin from *Bombax ceiba*. *Planta Medica*, *65*(4), pp. 331–334.
201. Yang, R., Zhou, Q., Wen, C., Hu, J., Li, H., Zhao, M. and Zhao, H., 2013. Mustard seed (S inapis A lba L inn) attenuates imiquimod-induced psoriasiform inflammation of BALB/c mice. *The Journal of Dermatology*, *40*(7), pp. 543–552.
202. Ono, H., Tesaki, S., Tanabe, S. and Watanabe, M., 1998. 6-Methylsulfinylhexyl isothiocyanate and its homologues as food-originated compounds with antibacterial activity against *Escherichia coli* and *Staphylococcus aureus*. *Bioscience, Biotechnology, and Biochemistry*, *62*(2), pp. 363–365.
203. Takasugi, M., Katsui, N. and Shirata, A., 1986. Isolation of three novel sulphur-containing phytoalexins from the chinese cabbage *Brassica campestris* L. ssp. pekinensis. *Journal of the Chemical Society, Chemical Communications*, *14*, pp. 1077–1078.
204. Hayes, J.D., Kelleher, M.O. and Eggleston, I.M., 2008. The cancer chemopreventive actions of phytochemicals derived from glucosinolates. *European Journal of Nutrition*, *47*(2), pp. 73–88.
205. Kumar, V., Thakur, A.K., Barothia, N.D. and Chatterjee, S.S., 2011. Therapeutic potentials of Brassica juncea: An overview. *TANG [HUMANITAS MEDICINE]*, *1*(1), pp. 2–1.
206. Saeidnia, S. and Gohari, A.R., 2012. Importance of *Brassica napus* as a medicinal food plant. *Journal of Medicinal Plants Research*, *6*(14), pp. 2700–2703.
207. Fahey, J.W., Zalcmann, A.T. and Talalay, P., 2001. The chemical diversity and distribution of glucosinolates and isothiocyanates among plants. *Phytochemistry*, *56*(1), pp. 5–51.
208. Brown, A.F., Yousef, G.G., Jeffery, E.H., Klein, B.P., Wallig, M.A., Kushad, M.M. and Juvik, J.A., 2002. Glucosinolate profiles in broccoli: variation in levels and implications in breeding for cancer chemoprotection. *Journal of the American Society for Horticultural Science*, *127*(5), pp. 807–813.
209. Bendimerad, N., Bendiab, S.A.T., Breme, K. and Fernandez, X., 2007. Essential oil composition of aerial parts of *Sinapis arvensis* L. from Algeria. *Journal of Essential Oil Research*, *19*(3), pp. 206–208.
210. Maurer, H.R., 2001. Bromelain: biochemistry, pharmacology and medical use. *Cellular and Molecular Life Sciences*, *58*(9), pp. 1234–1245.
211. Domingues, L.F., Giglioti, R., Feitosa, K.A., Fantatto, R.R., Rabelo, M.D., de Sena Oliveira, M.C., Bechara, G.H., de Oliveira, G.P., Junior, W.B. and de Souza Chagas, A.C., 2013. In vitro and in vivo evaluation of the activity of pineapple (*Ananas comosus*) on *Haemonchus contortus* in Santa Inês sheep. *Veterinary Parasitology*, *197*(1–2), pp. 263–270.
212. Shirwaikar, A., Rajendran, K. and Barik, R., 2006. Effect of aqueous bark extract of *Garuga pinnata* Roxb. in streptozotocin-nicotinamide induced type-II diabetes mellitus. *Journal of Ethnopharmacology*, *107*(2), pp. 285–290.
213. Ara, K., Kaisar, M.A., Rahman, M.S., Chowdhury, S.R., Islam, F. and Rashid, M.A., 2012. Antimicrobial Constituents from *Garuga pinnata* Roxb. *Latin American Journal of Pharmacy*, *31*(7), pp. 1071–1073.
214. Tomassini, L., Foddai, S., Serafini, M. and Cometa, M.F., 2000. Bis-iridoid glucosides from *Abelia chinensis*. *Journal of Natural Products*, *63*(7), pp. 998–999.
215. Chahar, M.K., Kumar, D.S., Lokesh, T. and Manohara, K.P., 2012. In-vivo antioxidant and immunomodulatory activity of mesuol isolated from *Mesua ferrea* L: seed oil. *International Immunopharmacology*, *13*(4), pp. 386

216. Jalalpure, S.S., Mandavkar, Y.D., Khalure, P.R., Shinde, G.S., Shelar, P.A. and Shah, A.S., 2011. Antiarthritic activity of various extracts of *Mesua ferrea* L: seed. *Journal of Ethnopharmacology*, *138*(3), pp. 700–704.
217. Mazumder, R., Dastidar, S.G., Basu, S.P. and Mazumder, A., 2005. Effect of *Mesua ferrea* L. flower extract on Salmonella. *Indian Journal of Experimental Biology*, *43*(6), pp. 566–568.
218. Govindarajan, R., Vijayakumar, M., Singh, M., Rao, C.V., Shirwaikar, A., Rawat, A.K.S. and Pushpangadan, P., 2006. Antiulcer and antimicrobial activity of *Anogeissus latifolia*. *Journal of Ethnopharmacology*, *106*(1), pp. 57–61.
219. Navale, A.M. and Paranjape, A., 2018. Antidiabetic and renoprotective effect of *Anogeissus acuminata* leaf extract on experimentally induced diabetic nephropathy. *Journal of Basic and Clinical Physiology and Pharmacology*, *29*(4), pp. 359–364.
220. Rimando, A.M., Pezzuto, J.M., Farnsworth, N.R., Santisuk, T., Reutrakul, V. and Kawanishi, K., 1994. New lignans from *Anogeissus acuminata* with HIV-1 reverse transcriptase inhibitory activity. *Journal of Natural Products*, *57*(7), pp. 896–904.
221. Govindarajan, R., Vijayakumar, M., Singh, M., Rao, C.V., Shirwaikar, A., Rawat, A.K.S. and Pushpangadan, P., 2006. Antiulcer and antimicrobial activity of *Anogeissus latifolia*. *Journal of Ethnopharmacology*, *106*(1), pp. 57–61.
222. Rahman, M.S., Rahman, M.Z., Uddin, A.A. and Rashid, M.A., 2007. Steriod and Triterpenoid from *Anogeissus latifolia*. *Dhaka University Journal of Pharmaceutical Sciences*, *6*(1), pp. 47–50.
223. Dwivedi, S., 2007. *Terminalia arjuna* Wight & Arn.—a useful drug for cardiovascular disorders. *Journal of Ethnopharmacology*, *114*(2), pp. 114–129.
224. Pettit, G.R., Hoard, M.S., Doubek, D.L., Schmidt, J.M., Pettit, R.K., Tackett, L.P. and Chapuis, J.C., 1996. Antineoplastic agents 338. The cancer cell growth inhibitory. Constituents of *Terminalia arjuna*. *Journal of Ethnopharmacology*, *53*(2), pp. 57–63.
225. Devi, R.S., Narayan, S., Vani, G. and Devi, C.S.S., 2007. Gastroprotective effect of *Terminalia arjuna* bark on diclofenac sodium induced gastric ulcer. *Chemico-Biological Interactions*, *167*(1), pp. 71–83.
226. Biswas, M., Biswas, K., Karan, T.K., Bhattacharya, S., Ghosh, A.K. and Haldar, P.K., 2011. Evaluation of analgesic and anti-inflammatory activities of *Terminalia arjuna* leaf. *Journal of Phytology*, *3*(1), pp. 33–38.
227. Motamarri, S.N., Karthikeyan, M., Kannan, M. and Rajasekar, S., 2012. *Terminalia belerica* Roxb.—a phytopharmacological review. *International Journal of Research in Pharmaceutical*, *3*, pp. 96–99.
228. Tariq, M., Hussain, S.J., Asif, M. and Jahan, M., 1977. Protective effect of fruit extracts of *Emblica officinalis* (Gaertn). & *Terminalia belerica* (Roxb.) in experimental myocardial necrosis in rats. *Indian Journal of Experimental Biology*, *15*(6), p. 485.
229. Kumar, G.P.S., Arulselvan, P., Kumar, D.S. and Subramanian, S.P., 2006. Anti-diabetic activity of fruits of *Terminalia chebula* on streptozotocin induced diabetic rats. *Journal of Health Science*, *52*(3), pp. 283–291.
230. Suguna, L., Singh, S., Sivakumar, P., Sampath, P. and Chandrakasan, G., 2002. Influence of *Terminalia chebula* on dermal wound healing in rats. *Phytotherapy Research*, *16*(3), pp. 227–231.
231. Kumar, C.U., Pokuri, V.K. and Pingali, U., 2015. Evaluation of the analgesic activity of standardized aqueous extract of *Terminalia chebula* in healthy human participants using hot air pain model. *Journal of Clinical and Diagnostic Research*, *9*(5), p. FC01.
232. Jadon, A., Bhadauria, M. and Shukla, S., 2007. Protective effect of *Terminalia belerica* Roxb. and gallic acid against carbon tetrachloride induced damage in albino rats. *Journal of Ethnopharmacology*, *109*(2), pp. 214–218.

233. Suryaprakash, D.V., Sreesatya, N., Avanigadda, S. and Vangalapati, M., 2012. Pharmacological review on *Terminalia chebula*. *International Journal of Research in Pharmaceutical and Biomedical Sciences*, *3*(2), pp. 679–683.
234. Saleem, A., Husheem, M., Härkönen, P. and Pihlaja, K., 2002. Inhibition of cancer cell growth by crude extract and the phenolics of *Terminalia chebula* retz. fruit. *Journal of Ethnopharmacology*, *81*(3), pp. 327–336.
235. Reddy, D.B., Reddy, T.C.M., Jyotsna, G., Sharan, S., Priya, N., Lakshmipathi, V. and Reddanna, P., 2009. Chebulagic acid, a COX–LOX dual inhibitor isolated from the fruits of *Terminalia chebula* Retz., induces apoptosis in COLO-205 cell line. *Journal of Ethnopharmacology*, *124*(3), pp. 506–512.
236. Reddy, D.B. and Reddanna, P., 2009. Chebulagic acid (CA) attenuates LPS-induced inflammation by suppressing NF-κB and MAPK activation in RAW 264.7 macrophages. *Biochemical and Biophysical Research Communications*, *381*(1), pp. 112–117.
237. Rao, N.K. and Nammi, S., 2006. Antidiabetic and renoprotective effects of the chloroform extract of *Terminalia chebula* Retz: seeds in streptozotocin-induced diabetic rats. *BMC Complementary and Alternative Medicine*, *6*(1), p. 17.
238. Pavithra, P.S., Sreevidya, N. and Verma, R.S., 2009. Antibacterial and antioxidant activity of methanol extract of *Evolvulus nummularius*. *Indian Journal of Pharmacology*, *41*(5), p. 233.
239. Sajak, A.A.B., Mediani, A., Dom, N.S.M., Machap, C., Hamid, M., Ismail, A., Khatib, A. and Abas, F., 2017. Effect of *Ipomoea aquatica* ethanolic extract in streptozotocin (STZ) induced diabetic rats via 1H NMR-based metabolomics approach. *Phytomedicine*, *36*, pp. 201–209.
240. Sokeng, S.D., Rokeya, B., Hannan, J.M.A., Junaida, K., Zitech, P., Ali, L., Ngounou, G., Lontsi, D. and Kamtchouing, P., 2007. Inhibitory effect of *Ipomoea aquatica* extracts on glucose absorption using a perfused rat intestinal preparation. *Fitoterapia*, *78*(7–8), pp. 526–529.
241. Malalavidhane, T.S., Wickramasinghe, S.N. and Jansz, E.R., 2000. Oral hypoglycaemic activity of *Ipomoea aquatica*. *Journal of Ethnopharmacology*, *72*(1–2), pp. 293–298.
242. Abdelhadi, A.A., Elkheir, Y.M., Hassan, T. and Mustafa, A.A., 1986. Neuromuscular blocking activity of a crude aqueous extract of *Ipomoea fistulosa*. *Clinical and Experimental Pharmacology and Physiology*, *13*(2), pp. 169–171.
243. Khan, N.M.M.U. and Hossain, M.S., 2015. Scopoletin and β-sitosterol glucoside from roots of *Ipomoea digitata*. *Journal of Pharmacognosy and Phytochemistry*, *4*(2), pp. 5–7.
244. Matin, M.A., Tewari, J.P. and Kalani, D.K., 1969. Pharmacological effects of paniculatin—a glycoside isolated from *Ipomoea digitata* linn. *Journal of Pharmaceutical Sciences*, *58*(6), pp. 757–759.
245. Pandey, A.K., Gupta, P.P. and Lal, V.K., 2013. Preclinical evaluation of hypoglycemic activity of *Ipomoea digitata* tuber in streptozotocin-induced diabetic rats. *Journal of Basic and Clinical Physiology and Pharmacology*, *24*(1), pp. 35–39.
246. Sohgaura, A.K., Bigoniya, P. and Shrivastava, B., 2018. In vitro antilithiatic potential of *Kalanchoe pinnata*, *Emblica officinalis*, Bambusa nutans, and *Cynodon dactylon*. *Journal of Pharmacy and Bioallied Sciences*, *10*(2), p. 83.
247. de Araújo, E.R.D., Guerra, G.C.B., de Souza Araújo, D.F., de Araújo, A.A., Fernandes, J.M., de Araújo Júnior, R.F., da Silva, V.C., de Carvalho, T.G., de Santis Ferreira, L. Zucolotto, S.M., 2018. Gastroprotective and antioxidant activity of *Kalanchoe brasiliensis* and *Kalanchoe pinnata* leaf juices against indomethacin and ethanol-induced gastric lesions in rats. *International Journal of Molecular Sciences*, *19*, p. 1265.
248. Mora-Perez, A. and Hernández-Medel, M.D.R., 2016. Anticonvulsant activity of methanolic extract from *Kalanchoe pinnata* (Lam.) stems and roots in mice: a comparison to diazepam. *Neurología (English Edition)*, *31*(3), pp. 161–168.

249. Bopda, O.S.M., Longo, F., Bella, T.N., Edzah, P.M.O., Taïwe, G.S., Bilanda, D.C., Tom, E.N.L., Kamtchouing, P. and Dimo, T., 2014. Antihypertensive activities of the aqueous extract of *Kalanchoe pinnata* (Crassulaceae) in high salt-loaded rats. *Journal of Ethnopharmacology*, *153*(2), pp. 400–407.
250. Mosaddik, M.A. and Haque, M.E., 2003. Cytotoxicity and antimicrobial activity of goniothalamin isolated from *Bryonopsis laciniosa*. *Phytotherapy Research*, *17*(10), pp. 1155–1157.
251. Kabir, K.E., Khan, A.R. and Mosaddik, M.A., 2003. Goniothalamin–a potent mosquito larvicide from *Bryonopsis laciniosa* L. *Journal of Applied Entomology*, *127*(2), pp. 112–115.
252. Niazi, J., Singh, P., Bansal, Y. and Goel, R.K., 2009. Anti-inflammatory, analgesic and antipyretic activity of aqueous extract of fresh leaves of *Coccinia indica*. *Inflammopharmacology*, *17*(4), pp. 239–244.
253. Munasinghe, M.A.A.K., Abeysena, C., Yaddehige, I.S., Vidanapathirana, T. and Piyumal, K.P.B., 2011. Blood sugar lowering effect of *Coccinia grandis* (L.) J. Voigt: path for a new drug for diabetes mellitus. *Experimental Diabetes Research*, *2011*, p. 4.
254. Farrukh, U., Shareef, H., Mahmud, S., Ali, S.A. and Rizwani, G.H., 2008. Antibacterial activities of *Coccinia grandis* L. *Pakistan Journal of Botany*, *40*(3), pp. 1259–1262.
255. Kumar, P.V., Sivaraj, A., Elumalai, E.K. and Kumar, B.S., 2009. Carbon tetrachloride-induced hepatotoxicity in rats-protective role of aqueous leaf extracts of *Coccinia grandis*. *International Journal of PharmTech Research*, *1*(4), pp. 1612–1615.
256. Grover, J.K. and Yadav, S.P., 2004. Pharmacological actions and potential uses of *Momordica charantia*: a review. *Journal of Ethnopharmacology*, *93*(1), pp. 123–132.
257. Ali, L., Azad Khan, A.K., Rouf Mamun, M.I., Mosihuzzaman, M., Nahar, N., Nur-E-Alam, M. and Rokeya, B., 1993. Studies on hypoglycemic effects of fruit pulp, seed, and whole plant of *Momordica charantia* on normal and diabetic model rats. *Planta Medica*, *59*(5), pp. 408–412.
258. Ray, R.B., Raychoudhuri, A., Steele, R. and Nerurkar, P., 2010. Bitter melon (*Momordica charantia*) extract inhibits breast cancer cell proliferation by modulating cell cycle regulatory genes and promotes apoptosis. *Cancer Research*, *70*(5), pp. 1925–1931.
259. Akter, M., Mitu, I.Z., Proma, J.J., Rahman, S.M., Islam, M.R., Rahman, S. and Rahmatullah, M., 2014. Antihyperglycemic and antinociceptive activity evaluation of methanolic extract of *Trichosanthes anguina* fruits in Swiss albino mice. *Advances in Natural and Applied Sciences*, *8*(8), pp. 70–75.
260. Peerzada, A.M., Ali, H.H., Naeem, M., Latif, M., Bukhari, A.H. and Tanveer, A., 2015. *Cyperus rotundus* L.: traditional uses, phytochemistry, and pharmacological activities. *Journal of Ethnopharmacology*, *174*, pp. 540–560.
261. Huang, B., He, D., Chen, G., Ran, X., Guo, W., Kan, X., Wang, W., Liu, D., Fu, S. and Liu, J., 2018. α-Cyperone inhibits LPS-induced inflammation in BV-2 cells through activation of Akt/Nrf2/HO-1 and suppression of the NF-κB pathway. *Food & Function*, *9*(5), pp. 2735–2743.
262. Xu, H.B., Ma, Y.B., Huang, X.Y., Geng, C.A., Wang, H., Zhao, Y., Yang, T.H., Chen, X.L., Yang, C.Y., Zhang, X.M. and Chen, J.J., 2015. Bioactivity-guided isolation of anti-hepatitis B virus active sesquiterpenoids from the traditional Chinese medicine: Rhizomes of *Cyperus rotundus*. *Journal of Ethnopharmacology*, *171*, pp. 131–140.
263. Ahn, J.H., Lee, T.W., Kim, K.H., Byun, H., Ryu, B., Lee, K.T., Jang, D.S. and Choi, J.H., 2015. 6-Acetoxy cyperene, a patchoulane-type sesquiterpene isolated from *Cyperus rotundus* rhizomes induces caspase-dependent apoptosis in human ovarian cancer cells. *Phytotherapy Research*, *29*(9), pp. 1330–1338.

264. Singh, P.A., Brindavanam, N.B., Kimothi, G.P. and Aeri, V., 2016. Evaluation of in vivo anti-inflammatory and analgesic activity of *Dillenia indica* f. elongata (Miq.) Miq. and *Shorea robusta* stem bark extracts. *Asian Pacific Journal of Tropical Disease*, *6*(1), pp. 75–81.
265. Kaur, N., Kishore, L. Singh, R., 2016. Antidiabetic effect of new chromane isolated from *Dillenia indica* L: leaves in streptozotocin induced diabetic rats. *Journal of Functional Foods*, *22*, pp. 547–555.
266. Mbiantcha, M., Kamanyi, A., Teponno, R.B., Tapondjou, A.L., Watcho, P. and Nguelefack, T.B., 2011. Analgesic and anti-inflammatory properties of extracts from the bulbils of *Dioscorea bulbifera* L. var sativa (Dioscoreaceae) in mice and rats. *Evidence-Based Complementary and Alternative Medicine*, *2011*, p. 9.
267. Kuete, V., BetrandTeponno, R., Mbaveng, A.T., Tapondjou, L.A., Meyer, J.J.M., Barboni, L. Lall, N., 2012. Antibacterial activities of the extracts, fractions and compounds from *Dioscorea bulbifera*. *BMC Complementary and Alternative Medicine*, *12*(1), p. 228.
268. Gao, H., Kuroyanagi, M., Wu, Lanti., Kawahara, N., Yasuno, T. and Nakamura, Y., 2002. Antitumor-promoting constituents from *Dioscorea bulbifera* L. in JB6 mouse epidermal cells. *Biological and Pharmaceutical Bulletin*, *25*(9), pp. 1241–1243.
269. Liu, H., Chou, G.X., Wang, J.M., Ji, L.L. Wang, Z.T., 2011. Steroidal saponins from the rhizomes of *Dioscorea bulbifera* and their cytotoxic activity. *Planta Medica*, *77*(8), pp. 845–848.
270. Teponno, R.B., Tapondjou, A.L., Gatsing, D., Djoukeng, J.D., Abou-Mansour, E., Tabacchi, R., Tane, P., Stoekli-Evans, H. and Lontsi, D., 2006. Bafoudiosbulbins A, and B, two anti-salmonellal clerodane diterpenoids from *Dioscorea bulbifera* L. var sativa. *Phytochemistry*, *67*(17), pp. 1957–1963.
271. Adesanya, S.A., Ogundana, S.K. and Roberts, M.F., 1989. Dihydrostilbene phytoalexins from *Dioscorea bulbifera* and D. dumentorum. *Phytochemistry*, *28*(3), pp. 773–774.
272. Shriram, V., Jahagirdar, S., Latha, C., Kumar, V., Puranik, V., Rojatkar, S., Dhakephalkar, P.K. and Shitole, M.G., 2008. A potential plasmid-curing agent, 8-epidiosbulbin E acetate, from *Dioscorea bulbifera* L. against multidrug-resistant bacteria. *International Journal of Antimicrobial Agents*, *32*(5), pp. 405–410.
273. Adeleye, A. and Ikotun, T., 1989. Antifungal activity of dihydrodioscorine extracted from a wild variety of *Dioscorea bulbifera* L. *Journal of Basic Microbiology*, *29*(5), pp. 265–267.
274. Chen, L., Su, J., Li, L., Li, B. and Li, W., 2011. A new source of natural D-borneol and its characteristic. *Journal of Medicinal Plants Research*, *5*(15), pp. 3440–3447.
275. Yaseen Khan, M., Ali, S.A. and Pundarikakshudu, K., 2016. Wound healing activity of extracts derived from *Shorea robusta* resin. *Pharmaceutical Biology*, *54*(3), pp. 542–548.
276. Santhoshkumar, M.U.T.H.U., Anusuya, N. and Bhuvaneswari, P., 2013. Antiulcerogenic effect of resin from *Shorea robusta* Gaertn. f. on experimentally induced ulcer models. *International Journal Pharmacy and Pharmaceutical Science*, *5*(1), pp. 269–272.
277. Chattopadhyay, D., Ojha, D., Mukherjee, H., Bag, P., Vaidya, S.P. and Dutta, S., 2018. Validation of a traditional preparation against multi-drug resistant Salmonella Typhi and its protective efficacy in S. Typhimurium infected mice. *Biomedicine & Pharmacotherapy*, *99*, pp. 286–289.
278. Dewanjee, S., Das, A.K., Sahu, R. and Gangopadhyay, M., 2009. Antidiabetic activity of *Diospyros peregrina* fruit: effect on hyperglycemia, hyperlipidemia and augmented oxidative stress in experimental type 2 diabetes. *Food and Chemical Toxicology*, *47*(10), pp. 2679–2685.

279. Uddin, S.J., Rouf, R., Shilpi, J.A., Alamgir, M., Nahar, L. and Sarker, S.D., 2008. Screening of some Bangladeshi medicinal plants for in vitro antibacterial activity. *Oriental Pharmacy and Experimental Medicine*, *8*(3), pp. 316–321.
280. Siraj, M.A., Shilpi, J.A., Hossain, M.G., Uddin, S.J., Islam, M.K., Jahan, I.A. and Hossain, H., 2016. Anti-inflammatory and antioxidant activity of *Acalypha hispida* leaf and analysis of its major bioactive polyphenols by HPLC. *Advanced Pharmaceutical Bulletin*, *6*(2), p. 275.
281. Reiersen, B., Kiremire, B.T., Byamukama, R. and Andersen, Ø.M., 2003. Anthocyanins acylated with gallic acid from chenille plant, *Acalypha hispida*. *Phytochemistry*, *64*(4), pp. 867–871.
282. Jagatheeswari, D., Deepa, J., Ali, H.S.J. and Ranganathan, P., 2013. *Acalypha indica* L-An important medicinal plant: a review of its traditional uses and pharmacological properties. *International Journal of Research in Botany*, *3*(1), pp. 19–22.
283. Ganeshkumar, M., Ponrasu, T., Krithika, R., Iyappan, K., Gayathri, V.S. and Suguna, L., 2012. Topical application of *Acalypha indica* accelerates rat cutaneous wound healing by up-regulating the expression of Type I and III collagen. *Journal of Ethnopharmacology*, *142*(1), pp. 14–22.
284. Gupta, R., Thakur, B., Singh, P., Singh, H.B., Sharma, V.D., Katoch, V.M. and Chauhan, S.V.S., 2010. Anti-tuberculosis activity of selected medicinal plants against multi-drug resistant *Mycobacterium tuberculosis* isolates. *Indian Journal of Medical Research*, *131*(6), p. 809.
285. Govindarajan, M., Jebanesan, A., Reetha, D., Amsath, R., Pushpanathan, T. and Samidurai, K., 2008. Antibacterial activity of *Acalypha indica* L. *European Review for Medical and Pharmacological Sciences*, *12*(5), pp. 299–302.
286. Kwan, Y.P., Saito, T., Ibrahim, D., Al-Hassan, F.M.S., Ein Oon, C., Chen, Y., Jothy, S.L., Kanwar, J.R. and Sasidharan, S., 2016. Evaluation of the cytotoxicity, cell-cycle arrest, and apoptotic induction by *Euphorbia hirta* in MCF-7 breast cancer cells. *Pharmaceutical Biology*, *54*(7), pp. 1223–1236.
287. Singh, G.D., Kaiser, P., Youssouf, M.S., Singh, S., Khajuria, A., Koul, A., Bani, S., Kapahi, B.K., Satti, N.K., Suri, K.A. and Johri, R.K., 2006. Inhibition of early and late phase allergic reactions by *Euphorbia hirta* L. *Phytotherapy Research*, *20*(4), pp. 316–321.
288. Perumal, S., Mahmud, R. and Ramanathan, S., 2015. Anti-infective potential of caffeic acid and epicatechin 3-gallate isolated from methanol extract of *Euphorbia hirta* (L.) against *Pseudomonas aeruginosa*. *Natural Product Research*, *29*(18), pp. 1766–1769.
289. Annapurna, J., Chowdary, I.P., Lalitha, G., Ramakrishna, S.V. and Iyengar, D.S., 2004. Antimicrobial activity of *Euphorbia nivulia* leaf extract. *Pharmaceutical Biology*, *42*(2), pp. 91–93.
290. Ravikanth, V., Reddy, V.L.N., Reddy, A.V., Ravinder, K., Rao, T.P., Ram, T.S., Kumar, K.A., Vamanarao, D.P. and Venkateswarlu, Y., 2003. Three new ingol diterpenes from *Euphorbia nivulia*: evaluation of cytotoxic activity. *Chemical and Pharmaceutical Bulletin*, *51*(4), pp. 431–434.
291. Mujumdar, A.M. and Misar, A.V., 2004. Anti-inflammatory activity of *Jatropha curcas* roots in mice and rats. *Journal of Ethnopharmacology*, *90*(1), pp. 11–15.
292. Mujumdar, A.M., Misar, A.V., Salaskar, M.V. and Upadhye, A.S., 2001. Antidiarrhoeal effect of isolated fraction (JC) of *Jatropha curcas* roots in mice. *Journal of Natural Remedies*, *1*(2), pp. 89–93.
293. Devappa, R.K., Rajesh, S.K., Kumar, V., Makkar, H.P. and Becker, K., 2012. Activities of *Jatropha curcas* phorbol esters in various bioassays. *Ecotoxicology and Environmental Safety*, *78*, pp. 57–62.

294. Félix-Silva, J., Giordani, R.B., Silva-Jr, A.A.D., Zucolotto, S.M. and Fernandes-Pedrosa, M.D.F., 2014. Jatropha gossypiifolia L.(Euphorbiaceae): a review of traditional uses, phytochemistry, pharmacology, and toxicology of this medicinal plant. *Evidence-Based Complementary and Alternative Medicine, 2014*, p. 32.
295. Ravindranath, N., Venkataiah, B., Ramesh, C., Jayaprakash, P. and Das, B., 2003. Jatrophenone, a novel macrocyclic bioactive diterpene from *Jatropha gossypifolia*. *Chemical and Pharmaceutical Bulletin*, *51*(7), pp. 870–871.
296. Nagaharika, Y. and Rasheed, S., 2013. Anti-inflammatory activity of leaves of *Jatropha gossypifolia* L. by HRBC membrane stabilization method. *Journal of Acute Disease*, *2*(2), pp. 156–158.
297. Bijesh, K. and Sebastian, D., 2013. Isolation and characterization of antibacterial compounds from *Macaranga peltata* against clinical isolates of *Staphylococcus aureus*. *International Journal of Biological & Pharmaceutical Research*, *4*, pp. 1196–1203.
298. Meenakshi Verma, N.K., Thakar, M., Subrahmanyam, V.M., Rao, V. and Dhanaraj, S.A., 2013. Investigation of anti bacterial and anti fugal potentials of *Macaranga peltata*. *International Journal of Current Research and Review*, *5*(7), pp. 26–32.
299. Kulkarni, R.R., Tupe, S.G., Gample, S.P., Chandgude, M.G., Sarkar, D., Deshpande, M.V. and Joshi, S.P., 2014. Antifungal dimeric chalcone derivative kamalachalcone E from Mallotus philippinensis. *Natural Product Research*, *28*(4), pp. 245–250.
300. Daikonya, A., Katsuki, S., Wu, J.B. and Kitanaka, S., 2002. Anti-allergic agents from natural sources (4): anti-allergic activity of new phloroglucinol derivatives from Mallotus philippensis (Euphorbiaceae). *Chemical and Pharmaceutical Bulletin*, *50*(12), pp. 1566–1569.
301. Tanaka, R., Nakata, T., Yamaguchi, C., Wada, S.I., Yamada, T. and Tokuda, H., 2008. Potential anti-tumor-promoting activity of 3α-Hydroxy-D: A-friedooleanan-2-one from the stem bark of Mallotus philippensis. *Planta Medica*, *74*(4), pp. 413–416.
302. Abreu, P., Matthew, S., González, T., Costa, D., Segundo, M.A. and Fernandes, E., 2006. Anti-inflammatory and antioxidant activity of a medicinal tincture from *Pedilanthus tithymaloides*. *Life Sciences*, *78*(14), pp. 1578–1585.
303. Ghosh, S., Samanta, A., Mandal, N.B., Bannerjee, S. and Chattopadhyay, D., 2012. Evaluation of the wound healing activity of methanol extract of *Pedilanthus tithymaloides* (L.) Poit leaf and its isolated active constituents in topical formulation. *Journal of Ethnopharmacology*, *142*(3), pp. 714–722.
304. Mongkolvisut, W. and Sutthivaiyakit, S., 2007. Antimalarial and antituberculous poly-O-acylated jatrophane diterpenoids from *Pedilanthus tithymaloides*. *The Journal of Natural Products*, *70*(9), pp. 1434–1438.
305. Jena, J. and Gupta, A.K., 2012. *Ricinus communis* Linn: a phytopharmacological review. International *Journal of Pharmacy and Pharmaceutical Sciences*, *4*(4), pp. 25–29.
306. Ogunniyi, D.S., 2006. Castor oil: a vital industrial raw material. *Bioresource Technology*, *97*(9), pp. 1086
307. Ilavarasan, R., Mallika, M. and Venkataraman, S., 2006. Anti-inflammatory and free radical scavenging activity of *Ricinus communis* root extract. *Journal of Ethnopharmacology*, *103*(3), pp. 478–480.
308. Jeyaseelan, E.C. and Jashothan, P.J., 2012. In vitro control of *Staphylococcus aureus* (NCTC 6571) and *Escherichia coli* (ATCC 25922) by *Ricinus communis* L. *Asian Pacific Journal of Tropical Biomedicine*, *2*(9), pp. 717–721.
309. Tewtrakul, S., Subhadhirasakul, S., Cheenpracha, S., Yodsaoue, O., Ponglimanont, C. and Karalai, C., 2011. Anti-inflammatory principles of *Suregada multiflora* against nitric oxide and prostaglandin E2 releases. *Journal of Ethnopharmacology*, *133*(1), pp. 63–66.

310. Cheenpracha, S., Yodsaoue, O., Karalai, C., Ponglimanont, C., Subhadhirasakul, S., Tewtrakul, S. and Kanjana-opas, A., 2006. Potential anti-allergic ent-kaurene diterpenes from the bark of *Suregada multiflora*. *Phytochemistry*, *67*(24), pp. 2630–2634.
311. Dhara, A.K., Suba, V., Sen, T., Pal, S. and Chaudhuri, A.N., 2000. Preliminary studies on the anti-inflammatory and analgesic activity of the methanolic fraction of the root extract of *Tragia involucrata* L. *Journal of Ethnopharmacology*, *72*(1–2), pp. 265–268.
312. Dhara, A.K., Pal, S. and Nag Chaudhuri, A.K., 2002. Psychopharmacological studies on *Tragia involucrata* root extract. *Phytotherapy Research*, *16*(4), pp. 326–330.
313. Farook, S.M. and Atlee, W.C., 2011. Antidiabetic and hypolipidemic potential of *Tragia involucrata* L. In streptozotocin-nicotinamide induced type II diabetic rats. *International Journal of Pharmacy and Pharmaceutical Sciences*, *3*(4), pp. 103–109.
314. Samy, R.P., Gopalakrishnakone, P., Sarumathi, M. and Ignacimuthu, S., 2006. Wound healing potential of *Tragia involucrata* extract in rats. *Fitoterapia*, *77*(4), pp. 300–302.
315. Nakamura, N. and Hattori, M., 1998. Saponins and C-glycosyl flavones from the seeds of *Abrus precatorius*. *Chemical and Pharmaceutical Bulletin*, *46*(6), pp. 982–987.
316. Zore, G.B., Awad, V., Thakre, A.D., Halde, U.K., Meshram, N.S., Surwase, B.S. and Mohan Karuppayil, S., 2007. Activity-directed fractionation and isolation of 4 antibacterial compounds from *Abrus precatorius* L: the roots. *Natural Product Research*, *21*(10), pp. 933–940.
317. Amer, S.A.A., Reda, A.S. and Dimetry, N.Z., 1989. Activity of *Abrus precatorius* L. extracts against the two-spotted spider mite Tetranychus urticae Koch (Acari: Tetranychidae). *Acarologia*, *30*(3), pp. 209–215.
318. Hata, Y., Raith, M., Ebrahimi, S.N., Zimmermann, S., Mokoka, T., Naidoo, D., Fouche, G., Maharaj, V., Kaiser, M., Brun, R. and Hamburger, M., 2013. Antiprotozoal isoflavan quinones from *Abrus precatorius* ssp. africanus. *Planta Medica*, *79*(6), pp. 492–498.
319. Dickers, K.J., Bradberry, S.M., Rice, P., Griffiths, G.D. and Vale, J.A., 2003. Abrin poisoning. *Toxicological Reviews*, *22*(3), pp. 137–142.
320. Clark, D.T., Gazi, M.I., Cox, S.W., Eley, B.M. and Tinsley, G.F., 1993. The effects of Acacia arabica gum on the in vitro growth and protease activities of periodontopathic bacteria. *Journal of Clinical Periodontology*, *20*(4), pp. 238–243.
321. Gazi, M.I., 1991. The finding of antiplaque features in Acacia Arabica type of chewing gum. *Journal of Clinical Periodontology*, *18*(1), pp. 75–77.
322. Bhanu, K.U., Rajadurai, S. and Nayudamma, Y., 1964. Studies on the tannins of babul, Acacia arabica, bark. *Australian Journal of Chemistry*, *17*(7), pp. 803–809.
323. Micucci, M., Gotti, R., Corazza, I., Tocci, G., Chiarini, A., De Giorgio, M., Camarda, L., Frosini, M., Marzetti, C., Cevenini, M. and Budriesi, R., 2017. Newer insights into the antidiarrheal effects of *Acacia catechu* Willd. extract in guinea pig. *Journal of Medicinal Food*, *20*(6), pp. 592–600.
324. Shahid, A., Ali, R., Ali, N., Kazim Hasan, S., Barnwal, P., Mohammad Afzal, S., Vafa, A. and Sultana, S., 2017. Methanolic bark extract of *Acacia catechu* ameliorates benzo (a) pyrene induced lung toxicity by abrogation of oxidative stress, inflammation, and apoptosis in mice. *Environmental Toxicology*, *32*(5), pp. 1566–1577.
325. Sham, J.S.K., Chiu, K.W. and Pang, P.K.T., 1984. Hypotensive action of *Acacia catechu*. *Planta Medica*, *50*(2), pp. 177–180.
326. Sun, Y.N., Li, W., Song, S.B., Yan, X.T., Zhao, Y., Jo, A.R., Kang, J.S. and Young Ho, K., 2016. A new phenolic derivative with soluble epoxide hydrolase and nuclear factor-kappaB inhibitory activity from the aqueous extract of *Acacia catechu*. *Natural Product Research*, *30*(18), pp. 2085–2092.
327. Ojha, D., Singh, G. and Upadhyaya, Y.N., 1969. Clinical evaluation of *Acacia catechu*, Willd.(Khadira) in the treatment of lepromatous leprosy. *International Journal of Leprosy and Other Mycobacterial Diseases: Official Organ of the International Leprosy Association*, *37*(3), pp. 302–307.

328. Gafur, M.A., Obata, T., Kiuchi, F. and Tsuda, Y., 1997. Acacia concinna saponins. I. Structures of prosapogenols, concinnosides AF, isolated from the alkaline hydrolysate of the highly polar saponin fraction. *Chemical and Pharmaceutical Bulletin*, *45*(4), pp. 620–625.
329. Kukhetpitakwong, R., Hahnvajanawong, C., Homchampa, P., Leelavatcharamas, V., Satra, J. and Khunkitti, W., 2006. Immunological adjuvant activities of saponin extracts from the pods of Acacia concinna. *International Immunopharmacology*, *6*(11), pp. 1729–1735.
330. Priya, P. and Veerakumari, L., 2017. Morphological and histological analysis of Cotylophoron cotylophorum treated with Acacia concinna. *Tropical Parasitology*, *7*(2), p. 92.
331. Sotohy, S.A., Müller, W. and Ismail, A.A., 1995. "In vitro" effect of Egyptian tannin-containing plants and their extracts on the survival of pathogenic bacteria. *DTW. Deutsche tierarztliche Wochenschrift*, *102*(9), pp. 344–348.
332. Sánchez, E., Heredia, N., Camacho-Corona, M.D.R. and García, S., 2013. Isolation, characterization and mode of antimicrobial action against Vibrio cholerae of methyl gallate isolated from *Acacia farnesiana*. *Journal of Applied Microbiology*, *115*(6), pp. 1307–1316.
333. Dholvitayakhun, A., Cushnie, T.T. and Trachoo, N., 2012. Antibacterial activity of three medicinal Thai plants against *Campylobacter jejuni* and other foodborne pathogens. *Natural Product Research*, *26*(4), pp. 356–363.
334. Olajide, O.A., Echianu, C.A., Adedapo, A.D. and Makinde, J.M., 2004. Anti-inflammatory studies on *Adenanthera pavonina* seed extract. *Inflammopharmacology*, *12*(2), pp. 196–201.
335. Babu, N.P., Pandikumar, P. and Ignacimuthu, S., 2009. Anti-inflammatory activity of *Albizia lebbeck* Benth., an ethnomedicinal plant, in acute and chronic animal models of inflammation. *Journal of Ethnopharmacology*, *125*(2), pp. 356–360.
336. Kumar, D., Kumar, S., Kohli, S., Arya, R. and Gupta, J., 2011. Antidiabetic activity of methanolic bark extract of *Albizia odoratissima* Benth. in alloxan induced diabetic albino mice. *Asian Pacific Journal of Tropical Medicine*, *4*(11), pp. 900–903.
337. Panthong, P., Bunluepuech, K., Boonnak, N., Chaniad, P., Pianwanit, S., Wattanapiromsakul, C. and Tewtrakul, S., 2015. Anti-HIV-1 integrase activity and molecular docking of compounds from *Albizia procera* bark. *Pharmaceutical Biology*, *53*(12), pp. 1861–1866.
338. Khatoon, M.M., Khatun, M.H., Islam, M.E. and Parvin, M.S., 2014. Analgesic, antibacterial and central nervous system depressant activities of *Albizia procera* leaves. *Asian Pacific Journal of Tropical Biomedicine*, *4*(4), pp. 279–284.
339. Somani, R., Kasture, S. and Singhai, A.K., 2006. Antidiabetic potential of *Butea monosperma* in rats. *Fitoterapia*, *77*(2), pp. 86–90.
340. Prashanth, D., Asha, M.K., Amit, A. and Padmaja, R., 2001. Anthelmintic activity of *Butea monosperma*. *Fitoterapia*, *72*(4), pp. 421–422.
341. Wagner, H., Geyer, B., Fiebig, M., Kiso, Y. and Hikino, H., 1986. Isoputrin and butrin, the antihepatotoxic principles of *Butea monosperma* flowers1. *Planta Medica*, *52*(2), pp. 77–79.
342. Bhargava, S.K., 1986. Estrogenic and postcoital anticonceptive activity in rats of butin isolated from *Butea monosperma* seed. *Journal of Ethnopharmacology*, *18*(1), pp. 95–101.
343. Yadava, R.N. and Tiwari, L., 2005. A potential antiviral flavone glycoside from the seeds of *Butea monosperma* O. Kuntze. *Journal of Asian Natural Products Research*, *7*(2), pp. 185–188.
344. Chokchaisiri, R., Suaisom, C., Sriphota, S., Chindaduang, A., Chuprajob, T. and Suksamrarn, A., 2009. Bioactive flavonoids of the flowers of *Butea monosperma*. *Chemical and Pharmaceutical Bulletin*, *57*(4), pp. 428–432.

345. Bandara, B.R., Kumar, N.S. and Samaranayake, K.S., 1989. An antifungal constituent from the stem bark of *Butea monosperma*. *Journal of Ethnopharmacology*, *25*(1), pp. 73–75.
346. Ata, A., Gale, E.M. and Samarasekera, R., 2009. Bioactive chemical constituents of *Caesalpinia bonduc*. *Phytochemistry Letters*, *2*(3), pp. 106–109.
347. Arif, T., Mandal, T.K., Kumar, N., Bhosale, J.D., Hole, A., Sharma, G.L., Padhi, M.M., Lavekar, G.S. and Dabur, R., 2009. In vitro and in vivo antimicrobial activities of seeds of *Caesalpinia bonduc* (Lin.) Roxb. *Journal of Ethnopharmacology*, *123*(1), pp. 177–180.
348. Simin, K., Khaliq-uz-Zaman, S.M. and Ahmad, V.U., 2001. Antimicrobial activity of seed extracts and bondenolide from *Caesalpinia bonduc* (L.) Roxb. *Phytotherapy Research*, *15*(5), pp. 437–440.
349. Kannur, D.M., Hukkeri, V.I. and Akki, K.S., 2006. Adaptogenic activity of *Caesalpinia bonduc* seed extracts in rats. *Journal of Ethnopharmacology*, *108*(3), pp. 327–331.
350. Kishalay, J., Kausik, C., Bera, T.K., Soumyajit, M., Debasis, D., Ali, K.M. and Debidas, G., 2010. Antihyperglycemic and antihyperlipidemic effects of hydro-methanolic extract of seed of *Caesalpinia bonduc* in streptozotocin induced diabetic male albino rat. *International Journal of PharmTech Research*, *2*(4), pp. 2234–2242.
351. Archana, P., Tandan, S.K., Chandra, S. and Lal, J., 2005. Antipyretic and analgesic activities of *Caesalpinia bonducella* seed kernel extract. *Phytotherapy Research*, *19*(5), pp. 376–381.
352. Kundu, R., Dasgupta, S., Biswas, A., Bhattacharya, A., Pal, B.C., Bandyopadhyay, D., Bhattacharya, S. and Bhattacharya, S., 2008. *Cajanus cajan* L.(Leguminosae) prevents alcohol-induced rat liver damage and augments cytoprotective function. *Journal of Ethnopharmacology*, *118*(3), pp. 440–447.
353. Capasso, F. and Gaginella, T.S., 2012. *Laxatives: A Practical Guide*. Springer Science & Business Media: Berlin, Germany.
354. Sony, P., Kalyani, M., Jeyakumari, D., Kannan, I. and Sukumar, R.G., 2018. In vitro antifungal activity of *Cassia fistula* extracts against fluconazole resistant strains of Candida species from HIV patients. *Journal de mycologie medicale*, *28*(1), pp. 193–200.
355. Irshad, M.D., Ahmad, A., Zafaryab, M.D., Ahmad, F., Manzoor, N., Singh, M. and Rizvi, M.M., 2013. Composition of *Cassia fistula* oil and its antifungal activity by disrupting ergosterol biosynthesis. *Natural Product Communications*, *8*(2), pp. 261–264.
356. Duraipandiyan, V. and Ignacimuthu, S., 2007. Antibacterial and antifungal activity of *Cassia fistula* L.: an ethnomedicinal plant. *Journal of Ethnopharmacology*, *112*(3), pp. 590–594.
357. Irshad, M., Shreaz, S., Manzoor, N., Khan, L.A. and Rizvi, M.M.A., 2011. Anticandidal activity of *Cassia fistula* and its effect on ergosterol biosynthesis. *Pharmaceutical Biology*, *49*(7), pp. 727–733.
358. Verma, H.P. and Singh, S.K., 2018. An insight into the possible mechanisms of anti-spermatogenic action of *Dalbergia sissoo* in male mice. *Andrologia*, *50*(3), p.e12917.
359. Karvande, A., Khedgikar, V., Kushwaha, P., Ahmad, N., Kothari, P., Verma, A., Kumar, P., Nagar, G.K., Mishra, P.R., Maurya, R. and Trivedi, R., 2017. Heartwood extract from *Dalbergia sissoo* promotes fracture healing and its application in ovariectomy-induced osteoporotic rats. *Journal of Pharmacy and Pharmacology*, *69*(10), pp. 1381–1397.
360. Gautam, J., Kumar, P., Kushwaha, P., Khedgikar, V., Choudhary, D., Singh, D., Maurya, R. and Trivedi, R., 2015. Neoflavonoid dalbergiphenol from heartwood of *Dalbergia sissoo* acts as bone savior in an estrogen withdrawal model for osteoporosis. *Menopause*, *22*(11), pp. 1246–1255.

361. Kim, E., Yoon, K.D., Lee, W.S., Yang, W.S., Kim, S.H., Sung, N.Y., Baek, K.S., Kim, Y., Htwe, K.M., Kim, Y.D. and Hong, S., 2014. Syk/Src-targeted anti-inflammatory activity of *Codariocalyx motorius* ethanolic extract. *Journal of Ethnopharmacology, 155*(1), pp. 185–193.
362. Uma, C., Suganya, N., Vanitha, P., Bhakkiyalakshmi, E., Suriyanarayanan, S., John, K.M., Sivasubramanian, S., Gunasekaran, P. and Ramkumar, K.M., 2014. Antihyperglycemic effect of *Codariocalyx motorius* modulated carbohydrate metabolic enzyme activities in streptozotocin-induced diabetic rats. *Journal of Functional Foods, 11*, pp. 517–527.
363. Kumar, P.N., Reddy, S.K., Babu, P.S. and Karthikeyan, R., 2010. Coagulant activity of the aqueous leaf extracts of *Erythrina variegata* L. *Pharmacological Research, 2*(2), pp. 74–76.
364. Tanaka, H., Sato, M., Fujiwara, S., Hirata, M., Etoh, H. and Takeuchi, H., 2002. Antibacterial activity of isoflavonoids isolated from *Erythrina variegata* against methicillin-resistant *Staphylococcus aureus*. *Letters in Applied Microbiology, 35*(6), pp. 494–498.
365. Ferreira, E.S., Hulme, A.N., McNab, H. and Quye, A., 2004. The natural constituents of historical textile dyes. *Chemical Society Reviews, 33*(6), pp. 329–336.
366. Meenakshisundaram, A., Harikrishnan, T.J. and Anna, T., 2016. Anthelmintic activity of *Indigofera tinctoria* against gastrointestinal nematodes of sheep. *Veterinary World, 9*(1), p. 101.
367. Garbhapu, A., Yalavarthi, P. and Koganti, P., 2011. Effect of ethanolic extract of *Indigofera tinctoria* on chemically-induced seizures and brain GABA levels in albino rats. *Iranian Journal of Basic Medical Sciences, 14*(4), p. 318.
368. Kunikata, T., Tatefuji, T., Aga, H., Iwaki, K., Ikeda, M. and Kurimoto, M., 2000. Indirubin inhibits inflammatory reactions in delayed-type hypersensitivity. *European Journal of Pharmacology, 410*(1), pp. 93–100.
369. Im, A.R., Kim, Y.H., Kim, Y.H., Yang, W.K., Kim, S.H. and Song, K.H., 2017. Dolichos lablab protects against nonalcoholic fatty liver disease in mice fed high-fat diets. *Journal of Medicinal Food, 20*(12), pp. 1222–1232.
370. Vigneshwaran, V., Thirusangu, P., Vijay Avin, B.R., Krishna, V., Pramod, S.N. and Prabhakar, B.T., 2017. Immunomodulatory glc/man-directed Dolichos lablab lectin (DLL) evokes anti-tumour response in vivo by counteracting angiogenic gene expressions. *Clinical & Experimental Immunology, 189*(1), pp. 21–35.
371. Shukla, K.K., Mahdi, A.A., Ahmad, M.K., Shankhwar, S.N., Rajender, S. and Jaiswar, S.P., 2009. *Mucuna pruriens* improves male fertility by its action on the hypothalamus–pituitary–gonadal axis. *Fertility and Sterility, 92*(6), pp. 1934–1940.
372. Champatisingh, D., Sahu, P.K., Pal, A. and Nanda, G.S., 2011. Anticataleptic and antiepileptic activity of ethanolic extract of leaves of *Mucuna pruriens*: a study on role of dopaminergic system in epilepsy in albino rats. *Indian Journal of Pharmacology, 43*(2), p. 197.
373. Bhaskar, A., Vidhya, V.G. and Ramya, M., 2008. Hypoglycemic effect of *Mucuna pruriens* seed extract on normal and streptozotocin-diabetic rats. *Fitoterapia, 79*(7–8), pp. 539–543.
374. Cho, J.Y., Park, J., Kim, P.S., Yoo, E.S., Baik, K.U. and Park, M.H., 2001. Savinin, a lignan from *Pterocarpus santalinus* inhibits tumor necrosis factor-α production and T cell proliferation. *Biological and Pharmaceutical Bulletin, 24*(2), pp. 167–171.
375. Kondeti, V.K., Badri, K.R., Maddirala, D.R., Thur, S.K.M., Fatima, S.S., Kasetti, R.B. and Rao, C.A., 2010. Effect of *Pterocarpus santalinus* bark, on blood glucose, serum lipids, plasma insulin and hepatic carbohydrate metabolic enzymes in streptozotocin-induced diabetic rats. *Food and Chemical Toxicology, 48*(5), pp. 1281–1287.

376. Kondeti, V.K., Badri, K.R., Maddirala, D.R., Thur, S.K.M., Fatima, S.S., Kasetti, R.B. and Rao, C.A., 2010. Effect of *Pterocarpus santalinus* bark, on blood glucose, serum lipids, plasma insulin and hepatic carbohydrate metabolic enzymes in streptozotocin-induced diabetic rats. *Food and Chemical Toxicology, 48*(5), pp. 1281–1287.
377. Kwon, H.J., Hong, Y.K., Kim, K.H., Han, C.H., Cho, S.H., Choi, J.S. and Kim, B.W., 2006. Methanolic extract of *Pterocarpus santalinus* induces apoptosis in HeLa cells. *Journal of Ethnopharmacology, 105*(1–2), pp. 229–234.
378. Pradhan, P., Joseph, L., Gupta, V., Chulet, R., Arya, H., Verma, R. and Bajpai, A., 2009. *Saraca asoca* (Ashoka): a review. *Journal of Chemical and Pharmaceutical Research, 1*(1), pp. 62–71.
379. Swamy, A.V., Patel, U.M., Koti, B.C., Gadad, P.C., Patel, N.L. and Thippeswamy, A.H.M., 2013. Cardioprotective effect of *Saraca indica* against cyclophosphamide induced cardiotoxicity in rats: a biochemical, electrocardiographic and histopathological study. *Indian Journal of Pharmacology, 45*(1), p. 44.
380. Shahid, M., Shahzad, A., Malik, A. and Anis, M., 2007. Antibacterial activity of aerial parts as well as in vitro raised calli of the medicinal plant *Saraca asoca* (Roxb.) de Wilde. *Canadian Journal of Microbiology, 53*(1), pp. 75–81.
381. Sreelatha, S., Padma, P.R. and Umasankari, E., 2011. Evaluation of anticancer activity of ethanol extract of *Sesbania grandiflora* (Agati Sesban) against Ehrlich ascites carcinoma in Swiss albino mice. *Journal of Ethnopharmacology, 134*(3), pp. 984–987.
382. Khatri, A., Garg, A. and Agrawal, S.S., 2009. Evaluation of hepatoprotective activity of aerial parts of *Tephrosia purpurea* L. and stem bark of *Tecomella undulata*. *Journal of Ethnopharmacology, 122*(1), pp. 1–5.
383. Kavitha, K. and Manoharan, S., 2006. Anticarcinogenic and antilipidperoxidative effects of *Tephrosia purpurea* (L.) Pers. in 7, 12-dimethylbenz (a) anthracene (DMBA) induced hamster buccal pouch carcinoma. *Indian Journal of Pharmacology, 38*(3), p. 185.
384. Bhadada, S.V. and Goyal, R.K., 2016. Effect of flavonoid rich fraction of *Tephrosia purpurea* (L.) Pers. on complications associated with streptozotocin-induced type I diabetes mellitus. *Indian Journal of Experimental Biology, 54*(7), pp. 457–466.
385. Park, K.W., Kundu, J., Chae, I.G., Bachar, S.C., Bae, J.W. and Chun, K.S., 2014. Methanol extract of *Flacourtia indica* aerial parts induces apoptosis via generation of ROS and activation of caspases in human colon cancer HCT116 cells. *Asian Pacific Journal of Cancer Prevention, 15*(17), pp. 7291–7296.
386. Kaou, A.M., Mahiou-Leddet, V., Canlet, C., Debrauwer, L., Hutter, S., Laget, M., Faure, R., Azas, N. and Ollivier, E., 2010. Antimalarial compounds from the aerial parts of *Flacourtia indica* (Flacourtiaceae). *Journal of Ethnopharmacology, 130*(2), pp. 272–274.
387. Singh, A.K. and Singh, J., 2010. Evaluation of anti-diabetic potential of leaves and stem of *Flacourtia jangomas* in streptozotocin-induced diabetic rats. *Indian Journal of Pharmacology, 42*(5), p. 301.
388. Levy, L., 1975. The activity of chaulmoogra acids against Mycobacterium leprae. *American Review of Respiratory Disease, 111*(5), pp. 703–705.
389. Devi, W.R., Singh, S.B. and Singh, C.B., 2013. Antioxidant and anti-dermatophytic properties leaf and stem bark of *Xylosma longifolium* clos. *BMC Complementary and Alternative Medicine, 13*(1), p. 155.
390. Kumar, V. and Van Staden, J., 2016. A review of *Swertia chirayita* (Gentianaceae) as a traditional medicinal plant. *Frontiers in Pharmacology, 6*, p. 308.
391. Karan, M., Vasisht, K. and Handa, S.S., 1999. Antihepatotoxic activity of *Swertia chirata* on carbon tetrachloride induced hepatotoxicity in rats. *Phytotherapy Research, 13*(1), pp. 24–30.

392. Bajpai, M.B., Asthana, R.K., Sharma, N.K., Chatterjee, S.K. and Mukherjee, S.K., 19Tribes living in lalpur, Natore District, Ari. Hypoglycemic effect of swerchirin from the hexane fraction of *Swertia chirayita. Planta Medica, 57*(2), pp. 102–104.
393. Iqbal, Z., Lateef, M., Khan, M.N., Jabbar, A. and Akhtar, M.S., 2006. Anthelmintic activity of *Swertia chirata* against gastrointestinal nematodes of sheep. *Fitoterapia, 77*(6), pp. 463–465.
394. Rafatullah, S., Tariq, M., Mossa, J.S., Al-Yahya, M.A., Al-Said, M.S. and Ageel, A.M., 1993. Protective effect of *Swertia chirata* against indomethacin and other ulcerogenic agent-induced gastric ulcers. *Drugs Under Experimental and Clinical Research, 19*(2), pp. 69–73.
395. Sahoo, A.K. and Kanhar, S., 2017. Antioxidant and antiulcer potential of *Hydrolea zeylanica* (L.) Vahl against gastric ulcers in rats. *International Journal of Complementary and Alternative Medicine, 10*(1), p. 00324.
396. Qureshi, M.S., Kumar, G.S. and Reddy, A.V., 2015. Exploring the in-vitro anthelmintic potential of *Hydrolea zeylanica*-An aquatic medicinal herb. *Research Journal of Pharmacy and Technology, 8*(12), p. 1645.
397. Xiaoyu, S., Isogai, A., Furihata, K., Handong, S. and Suzuki, A., 1993. Neo-clerodane diterpenoids from *Ajuga macrosperma* and *Ajuga pantantha. Phytochemistry, 34*(4), pp. 1091–1094.
398. Shen, X., Isogai, A., Furihata, K., Sun, H. and Suzuki, A., 1993. Neo-clerodane diterpenoids from *Ajuga macrosperma. Phytochemistry, 33*(4), pp. 887–889.
399. Kulkarni, R.R., Shurpali, K., Gawde, R.L., Sarkar, D., Puranik, V.G. and Joshi, S.P., 2012. Phyllocladane diterpenes from *Anisomeles heyneana. Journal of Asian Natural Products Research, 14*(12), pp. 1162–1168.
400. Rao, Y.K., Fang, S.H., Hsieh, S.C., Yeh, T.H. and Tzeng, Y.M., 2009. The constituents of *Anisomeles indica* and their anti-inflammatory activities. *Journal of Ethnopharmacology, 121*(2), pp. 292–296.
401. Tu, Y.X., Wang, S.B., Fu, L.Q., Li, S.S., Guo, Q.P., Wu, Y., Mou, X.Z. and Tong, X.M., 2018. Ovatodiolide targets chronic myeloid leukemia stem cells by epigenetically upregulating hsa-miR-155, suppressing the BCR-ABL fusion gene and dysregulating the PI3K/AKT/mTOR pathway. *Oncotarget, 9*(3), p. 3267.
402. Moreira, A.C.P., Lima, E.D.O., Wanderley, P.A., Carmo, E.S. and Souza, E.L.D., 2010. Chemical composition and antifungal activity of *Hyptis suaveolens* (L.) poit leaves essential oil against *Aspergillus* species. *Brazilian Journal of Microbiology, 41*(1), pp. 28–33.
403. Santos, T.C., Marques, M.S., Menezes, I.A., Dias, K.S., Silva, A.B., Mello, I.C., Carvalho, A.C., Cavalcanti, S.C., Antoniolli, Â.R. and Marçal, R.M., 2007. Antinociceptive effect and acute toxicity of the *Hyptis suaveolens* leaves aqueous extract on mice. *Fitoterapia, 78*(5), pp. 333–336.
404. Vera-Arzave, C., Antonio, L.C., Arrieta, J., Cruz-Hernández, G., Velázquez-Méndez, A.M., Reyes-Ramírez, A. and Sánchez-Mendoza, M.E., 2012. Gastroprotection of suaveolol, isolated from *Hyptis suaveolens*, against ethanol-induced gastric lesions in Wistar rats: role of prostaglandins, nitric oxide and sulfhydryls. *Molecules, 17*(8), pp. 8917–8927.
405. Grassi, P., Reyes, T.S.U., Sosa, S., Tubaro, A., Hofer, O. and Zitterl-Eglseer, K., 2006. Anti-inflammatory activity of two diterpenes of *Hyptis suaveolens* from El Salvador. *Zeitschrift für Naturforschung C, 61*(3–4), pp. 165–170.
406. Oliveira, A.S., Cercato, L.M., de Santana Souza, M.T., de Oliveira Melo, A.J., dos Santos Lima, B., Duarte, M.C., de Souza Araujo, A.A., e Silva, A.M.D.O. and Camargo, E.A., 2017. The ethanol extract of *Leonurus sibiricus* L. induces antioxidant, antinociceptive and topical anti-inflammatory effects. *Journal of Ethnopharmacology, 206*, pp. 144–151.

407. Islam, M.A., Ahmed, F., Das, A.K. and Bachar, S.C., 2005. Analgesic and anti-inflammatory activity of *Leonurus sibiricus*. *Fitoterapia*, *76*(3–4), pp. 359–362.
408. Kittl, M., Beyreis, M., Tumurkhuu, M., Fürst, J., Helm, K., Pitschmann, A., Gaisberger, M., Glasl, S., Ritter, M. and Jakab, M., 2016. Quercetin stimulates insulin secretiinon and reduces the viability of rat INS-1 beta-cells. *Cellular Physiology and Biochemistry*, *39*(1), pp. 278–293.
409. Gopi, K., Renu, K. and Jayaraman, G., 2014. Inhibition of Naja naja venom enzymes by the methanolic extract of *Leucas aspera* and its chemical profile by GC–MS. *Toxicology Reports*, *1*, pp. 667–673.
410. Srinivas, K., Rao, M.E.B. and Rao, S.S., 2000. Anti-inflammatory activity of *Heliotropium indicum* L. *Leucas aspera* spreng. in albino rats. *Indian Journal of Pharmacology*, *32*(1), pp. 37–38.
411. Rahman, M.S., Sadhu, S.K. and Hasan, C.M., 2007. Preliminary antinociceptive, antioxidant and cytotoxic activities of *Leucas aspera* root. *Fitoterapia*, *78*(7–8), pp. 552–555.
412. Yamani, H.A., Pang, E.C., Mantri, N. and Deighton, M.A., 2016. Antimicrobial activity of Tulsi (*Ocimum tenuiflorum*) essential oil and their major constituents against three species of bacteria. *Frontiers in Microbiology*, *7*, p. 681.
413. Chaudhary, S.K., Mukherjee, P.K., Maity, N., Nema, N.K., Bhadra, S. and Saha, B.P., 2014. Ocimum sanctum L. a potential angiotensin converting enzyme (ACE) inhibitor useful in hypertension. *Indian Journal of Natural Products and Resources*, *5*(1), pp. 83–87.
414. Shynu, M., Saini, M., Sharma, B., Gupta, L.K. and Gupta, P.K., 2006. *Ocimum tenuiflorum* possesses antiviral activity against bovine herpes virus-1. *Indian Journal of Virology*, *17*, p. 28.
415. Shynu, M., Gupta, P.K., Sharma, B. and Saini, M., 2007. Immunomodulatory potential of *Ocimum tenuiflorum* extracts in bovine peripheral blood mononuclear cells in vitro. *Journal of Immunology and Immunopathology*, *9*(1 and 2), pp. 31–36.
416. Jung, H.J., Song, Y.S., Lim, C.J. and Park, E.H., 2009. Anti-inflammatory, anti-angiogenic and anti-nociceptive activities of an ethanol extract of *Salvia plebeia* R. Brown. *Journal of Ethnopharmacology*, *126*(2), pp. 355–360.
417. Nugroho, A., Kim, M.H., Choi, J., Baek, N.I. and Park, H.J., 2012. In vivo sedative and gastroprotective activities of *Salvia plebeia* extract and its composition of polyphenols. *Archives of Pharmacal Research*, *35*(8), pp. 1403–1411.
418. Qu, X.J., Xia, X., Wang, Y.S., Song, M.J., Liu, L.L., Xie, Y.Y., Cheng, Y.N., Liu, X.J., Qiu, L.L., Xiang, L. and Gao, J.J., 2009. Protective effects of *Salvia plebeia* compound homoplantaginin on hepatocyte injury. *Food and Chemical Toxicology*, *47*(7), pp. 1710–1715.
419. Eswaran, M.B., Surendran, S., Vijayakumar, M., Ojha, S.K., Rawat, A.K.S. and Rao, C.V., 2010. Gastroprotective activity of *Cinnamomum tamala* leaves on experimental gastric ulcers in rats. *Journal of Ethnopharmacology*, *128*(2), pp. 537–540.
420. Bisht, S. and Sisodia, S.S., 2011. Assessment of antidiabetic potential of *Cinnamomum tamala* leaves extract in streptozotocin induced diabetic rats. *Indian Journal of Pharmacology*, *43*(5), p. 582.
421. Mishra, A.K., Singh, B.K. and Pandey, A.K., 2010. In vitro-antibacterial activity and phytochemical profiles of *Cinnamomum tamala* (Tejpat) leaf extracts and oil. *Reviews in Infection*, *1*(3), pp. 134–139.
422. Zhang, X., Jin, Y., Wu, Y., Zhang, C., Jin, D., Zheng, Q. and Li, Y., 2018. Anti-hyperglycemic and anti-hyperlipidemia effects of the alkaloid-rich extract from barks of *Litsea glutinosa* in ob/ob mice. *Scientific Reports*, *8*(1), p. 12646.
423. Mandal, S.C., Kumar, C.A., Majumder, A., Majumder, R. and Maity, B.C., 2000. Antibacterial activity of *Litsea glutinosa* bark. *Fitoterapia*, *71*(4), pp. 439–441.
424. Pal, B.C., Chaudhuri, T., Yoshikawa, K. and Arihara, S., 1994. Saponins from *Barringtonia acutangula*. *Phytochemistry*, *35*(5), pp. 1315–1318.

425. Rahman, M.M., Polfreman, D., MacGeachan, J. and Gray, A.I., 2005. Antimicrobial activities of *Barringtonia acutangula. Phytotherapy Research*, *19*(6), pp. 543–545.
426. Imam, M.Z., Sultana, S. and Akter, S., 2012. Antinociceptive, antidiarrheal, and neuropharmacological activities of *Barringtonia acutangula. Pharmaceutical Biology*, *50*(9), pp. 1078–1084.
427. Patil, K.R. and Patil, C.R., 2017. Anti-inflammatory activity of bartogenic acid containing fraction of fruits of *Barringtonia racemosa* Roxb. in acute and chronic animal models of inflammation. *Journal of Traditional and Complementary Medicine*, *7*(1), pp. 86–93.
428. Deraniyagala, S.A., Ratnasooriya, W.D. and Goonasekara, C.L., 2003. Antinociceptive effect and toxicological study of the aqueous bark extract of *Barringtonia racemosa* on rats. *Journal of Ethnopharmacology*, *86*(1), pp. 21–26.
429. Khan, S., Jabbar, A., Hasan, C.M. and Rashid, M.A., 2001. Antibacterial activity of *Barringtonia racemosa. Fitoterapia*, *72*(2), pp. 162–164.
430. Thomas, T.J., Panikkar, B., Subramoniam, A., Nair, M.K. and Panikkar, K.R., 2002. Antitumour property and toxicity of *Barringtonia racemosa* Roxb seed extract in mice. *Journal of Ethnopharmacology*, *82*(2–3), pp. 223–227.
431. Emran, T.B., Rahman, M.A., Hosen, S.Z., Rahman, M.M., Islam, A.M.T., Chowdhury, M.A.U. and Uddin, M.E., 2012. Analgesic activity of *Leea indica* (Burm. f.) Merr. *Phytopharmacology*, *3*(1), pp. 150–157.
432. Dewanjee, S., Dua, T.K. and Sahu, R., 2013. Potential anti-inflammatory effect of *Leea macrophylla* Roxb. leaves: a wild edible plant. *Food and Chemical Toxicology*, *59*, pp. 514–520.
433. Nizami, A.N., Rahman, M.A., Ahmed, N.U. and Islam, M.S., 2012. Whole *Leea macrophylla* ethanolic extract normalizes kidney deposits and recovers renal impairments in an ethylene glycol–induced urolithiasis model of rats. *Asian Pacific Journal of Tropical Medicine*, *5*(7), pp. 533–538.
434. Joshi, A., Prasad, S.K., Joshi, V.K. and Hemalatha, S., 2016. Phytochemical standardization, antioxidant, and antibacterial evaluations of *Leea macrophylla*: a wild edible plant. *Journal of Food and Drug Analysis*, *24*(2), pp. 324–331.
435. Dorsch, W., 1997. *Allium cepa* L (Onion): Part 2 chemistry, analysis and pharmacology. *Phytomedicine*, *3*(4), p. 3
436. Ari Ghorani, V., Marefati, N., Shakeri, F., Rezaee, R., Boskabady, M. and Boskabady, M.H., 2018. The effects of *Allium cepa* extract on tracheal responsiveness, lung inflammatory cells and phospholipase A2 level in Asthmatic rats. *Iranian Journal of Allergy, Asthma and Immunology*, *17*(3), pp. 221–231.
437. Abouzed, T.K., del Mar Contreras, M., Sadek, K.M., Shukry, M., Abdelhady, D.H., Gouda, W.M., Abdo, W., Nasr, N.E., Mekky, R.H., Segura-Carretero, A. and Kahilo, K.A.A., 2018. Red onion scales ameliorated streptozotocin-induced diabetes and diabetic nephropathy in Wistar rats in relation to their metabolite fingerprint. *Diabetes Research and Clinical Practice*, *140*, pp. 253–264.
438. Reuter, H.D., 1995. *Allium sativum* and allium ursinum: Part 2 pharmacology and medicinal application. *Phytomedicine*, *2*(1), pp. 73–91.
439. Koch, H.P. and Lawson, L.D., 1996. *Garlic: The Science and Therapeutic Application of Allium Sativum L. Related Species* (Vol. 683181475). Williams & Wilkins: Baltimore, MD, p. 329.
440. Kendler, B.S., 1987. Garlic (*Allium sativum)* and onion (*Allium cepa*): a review of their relationship to cardiovascular disease. *Preventive Medicine*, *16*(5), pp. 670–685.
441. Singh, V.K. and Singh, D.K., 2008. Pharmacological effects of garlic (*Allium sativum* L.). *Annual Review of Biomedical Sciences*, *10*.
442. Bozin, B., Mimica-Dukic, N., Samojlik, I., Goran, A. and Igic, R., 2008. Phenolics as antioxidants in garlic (*Allium sativum* L., Alliaceae). *Food Chemistry*, *111*(4), pp. 925–929.

443. Chung, L.Y., 2006. The antioxidant properties of garlic compounds: allyl cysteine, alliin, allicin, and allyl disulfide. *Journal of Medicinal Food*, *9*(2), pp. 205–213.
444. Singh, R. and Geetanjali, 2016. *Asparagus racemosus* : a review on its phytochemical and therapeutic potential. *Natural Product Research*, *30*(17), pp. 1896–1908.
445. Pahwa, P. and Goel, R.K., 2016. Ameliorative effect of *Asparagus racemosus* root extract against pentylenetetrazol-induced kindling and associated depression and memory deficit. *Epilepsy & Behavior*, *57*, pp. 196–201.
446. Battu, G.R. and Kumar, B.M., 2010. Anti-inflammatory activity of leaf extract of *Asparagus racemosus* Willd. *International Journal of Chemical Sciences*, *8*(2), pp. 1329–1338.
447. Hannan, J.M.A., Marenah, L., Ali, L., Rokeya, B., Flatt, P.R. and Abdel-Wahab, Y.H., 2007. Insulin secretory actions of extracts of *Asparagus racemosus* root in perfused pancreas, isolated islets and clonal pancreatic β-cells. *Journal of Endocrinology*, *192*(1), pp. 159–168.
448. Mandal, S.C., Nandy, A., Pal, M. and Saha, B.P., 2000. Evaluation of antibacterial activity of *Asparagus racemosus* Willd: the root. *Phytotherapy Research*, *14*(2), pp. 118–119.
449. Singh, D.K., Luqman, S. and Mathur, A.K., 2015. *Lawsonia inermis* L.–a commercially important primaeval dyeing and medicinal plant with diverse pharmacological activity: a review. *Industrial Crops and Products*, *65*, pp. 269–286.
450. Babu, P.D. and Subhasree, R.S., 2009. Antimicrobial activities of *Lawsonia inermis*—a review. *Academic Journal of Plant Sciences*, *2*(4), pp. 231–232.
451. Darvin, S.S., Esakkimuthu, S., Toppo, E., Balakrishna, K., Paulraj, M.G., Pandikumar, P., Ignacimuthu, S. and Al-Dhabi, N.A., 2018. Hepatoprotective effect of lawsone on rifampicin-isoniazid induced hepatotoxicity in in vitro and in vivo models. *Environmental Toxicology and Pharmacology*, *61*, pp. 87–94.
452. Alia, B.H., Bashir, A.K. and Tanira, M.O.M., 1995. Anti-inflammatory, antipyretic, and analgesic effects of *Lawsonia inermis* L. (henna) in rats. *Pharmacology*, *51*(6), pp. 356–363.
453. Hossain, S.J., Basar, M.H., Rokeya, B., Arif, K.M.T., Sultana, M.S. and Rahman, M.H., 2013. Evaluation of antioxidant, antidiabetic and antibacterial activities of the fruit of *Sonneratia apetala* (Buch.-Ham.). *Oriental Pharmacy and Experimental Medicine*, *13*(2), pp. 95–102.
454. Hsu, C.L., Fang, S.C., Huang, H.W. and Yen, G.C., 2015. Anti-inflammatory effects of triterpenes and steroid compounds isolated from the stem bark of *Hiptage benghalensis*. *Journal of Functional Foods*, *12*, pp. 420–427.
455. Ngente, L., Nachimuthu, S.K. and Guruswami, G., 2012. Insecticidal and repellent activity of *Hiptage benghalensis* L. Kruz (Malpighiaceae) against mosquito vectors. *Parasitology Research*, *111*(3), pp. 1007–1017.
456. Gul, M.Z., Bhakshu, L.M., Ahmad, F., Kondapi, A.K., Qureshi, I.A. and Ghazi, I.A., 2011. Evaluation of *Abelmoschus moschatus* extracts for antioxidant, free radical scavenging, antimicrobial and antiproliferative activities using in vitro assays. *BMC Complementary and Alternative Medicine*, *11*(1), p. 64.
457. Liu, I.M., Liou, S.S., Lan, T.W., Hsu, F.L. and Cheng, J.T., 2005. Myricetin as the active principle of *Abelmoschus moschatus* to lower plasma glucose in streptozotocin-induced diabetic rats. *Planta Medica*, *71*(7), pp. 617–621.
458. Porchezhian, E. and Ansari, S.H., 2005. Hepatoprotective activity of *Abutilon indicum* on experimental liver damage in rats. *Phytomedicine*, *12*(1–2), pp. 62–64.
459. Seetharam, Y.N., Chalageri, G. and Setty, S.R., 2002. Hypoglycemic activity of *Abutilon indicum* leaf extracts in rats. *Fitoterapia*, *73*(2), pp. 156–159.
460. Rahuman, A.A., Gopalakrishnan, G., Venkatesan, P. and Geetha, K., 2008. Isolation and identification of mosquito larvicidal compound from *Abutilon indicum* (L.) Sweet. *Parasitology Research*, *102*(5), pp. 981–988.

461. Sharma, P.V. and Ahmad, Z.A., 1989. Two sesquiterpene lactones from *Abutilon indicum*. *Phytochemistry*, *28*(12), p. 3525.
462. Annan, K. and Houghton, P.J., 2008. Antibacterial, antioxidant and fibroblast growth stimulation of aqueous extracts of *Ficus asperifolia* Miq. and *Gossypium arboreum* L., wound-healing plants of Ghana. *Journal of Ethnopharmacology*, *119*(1), pp. 141–144.
463. Waage, S.K. and Hedin, P.A., 1984. Biologically-active flavonoids from *Gossypium arboreum*. *Phytochemistry*, *23*(11), pp. 2509–2511.
464. Karou, D., Dicko, M.H., Sanon, S., Simpore, J. and Traore, A.S., 2003. Antimalarial activity of *Sida acuta* Burm. f.(Malvaceae) and *Pterocarpus erinaceus* Poir. (Fabaceae). *Journal of Ethnopharmacology*, *89*(2–3), pp. 291–294.
465. Ekpo, M.A. and Etim, P.C., 2009. Antimicrobial activity of ethanolic and aqueous extracts of *Sida acuta* on microorganisms from skin infections. *Journal of Medicinal Plants Research*, *3*(9), pp. 621–624.
466. Sreedevi, C.D., Latha, P.G., Ancy, P., Suja, S.R., Shyamal, S., Shine, V.J., Sini, S., Anuja, G.I. and Rajasekharan, S., 2009. Hepatoprotective studies on *Sida acuta* Burm. f. *Journal of Ethnopharmacology*, *124*(2), pp. 171–175.
467. Mazumder, U.K., Gupta, M., Manikandan, L. and Bhattacharya, S., 2001. Antibacterial activity of *Urena lobata* root. *Fitoterapia*, *72*(8), pp. 927–929.
468. Purnomo, Y., Soeatmadji, D.W., Sumitro, S.B. and Widodo, M.A., 2015. Anti-diabetic potential of *Urena lobata* leaf extract through inhibition of dipeptidyl peptidase IV activity. *Asian Pacific Journal of Tropical Biomedicine*, *5*(8), pp. 645–649.
469. Atawodi, S.E. and Atawodi, J.C., 2009. *Azadirachta indica* (neem): a plant of multiple biological and pharmacological activities. *Phytochemistry Reviews*, *8*(3), pp. 601–620.
470. Saleem, S., Muhammad, G., Hussain, M.A. and Bukhari, S.N.A., 2018. A comprehensive review of phytochemical profile, bioactives for pharmaceuticals, and pharmacological attributes of *Azadirachta indica*. *Phytotherapy Research*, *32*(7), pp. 1241–1272.
471. Ponnusamy, S., Haldar, S., Mulani, F., Zinjarde, S., Thulasiram, H. and RaviKumar, A., 2015. Gedunin and Azadiradione: human pancreatic alpha-amylase inhibiting limonoids from neem (*Azadirachta indica*) as anti-diabetic agents. *PLoS One*, *10*(10), p. e0140113.
472. Siddiqui, S., Faizi, S. and Siddiqui, B.S., 1992. Constituents of *Azadirachta indica*: isolation and structure elucidation of a new antibacterial tetranortriterpenoid, mahmoodin, and a new protolimonoid, naheedin. *Journal of Natural Products*, *55*(3), pp. 303–310.
473. Koul, O., Multani, J.S., Singh, G., Daniewski, W.M. and Berlozecki, S., 2003. 6β-Hydroxygedunin from *Azadirachta indica*. Its potentiation effects with some non-azadirachtin limonoids in neem against lepidopteran larvae. *Journal of Agricultural and Food Chemistry*, *51*(10), pp. 2937–2942.
474. Akihisa, T., Noto, T., Takahashi, A., Fujita, Y., Banno, N., Tokuda, H., Koike, K., Suzuki, T., Yasukawa, K. and Kimura, Y., 2009. Melanogenesis inhibitory, anti-inflammatory, and chemopreventive effects of limonoids from the seeds of *Azadirachta indicia* A. Juss. (neem). *Journal of Oleo Science*, *58*(11), pp. 581–594.
475. Rana, A., 2008. *Melia azedarach*: a phytopharmacological review. *Pharmacognosy Reviews*, *2*(3), p. 173.
476. Samudram, P., Vasuki, R., Rajeshwari, H., Geetha, A. and Moorthi, P.S., 2009. Antioxidant and antihepatotoxic activities of ethanolic crude extract of *Melia azedarach* and *Piper longum*. *Journal of Medicinal Plants Research*, *3*(12), pp. 1078–1083.
477. Borges, L.M.F., Ferri, P.H., Silva, W.J., Silva, W.C. and Silva, J.G., 2003. In vitro efficacy of extracts of *Melia azedarach* against the tick *Boophilus microplus*. *Medical and Veterinary Entomology*, *17*(2), pp. 228–231.
478. Carpinella, M.C., Giorda, L.M., Ferrayoli, C.G. and Palacios, S.M., 2003. Antifungal effects of different organic extracts from *Melia azedarach* L. on phytopathogenic fungi and their isolated active components. *Journal of Agricultural and Food Chemistry*, *51*(9), pp. 2506–2511.

479. Amresh, G., Zeashan, H., Gupta, R.J., Kant, R., Rao, C.V. and Singh, P.N., 2007. Gastroprotective effects of ethanolic extract from *Cissampelos pareira* in experimental animals. *Journal of Natural Medicines*, *61*(3), pp. 323–328.
480. Amresh, G., Singh, P.N. and Rao, C.V., 2007. Antinociceptive and antiarthritic activity of *Cissampelos pareira* roots. *Journal of Ethnopharmacology*, *111*(3), pp. 531–536.
481. Amresh, G., Reddy, G.D., Rao, C.V. and Singh, P.N., 2007. Evaluation of anti-inflammatory activity of *Cissampelos pareira* root in rats. *Journal of Ethnopharmacology*, *110*(3), pp. 526–531.
482. Ganapaty, S., Dash, G.K., Subburaju, T. and Suresh, P., 2002. Diuretic, laxative and toxicity studies of *Cocculus hirsutus* aerial parts. *Fitoterapia*, *73*(1), pp. 28–31.
483. Badole, S.L., Bodhankar, S.L., Patel, N.M. and Bhardwaj, S., 2009. Acute and chronic diuretic effect of ethanolic extract of leaves of *Cocculus hirsutus* (L.) Diles in normal rats. *Journal of Pharmacy and Pharmacology*, *61*(3), pp. 387–393.
484. Nayak, S. and Singhai, A.K., 2003. Antimicrobial activity of the roots of *Cocculus hirsutus*. *Ancient Science of Life*, *22*(3), p. 101.
485. Guinaudeau, H., Lin, L.Z., Ruangrungsi, N. and Cordell, G.A., 1993. Bisbenzylisoquinoline alkaloids from *Cyclea barbata*. *Journal of Natural Products*, *56*(11), pp. 1989–1992.
486. Lin, L.Z., Shieh, H.L., Angerhofer, C.K., Pezzuto, J.M., Cordell, G.A., Xue, L., Johnson, M.E. and Ruangrungsi, N., 1993. Cytotoxic and antimalarial bisbenzylisoquinoline alkaloids from *Cyclea barbata*. *Journal of Natural Products*, *56*(1), pp. 22–29.
487. Choi, H.S., Kim, H.S., Min, K.R., Kim, Y., Lim, H.K., Chang, Y.K. and Chung, M.W., 2000. Anti-inflammatory effects of fangchinoline and tetrandrine. *Journal of Ethnopharmacology*, *69*(2), pp. 173–179.
488. Lee, Y.S., Han, S.H., Lee, S.H., Kim, Y.G., Park, C.B., Kang, O.H., Keum, J.H., Kim, S.B., Mun, S.H., Seo, Y.S. and Myung, N.Y., 2012. The mechanism of antibacterial activity of tetrandrine against *Staphylococcus aureus*. *Foodborne Pathogens and Disease*, *9*(8), pp. 686–691.
489. Lee, Y.S., Han, S.H., Lee, S.H., Kim, Y.G., Park, C.B., Kang, O.H., Keum, J.H., Kim, S.B., Mun, S.H., Shin, D.W. and Kwon, D.Y., 2011. Synergistic effect of tetrandrine and ethidium bromide against methicillin-resistant *Staphylococcus aureus* (MRSA). *The Journal of Toxicological Sciences*, *36*(5), pp. 645–651.
490. Zhang, H., Wang, K., Zhang, G., Ho, H.I. and Gao, A., 2010. Synergistic anti-candidal activity of tetrandrine on ketoconazole: an experimental study. *Planta Medica*, *76*(1), pp. 53–61.
491. Li, S.X., Song, Y.J., Jiang, L., Zhao, Y.J., Guo, H., Li, D.M., Zhu, K.J. and Zhang, H., 2017. Synergistic effects of tetrandrine with posaconazole against *Aspergillus fumigatus*. *Microbial Drug Resistance*, *23*(6), pp. 674–681.
492. Hu, S., Dutt, J., Zhao, T. and Foster, C.S., 1997. Tetrandrine potently inhibits herpes simplex virus type-1-induced keratitis in BALB/c mice. *Ocular Immunology and Inflammation*, *5*(3), pp. 173–180.
493. Moniruzzaman, M., Hossain, M.S. and Bhattacharjee, P.S., 2016. Evaluation of antinociceptive activity of methanolic extract of leaves of *Stephania japonica* Linn. *Journal of Ethnopharmacology*, *186*, pp. 205–208.
494. Kang, H.S., Kim, Y.H., Lee, C.S., Lee, J.J., Choi, I. and Pyun, K.H., 1996. Anti-inflammatory effects of *Stephania tetrandra* S. Moore on interleukin-6 production and experimental inflammatory disease models. *Mediators of Inflammation*, *5*(4), pp. 280–291.
495. Hall, A.M. and Chang, C.J., 1997. Multidrug-resistance modulators from *Stephania japonica*. *Journal of Natural Products*, *60*(11), pp. 1193–1195.
496. Stanely, P., Prince, M. and Menon, V.P., 2000. Hypoglycaemic and other related actions of *Tinospora cordifolia* roots in alloxan-induced diabetic rats. *Journal of Ethnopharmacology*, *70*(1), pp. 9–15.

497. Kapil, A. and Sharma, S., 1997. Immunopotentiating compounds from *Tinospora cordifolia*. *Journal of Ethnopharmacology*, *58*(2), pp. 89–95.
498. Bishayi, B., Roychowdhury, S., Ghosh, S. and Sengupta, M., 2002. Hepatoprotective and immunomodulatory properties of *Tinospora cordifolia* in CCl_4 intoxicated mature albino rats. *The Journal of Toxicological Sciences*, *27*(3), pp. 139–146.
499. Traore, F., Faure, R., Ollivier, E., Gasquet, M., Azas, N., Debrauwer, L., Keita, A., Timon-David, P. and Balansard, G., 2000. Structure and antiprotozoal activity of triterpenoid saponins from *Glinus oppositifolius*. *Planta Medica*, *66*(4), pp. 368–371.
500. Martin-Puzon, J.J.R., Valle Jr., D.L. and Rivera, W.L., 2015. TLC profiles and antibacterial activity of *Glinus oppositifolius* L. Aug. DC. (Molluginaceae) leaf and stem extracts against bacterial pathogens. *Asian Pacific Journal of Tropical Disease*, *5*(7), pp. 569–574.
501. Shukla, R., Anand, K., Prabhu, K.M. and Murthy, P.S., 1995. Hypolipidemic effect of water extract of *Ficus bengalensis* in alloxan induced diabetes mellitus in rabbits. *Indian Journal of Clinical Biochemistry*, *10*(2), p. 119.
502. Murti, K. and Kumar, U., 2011. Antimicrobial activity of *Ficus benghalensis* and *Ficus racemosa* roots L. *American Journal of Microbiology*, *2*(1), pp. 21–24.
503. Murti, K., Kumar, U. and Panchal, M., 2011. Healing promoting potentials of roots of *Ficus benghalensis* L. in albino rats. *Asian Pacific Journal of Tropical Medicine*, *4*(11), pp. 921–924.
504. Mandal, S.C. and Kumar, C.A., 2002. Studies on anti-diarrhoeal activity of *Ficus hispida*. Leaf extract in rats. *Fitoterapia*, *73*(7–8), pp. 663–667.
505. Mandal, S.C., Saraswathi, B., Ashok Kumar, C.K., Mohana Lakshmi, S. and Maiti, B.C., 2000. Protective effect of leaf extract of *Ficus hispida* L. against paracetamol-induced hepatotoxicity in rats. *Phytotherapy Research*, *14*(6), pp. 457–459.
506. Yap, V.A., Loong, B.J., Ting, K.N., Hwei-San Loh, S., Yong, K.T., Low, Y.Y., Kam, T.S. and Lim, K.H., 2015. Hispidacine, an unusual 8, 4′-oxyneolignan-alkaloid with vasorelaxant activity, and hispiloscine, an antiproliferative phenanthroindolizidine alkaloid, from *Ficus hispida* L. *Phytochemistry*, *109*, pp. 96–102.
507. Jahan, I.A., Nahar, N., Mosihuzzaman, M., Rokeya, B., Ali, L., Azad Khan, A.K., Makhmur, T. and Iqbal Choudhary, M., 2009. Hypoglycaemic and antioxidant activities of *Ficus racemosa* L: fruits. *Natural Product Research*, *23*(4), pp. 399–408.
508. Keshari, A.K., Kumar, G., Kushwaha, P.S., Bhardwaj, M., Kumar, P., Rawat, A., Kumar, D., Prakash, A., Ghosh, B. and Saha, S., 2016. Isolated flavonoids from *Ficus racemosa* stem bark possess antidiabetic, hypolipidemic and protective effects in albino Wistar rats. *Journal of Ethnopharmacology*, *181*, pp. 252–262.
509. Ahmed, F. and Urooj, A., 2012. Cardioprotective activity of standardized extract of *Ficus racemosa* stem bark against doxorubicin-induced toxicity. *Pharmaceutical Biology*, *50*(4), pp. 468–473.
510. Pandit, R., Phadke, A. and Jagtap, A., 2010. Antidiabetic effect of *Ficus religiosa* extract in streptozotocin-induced diabetic rats. *Journal of Ethnopharmacology*, *128*(2), pp. 462–466.
511. Ghosh, M., Civra, A., Rittà, M., Cagno, V., Mavuduru, S.G., Awasthi, P., Lembo, D. and Donalisio, M., 2016. *Ficus religiosa* L: bark extracts inhibit infection by herpes simplex virus type 2 in vitro. *Archives of Virology*, *161*(12), pp. 3509–3514.
512. Cagno, V., Civra, A., Kumar, R., Pradhan, S., Donalisio, M., Sinha, B.N., Ghosh, M. and Lembo, D., 2015. *Ficus religiosa* L: bark extracts inhibit human rhinovirus and respiratory syncytial virus infection in vitro. *Journal of Ethnopharmacology*, *176*, pp. 252–257.
513. Choudhari, A.S., Suryavanshi, S.A. and Kaul-Ghanekar, R., 2013. The aqueous extract of *Ficus religiosa* induces cell cycle arrest in human cervical cancer cell lines SiHa (HPV-16 Positive) and apoptosis in HeLa (HPV-18 Positive). *PLoS One*, *8*(7), p. e70127.

514. Pistelli, L., Chiellini, E.E. and Morelli, I., 2000. Flavonoids from *Ficus pumila*. *Biochemical Systematics and Ecology*, *28*(3), pp. 287–289.
515. Ragasa, C.Y., Juan, E. and Rideout, J.A., 1999. A triterpene from *Ficus pumila*. *Journal of Asian Natural Products Research*, *1*(4), pp. 269–275.
516. Liao, C.R., Kao, C.P., Peng, W.H., Chang, Y.S., Lai, S.C. and Ho, Y.L., 2012. Analgesic and anti-inflammatory activities of methanol extract of *Ficus pumila* L. in mice. *Evidence-Based Complementary and Alternative Medicine*, *2012*, p. 340141.
517. Chatterjee, R.K., Fatma, N., Murthy, P.K., Sinha, P., Kulshrestha, D.K. and Dhawan, B.N., 1992. Macrofilaricidal activity of the stembark of *Streblus asper* and its major active constituents. *Drug Development Research*, *26*(1), pp. 67–78.
518. Wongkham, S., Laupattarakasaem, P., Pienthaweechai, K., Areejitranusorn, P., Wongkham, C. and Techanitiswad, T., 2001. Antimicrobial activity of *Streblus asper* leaf extract. *Phytotherapy Research*, *15*(2), pp. 119–121.
519. Sripanidkulchai, B., Junlatat, J., Wara-aswapati, N. and Hormdee, D., 2009. Anti-inflammatory effect of *Streblus asper* leaf extract in rats and its modulation on inflammation-associated genes expression in RAW 264.7 macrophage cells. *Journal of Ethnopharmacology*, *124*(3), pp. 566–570.
520. Kumar, R.S., Kar, B., Dolai, N., Bala, A. and Haldar, P.K., 2012. Evaluation of antihyperglycemic and antioxidant properties of *Streblus asper* Lour against streptozotocin–induced diabetes in rats. *Asian Pacific Journal of Tropical Disease*, *2*(2), pp. 139–143.
521. Ghasi, S., Nwobodo, E. and Ofili, J.O., 2000. Hypocholesterolemic effects of crude extract of leaf of *Moringa oleifera* Lam in high-fat diet fed Wistar rats. *Journal of Ethnopharmacology*, *69*(1), pp. 21–25.
522. Pari, L. and Kumar, N.A., 2002. Hepatoprotective activity of *Moringa oleifera* on anti-tubercular drug-induced liver damage in rats. *Journal of Medicinal Food*, *5*(3), pp. 171–177.
523. Faizi, S., Siddiqui, B.S., Saleem, R., Aftab, K., Shaheen, F. and Gilani, A.U.H., 1998. Hypotensive constituents from the pods of *Moringa oleifera*. *Planta Medica*, *64*(3), pp. 225–228.
524. Caceres, A., Cabrera, O., Morales, O., Mollinedo, P. and Mendia, P., 1991. Pharmacological properties of *Moringa oleifera*. 1: preliminary screening for antimicrobial activity. *Journal of Ethnopharmacology*, *33*(3), pp. 213–216.
525. Imam, M.Z. and Akter, S., 2011. *Musa paradisiaca* L. and *Musa sapientum* L.: a phytochemical and pharmacological review. *Journal of Applied Pharmaceutical Science*, *1*(5), pp. 14–20.
526. Silva, A.A.S., Morais, S.M., Falcão, M.J.C., Vieira, I.G.P., Ribeiro, L.M., Viana, S.M., Teixeira, M.J., Barreto, F.S., Carvalho, C.A., Cardoso, R.P.A. and Andrade-Junior, H.F., 2014. Activity of cycloartane-type triterpenes and sterols isolated from *Musa paradisiaca* fruit peel against *Leishmania infantum* chagasi. *Phytomedicine*, *21*(11), pp. 1419–1423.
527. Jawla, S., Kumar, Y. and Khan, M.S.Y., 2012. Antimicrobial and antihyperglycemic activities of *Musa paradisiaca* flowers. *Asian Pacific Journal of Tropical Biomedicine*, *2*(2), pp. S914–S918.
528. Gutiérrez, R.M.P., Mitchell, S. and Solis, R.V., 2008. *Psidium guajava*: a review of its traditional uses, phytochemistry and pharmacology. *Journal of Ethnopharmacology*, *117*(1), pp. 1–27.
529. Olajide, O.A., Awe, S.O. and Makinde, J.M., 1999. Pharmacological studies on the leaf of *Psidium guajava*. *Fitoterapia*, *70*(1), pp. 25–31.
530. Ojewole, J.A.O., 2005. Hypoglycemic and hypotensive effects of *Psidium guajava* Linn. (Myrtaceae) leaf aqueous extract. *Methods and Findings in Experimental and Clinical Pharmacology*, *27*(10), pp. 689–696.

531. Teixeira, C.C., Pinto, L.P., Kessler, F.H.P., Knijnik, L., Pinto, C.P., Gastaldo, G.J. and Fuchs, F.D., 1997. The effect of *Syzygium cumini* (L.) skeels on post-prandial blood glucose levels in non-diabetic rats and rats with streptozotocin-induced diabetes mellitus. *Journal of Ethnopharmacology*, *56*(3), pp. 209–213.
532. Oliveira, G.F.D., Furtado, N.A.J.C., Silva Filho, A.A.D., Martins, C.H.G., Bastos, J.K. and Cunha, W.R., 2007. Antimicrobial activity of *Syzygium cumini* leaves extract. *Brazilian Journal of Microbiology*, *38*(2), pp. 381–384.
533. Ramirez, R.O. and Roa Jr., C.C., 2003. The gastroprotective effect of tannins extracted from duhat (*Syzygium cumini* Skeels) bark on HCl/ethanol induced gastric mucosal injury in Sprague–Dawley rats. *Clinical Hemorheology and Microcirculation*, *29*(3–4), pp. 253–261.
534. Sutariya, B., Taneja, N. and Saraf, M., 2017. Betulinic acid, isolated from the leaves of *Syzygium cumini* (L.) Skeels, ameliorates the proteinuria in experimental membranous nephropathy through regulating Nrf2/NF-κB pathways. *Chemico-Biological Interactions*, *274*, pp. 124–137.
535. Dung, N.T., Bajpai, V.K., Yoon, J.I. and Kang, S.C., 2009. Anti-inflammatory effects of essential oil isolated from the buds of *Cleistocalyx operculatus* (Roxb.) Merr and Perry. *Food and Chemical Toxicology*, *47*(2), pp. 449–453.
536. Dash, B.K., Sen, M.K., Alam, K., Hossain, K., Islam, R., Banu, N.A., Rahman, S. and Jamal, A.M., 2013. Antibacterial activity of *Nymphaea nouchali* (Burm. f) flower. *Annals of Clinical Microbiology and Antimicrobials*, *12*(1), p. 27.
537. Parimala, M. and Shoba, F.G., 2014. In vitro antimicrobial activity and HPTLC analysis of hydroalcoholic seed extract of *Nymphaea nouchali* Burm. f. *BMC Complementary and Alternative Medicine*, *14*(1), p. 361.
538. Parimala, M. and Shoba, F.G., 2014. Evaluation of antidiabetic potential of *Nymphaea nouchali* Burm. f: seeds in STZ-induced diabetic rats. *International Journal of Pharmaceutical Sciences*, *6*(4), pp. 536–541.
539. Antonisamy, P., Subash-Babu, P., Alshatwi, A.A., Aravinthan, A., Ignacimuthu, S., Choi, K.C. and Kim, J.H., 2014. Gastroprotective effect of nymphayol isolated from *Nymphaea stellata* (Willd.) flowers: contribution of antioxidant, anti-inflammatory and anti-apoptotic activities. *Chemico-Biological Interactions*, *224*, pp. 157–163.
540. Debnath, S., Ghosh, S. and Hazra, B., 2013. Inhibitory effect of *Nymphaea pubescens* Willd. flower extract on carrageenan-induced inflammation and CCl_4-induced hepatotoxicity in rats. *Food and Chemical Toxicology*, *59*, pp. 485–491.
541. Selvakumari, S. and Arcot, S., 2010. Andiabetic activity of *Nymphaea pubescens* Willd—a plant drug of aquatic flora interest. *Journal of Pharmacy Research*, *3*(12), pp. 3067–3069.
542. Sengar, N., Joshi, A., Prasad, S.K. and Hemalatha, S., 2015. Anti-inflammatory, analgesic and anti-pyretic activities of standardized root extract of *Jasminum sambac*. *Journal of Ethnopharmacology*, *160*, pp. 140–148.
543. AlRashdi, A.S., Salama, S.M., Alkiyumi, S.S., Abdulla, M.A., Hadi, A.H.A., Abdelwahab, S.I., Taha, M.M., Hussiani, J. and Asykin, N., 2012. Mechanisms of gastroprotective effects of ethanolic leaf extract of *Jasminum sambac* against HCl/ethanol-induced gastric mucosal injury in rats. *Evidence-Based Complementary and Alternative Medicine*, *2012*, p. 786426.
544. Khatune, N.A., Mosaddik, M.A. and Haque, M.E., 2001. Antibacterial activity and cytotoxicity of *Nyctanthes arbor-tristis* flowers. *Fitoterapia*, *72*(4), pp. 412–414.
545. Mishra, R.K., Mishra, V., Pandey, A., Tiwari, A.K., Pandey, H., Sharma, S., Pandey, A.C. and Dikshit, A., 2016. Exploration of anti-Malassezia potential of *Nyctanthes arbor-tristis* L. their application to combat the infection caused by Mala s1 a novel allergen. *BMC Complementary and Alternative Medicine*, *16*(1), p. 114.

546. Saxena, R.S., Gupta, B., Saxena, K.K., Srivastava, V.K. and Prasad, D.N., 1987. Analgesic, antipyretic and ulcerogenic activity of *Nyctanthes arbor tristis* leaf extract. *Journal of Ethnopharmacology*, *19*(2), pp. 193–200.
547. Saxena, R.S., Gupta, B., Saxena, K.K., Singh, R.C. and Prasad, D.M., 1984. Study of anti-inflammatory activity in the leaves of *Nyctanthes arbor tristis* L.—an Indian medicinal plant. *Journal of Ethnopharmacology*, *11*(3), pp. 319–330.
548. Chang, C.I., Kuo, C.C., Chang, J.Y. and Kuo, Y.H., 2004. Three new oleanane-type triterpenes from *Ludwigia octovalvis* with cytotoxic activity against two human cancer cell lines. *Journal of Natural Products*, *67*(1), pp. 91–93.
549. Yakob, H.K., Sulaiman, S.F. and Uyub, A.M., 2012. Antioxidant and antibacterial activity of *Ludwigia octovalvis* on *Escherichia coli* O157: H7 and some pathogenic bacteria. *World Applied Sciences Journal*, *16*, pp. 22–29.
550. Paul, P., Chowdhury, K., Nath, D. and Bhattacharjee, M.K., 2013. Antimicrobial efficacy of orchid extracts as potential inhibitors of antibiotic resistant strains of *Escherichia coli*. *Asian Journal of Pharmaceutical Clinical Research*, *6*(3), pp. 108–111.
551. Hoque, M.M., Khaleda, L. and Al-Forkan, M., 2016. Evaluation of pharmaceutical properties on microbial activities of some important medicinal orchids of Bangladesh. *Journal of Pharmacognosy and Phytochemistry*, *5*(2), p. 265.
552. Howlader, M.A., Alam, M., Ahmed, K., Khatun, F. and Apu, A.S., 2011. Antinociceptive and anti-inflammatory activity of the ethanolic extract of *Cymbidium aloifolium* (L.). *Pakistan Journal of Biological Sciences*, *14*(19), p. 909.
553. Juneja, R.K., Sharma, S.C. and Tandon, J.S., 1987. Two substituted bibenzyls and a dihydrophenanthrene from *Cymbidium aloifolium*. *Phytochemistry*, *26*(4), pp. 1123–1125.
554. Pang, D., You, L., Zhou, L., Li, T., Zheng, B. and Liu, R.H., 2017. *Averrhoa carambola* free phenolic extract ameliorates nonalcoholic hepatic steatosis by modulating mircoRNA-34a, mircoRNA-33 and AMPK pathways in leptin receptor-deficient db/db mice. *Food & Function*, *8*(12), pp. 4496–4507.
555. Rehman, A., Rehman, A. and Ahmad, I., 2015. Antibacterial, antifungal, and insecticidal potentials of *Oxalis corniculata* and its isolated compounds. *International Journal of Analytical Chemistry*, *2015*, p. 842468.
556. Sreejith, G., Jayasree, M., Latha, P.G., Suja, S.R., Shyamal, S., Shine, V.J., Anuja, G.I., Sini, S., Shikha, P., Krishnakumar, N.M. and Vilash, V., 2014. Hepatoprotective activity of *Oxalis corniculata* L. ethanolic extract against paracetamol induced hepatotoxicity in Wistar rats and its in vitro antioxidant effects. *Indian Journal of Experimental Biology*, *52*(2), pp. 147–152.
557. Sakat, S.S., Tupe, P. and Juvekar, A., 2012. Gastroprotective effect of *Oxalis corniculata* (whole plant) on experimentally induced gastric ulceration in Wistar rats. *Indian Journal of Pharmaceutical Sciences*, *74*(1), p. 48.
558. Tan, M.A., Takayama, H., Aimi, N., Kitajima, M., Franzblau, S.G. and Nonato, M.G., 2008. Antitubercular triterpenes and phytosterols from *Pandanus tectorius* Soland. var. laevis. *Journal of Natural Medicines*, *62*(2), pp. 232–235.
559. Zhang, X., Wu, C., Wu, H., Sheng, L., Su, Y., Zhang, X., Luan, H., Sun, G., Sun, X., Tian, Y. and Ji, Y., 2013. Anti-hyperlipidemic effects and potential mechanisms of action of the caffeoylquinic acid-rich *Pandanus tectorius* fruit extract in hamsters fed a high fat-diet. *PLoS One*, *8*(4), p. e61922.
560. Andersen, L., Adsersen, A. and Jaroszewski, J.W., 1998. Cyanogenesis of *Passiflora foetida*. *Phytochemistry*, *47*(6), pp. 1049–1050.
561. Echeverri, F., Cardona, G., Torres, F., Pelaez, C., Quiñones, W. and Renteria, E., 1991. Ermanin: an insect deterrent flavonoid from *Passiflora foetida* resin. *Phytochemistry*, *30*(1), pp. 153–155.

562. Sasikala, V., Saravanan, S. and Parimelazhagan, T., 2011. Analgesic and anti–inflammatory activities of *Passiflora foetida* L. *Asian Pacific Journal of Tropical Medicine*, *4*(8), pp. 600–603.
563. Kaennakam, S., Sichaem, J., Siripong, P. and Tip-Pyang, S., 2013. A new cytotoxic phenolic derivative from the roots of *Antidesma acidum*. *Natural Product Communications*, *8*(8), pp. 1111–1113.
564. Jayasinghe, L., Kumarihamy, B.M., Jayarathna, K.N., Udishani, N.G., Bandara, B.R., Hara, N. and Fujimoto, Y., 2003. Antifungal constituents of the stem bark of *Bridelia retusa*. *Phytochemistry*, *62*(4), pp. 637–641.
565. Mehare, I.D. and Hatapakki, B.C., 2003. Antiinflammatory activity of bark of *Bridelia retusa* Spreng. *Indian Journal of Pharmaceutical Sciences*, *65*(4), p. 410.
566. Kumar, T. and Jain, V., 2014. Antinociceptive and anti-inflammatory activities of *Bridelia retusa* methanolic fruit extract in experimental animals. *The Scientific World Journal*, *2014*, p. 890151.
567. Tatiya, A.U., Deore, U.V., Jain, P.G. and Surana, S.J., 2011. Hypoglycemic potential of *Bridelia retusa* bark in albino rats. *Asian Journal of Biological Sciences*, *4*(1), pp. 84–89.
568. Anjum, A., Haque, M.R., Rahman, M.S., Hasan, C.M., Haque, M.E. and Rashid, M.A., 2011. In vitro antibacterial, antifungal and cytotoxic activity of three Bangladeshi *Bridelia* species. *International Research of Pharmacy and Pharmacology*, *1*(7), pp. 149–154.
569. Talapatra, S.K., Bhattacharya, S., Maiti, B.C. and Talapatra, B., 1973. Structure of glochilocudiol: a new triterpenoid from *Glochidion multiloculare*: natural occurrence of dimedone. *Chemistry & Industry (London)*, *21*, pp. 1033–1034.
570. Kabir, S., Zahan, R., Chowdhury, A.M.S., Haque, M.R. and Rashid, M.A., 2015. Antitumor, analgesic and anti-inflammatory activities of *Glochidion multiloculare* (Rottler ex Willd) Voigt. *Bangladesh Pharmaceutical Journal*, *18*(2), pp. 142–148.
571. Azam, A.Z., Al Hasan, A., Uddin, M.G., Masud, M.M. and Hasan, C.M., 2012. Antimicrobial, antioxidant and cytotoxic activities of *Glochidion multiloculare* (Roxb. ex Willd.) Müll. Arg. (Euphorbiaceae). *Dhaka University Journal of Pharmaceutical Sciences*, *11*(2), pp. 117–120.
572. Gaire, B.P. and Subedi, L., 2014. Phytochemistry, pharmacology and medicinal properties of *Phyllanthus emblica* L. *Chinese Journal of Integrative Medicine*, pp. 1–8.
573. Perianayagam, J.B., Sharma, S.K., Joseph, A. and Christina, A.J.M., 2004. Evaluation of anti-pyretic and analgesic activity of *Emblica officinalis* Gaertn. *Journal of Ethnopharmacology*, *95*(1), pp. 83–85.
574. Usharani, P., Fatima, N. and Muralidhar, N., 2013. Effects of *Phyllanthus emblica* extract on endothelial dysfunction and biomarkers of oxidative stress in patients with type 2 diabetes mellitus: a randomized, double-blind, controlled study. *Diabetes, Metabolic Syndrome and Obesity: Targets and Therapy*, *6*, p. 275.
575. Mahata, S., Pandey, A., Shukla, S., Tyagi, A., Husain, S.A., Das, B.C. and Bharti, A.C., 2013. Anticancer activity of *Phyllanthus emblica* L. (Indian gooseberry): inhibition of transcription factor AP-1 and HPV gene expression in cervical cancer cells. *Nutrition and Cancer*, *65*(suppl 1), pp. 88–97.
576. Bagalkotkar, G., Sagineedu, S.R., Saad, M.S. and Stanslas, J., 2006. Phytochemicals from *Phyllanthus niruri* Linn. and their pharmacological properties: a review. *Journal of Pharmacy and Pharmacology*, *58*(12), pp. 1559–1570.
577. Syamasundar, K.V., Singh, B., Thakur, R.S., Husain, A., Yoshinobu, K. and Hiroshi, H., 1985. Antihepatotoxic principles of *Phyllanthus niruri* herbs. *Journal of Ethnopharmacology*, *14*(1), pp. 41–44.

578. Khanna, A.K., Rizvi, F. and Chander, R., 2002. Lipid lowering activity of *Phyllanthus niruri* in hyperlipemic rats. *Journal of Ethnopharmacology*, *82*(1), pp. 19–22.
579. Qian-Cutrone, J., Huang, S., Trimble, J., Li, H., Lin, P.F., Alam, M., Klohr, S.E. and Kadow, K.F., 1996. Niruriside, a new HIV REV/RRE binding inhibitor from *Phyllanthus niruri*. *Journal of Natural Products*, *59*(2), pp. 196–199.
580. Kumar, S., Kumar, D., Deshmukh, R.R., Lokhande, P.D., More, S.N. and Rangari, V.D., 2008. Antidiabetic potential of *Phyllanthus reticulatus* in alloxan-induced diabetic mice. *Fitoterapia*, *79*(1), pp. 21–23.
581. Saha, A., Masud, M.A., Bachar, S.C., Kundu, J.K., Datta, B.K., Nahar, L. and Sarker, S.D., 2007. The analgesic and anti-inflammatory activities of the extracts of *Phyllanthus reticulatus* in mice model. *Pharmaceutical Biology*, *45*(5), pp. 355–359.
582. Omulokoli, E., Khan, B. and Chhabra, S.C., 1997. Antiplasmodial activity of four Kenyan medicinal plants. *Journal of Ethnopharmacology*, *56*(2), pp. 133–137.
583. Hasan, I., Hussain, M.S., Millat, M.S., Sen, N., Rahman, M.A., Rahman, M.A., Islam, S. and Moghal, M.M.R., 2018. Ascertainment of pharmacological activities of *Allamanda neriifolia* Hook and *Aegialitis rotundifolia* Roxb used in Bangladesh: an in vitro study. *Journal of Traditional and Complementary Medicine*, *8*(1), pp. 107–112.
584. Sett, S., Hazra, J., Datta, S., Mitra, A. and Mitra, A.K., 2014. Screening the Indian Sundarban mangrove for antimicrobial activity. *International Journal of Science Innovations and Discoveries*, *4*, pp. 17–25.
585. Devarshi, P., Patil, S. and Kanase, A., 1991. Effect of *Plumbago zeylanica* root powder induced preimplantationary loss and abortion on uterine luminal proteins in albino rats. *Indian Journal of Experimental Biology*, *29*(6), pp. 521–522.
586. Checker, R., Sharma, D., Sandur, S.K., Khanam, S. and Poduval, T.B., 2009. Anti-inflammatory effects of plumbagin are mediated by inhibition of NF-kappaB activation in lymphocytes. *International Immunopharmacology*, *9*(7–8), pp. 949–958.
587. Kodati, D.R., Burra, S. and Kumar, G.P., 2011. Evaluation of wound healing activity of methanolic root extract of *Plumbago zeylanica* L. in wistar albino rats. *Asian Journal of Plant Science and Research*, *1*(2), pp. 26–34.
588. Kim, J.E., Kim, E.H. and Park, S.N., 2010. Antibacterial activity of *Persicaria hydropiper* extracts and its application for cosmetic material. *Korean Journal of Microbiology and Biotechnology*, *38*, pp. 112–115.
589. Lai, C.Y., Tsai, A.C., Chen, M.C., Chang, L.H., Sun, H.L., Chang, Y.L., Chen, C.C., Teng, C.M. and Pan, S.L., 2012. Aciculatin induces p53-dependent apoptosis via MDM2 depletion in human cancer cells in vitro and in vivo. *PLoS One*, *7*(8), p. e42192.
590. Atmani, F., Sadki, C., Aziz, M., Mimouni, M. and Hacht, B., 2009. *Cynodon dactylon* extract as a preventive and curative agent in experimentally induced nephrolithiasis. *Urological Research*, *37*(2), pp. 75–82.
591. Biswas, T.K., Pandit, S., Chakrabarti, S., Banerjee, S., Poyra, N. and Seal, T., 2017. Evaluation of *Cynodon dactylon* for wound healing activity. *Journal of Ethnopharmacology*, *197*, pp. 128–137.
592. Merish, S., Tamizhamuthu, M. and Thomas, M.W., 2014. Styptic and wound healing properties of siddha medicinal plants–a review. *International Journal of Pharma & Bio Sciences*, *5*(2), pp. 43–49.
593. Ahmed, S., Reza, M.S., Haider, S.S. and Jabbar, A., 1994. Antimicrobial activity of *Cynodon dactylon*. *Fitoterapia*, *65*, pp. 463–464.
594. Murali, K.S., Sivasubramanian, S., Vincent, S., Murugan, S.B., Giridaran, B., Dinesh, S., Gunasekaran, P., Krishnasamy, K. and Sathishkumar, R., 2015. Anti—chikungunya activity of luteolin and apigenin rich fraction from *Cynodon dactylon*. *Asian Pacific Journal of Tropical Medicine*, *8*(5), pp. 352–358.

595. De Melo, G.O., Muzitano, M.F., Legora-Machado, A., Almeida, T.A., De Oliveira, D.B., Kaiser, C.R., Koatz, V.L.G. and Costa, S.S., 2005. C-glycosylflavones from the aerial parts of *Eleusine indica* inhibit LPS-induced mouse lung inflammation. *Planta Medica*, *71*(4), pp. 362–363.
596. Luqman, S., Srivastava, S., Darokar, M.P. and Khanuja, S.P., 2005. Detection of antibacterial activity in spent roots of two genotypes of aromatic grass *Vetiveria zizanioides*. *Pharmaceutical Biology*, *43*(8), pp. 732–736.
597. Chou, S.T., Lai, C.P., Lin, C.C. and Shih, Y., 2012. Study of the chemical composition, antioxidant activity and anti-inflammatory activity of essential oil from *Vetiveria zizanioides*. *Food Chemistry*, *134*(1), pp. 262–268.
598. Khatun, A., Imam, M.Z. and Rana, M.S., 2015. Antinociceptive effect of methanol extract of leaves of *Persicaria hydropiper* in mice. *BMC Complementary and Alternative Medicine*, *15*(1), p. 63.
599. Yang, Y., Yu, T., Jang, H.J., Byeon, S.E., Song, S.Y., Lee, B.H., Rhee, M.H., Kim, T.W., Lee, J., Hong, S. and Cho, J.Y., 2012. In vitro and in vivo anti-inflammatory activities of *Polygonum hydropiper* methanol extract. *Journal of Ethnopharmacology*, *139*(2), pp. 616–625.
600. Zihad, S.N.K., Bhowmick, N., Uddin, S.J., Sifat, N., Rahman, M.S., Rouf, R., Islam, M.T., Dev, S., Hazni, H., Aziz, S. and Ali, E.S., 2018. Analgesic activity, chemical profiling and computational study on *Chrysopogon aciculatus*. *Frontiers in Pharmacology*, *9*, p. 1164.
601. Ekambaram, S.P., Perumal, S.S. and Pavadai, S., 2017. Anti-inflammatory effect of *Naravelia zeylanica* DC via suppression of inflammatory mediators in carrageenan-induced abdominal oedema in zebrafish model. *Inflammopharmacology*, *25*(1), pp. 147–158.
602. Shenoy, M.A., Shastry, C.S. and Gopkumar, P., 2009. Anti ulcer activity of *Naravelia zeylanica* leaves extract. *Journal of Pharmacy Research*, *2*(7), pp. 1218–1220.
603. Naika, H.R., Krishna, V., Harish, B.G., Ahamed, B.K. and Mahadevan, K.M., 2007. Antimicrobial activity of bioactive constituents isolated from the leaves of *Naravelia zeylanica* (L.) DC. *International Journal of Biomedical and Pharmaceutical Sciences*, *1*(2), pp. 153–159.
604. Talmale, S., Bhujade, A. and Patil, M., 2015. Anti-allergic and anti-inflammatory properties of *Zizyphus mauritiana* root bark. *Food & Function*, *6*(9), pp. 2975–2983.
605. Bhatia, A. and Mishra, T., 2010. Hypoglycemic activity of *Ziziphus mauritiana* aqueous ethanol seed extract in alloxan-induced diabetic mice. *Pharmaceutical Biology*, *48*(6), pp. 604–610.
606. Panseeta, P., Lomchoey, K., Prabpai, S., Kongsaeree, P., Suksamrarn, A., Ruchirawat, S. and Suksamrarn, S., 2011. Antiplasmodial and antimycobacterial cyclopeptide alkaloids from the root of *Ziziphus mauritiana*. *Phytochemistry*, *72*(9), pp. 909–915.
607. Khoubnasabjafari, M. and Jouyban, A., 2011. A review of phytochemistry and bioactivity of quince (*Cydonia oblonga* Mill.). *Journal of Medicinal Plants Research*, *5*(16), pp. 3577–3594.
608. Gholami, S., Hosseini, M.J., Jafari, L., Omidvar, F., Kamalinejad, M., Mashayekhi, V., Hosseini, S.H., Kardan, A., Pourahmad, J. and Eskandari, M.R., 2017. Mitochondria as a target for the cardioprotective effects of *Cydonia oblonga* Mill. and *Ficus carica* L. in doxorubicin-induced cardiotoxicity. *Drug Research*, *67*(6), pp. 358–365.
609. Riahi-Chebbi, I., Haoues, M., Essafi, M., Zakraoui, O., Fattouch, S., Karoui, H. and Essafi-Benkhadir, K., 2015. Quince peel polyphenolic extract blocks human colon adenocarcinoma LS174 cell growth and potentiates 5-fluorouracil efficacy. *Cancer Cell International*, *16*(1), p. 1.

610. Mirmohammadlu, M., Hosseini, S.H., Kamalinejad, M., Gavgani, M.E., Noubarani, M. and Eskandari, M.R., 2015. Hypolipidemic, hepatoprotective and renoprotective effects of *Cydonia oblonga* Mill. fruit in streptozotocin-induced diabetic rats. *Iranian Journal of Pharmaceutical Research*, *14*(4), p. 1207.
611. Kumar, R., Nair, V., Gupta, Y.K. and Singh, S., 2017. Anti-inflammatory and anti-arthritic activity of aqueous extract of *Rosa centifolia* in experimental rat models. *International Journal of Rheumatic Diseases*, *20*(9), pp. 1072–1078.
612. Sankar, R.A., Nikhila, C., Lakshmiprasanna, V.C., Mobeena, S.K., Karunakar, K. and Bharathi, N., 2011. Evaluation of anti-tussive activity of *Rosa centifolia*. *International Journal of Pharmaceutical Sciences and Research*, *2*(6), p. 1473.
613. Hirulkar, N.B., 2010. Antimicrobial activity of rose petals extract against some pathogenic bacteria. *International Journal of Pharmaceutical & Biological Archive*, *1*(5), pp. 478–484.
614. Iqbal, P.F., Bhat, A.R. and Azam, A., 2009. Antiamoebic coumarins from the root bark of *Adina cordifolia* and their new thiosemicarbazone derivatives. *European Journal of Medicinal Chemistry*, *44*(5), pp. 2252–2259.
615. Alam, M.A., Subhan, N., Chowdhury, S.A., Awal, M.A., Mostofa, M., Rashid, M.A., Hasan, C.M., Nahar, L. and Sarker, S.D., 2011. *Anthocephalus cadamba* (Roxb.) Miq., Rubiaceae, extract shows hypoglycemic effect and eases oxidative stress in alloxan-i nduced diabetic rats. *Revista Brasileira de Farmacognosia*, *21*(1).
616. Pandey, A. and Negi, P.S., 2018. Phytochemical composition, in vitro antioxidant activity and antibacterial mechanisms of *Neolamarckia cadamba* fruits extracts. *Natural Product Research*, *32*(10), pp. 1189–1192.
617. Qureshi, A.K., Mukhtar, M.R., Hirasawa, Y., Hosoya, T., Nugroho, A.E., Morita, H., Shirota, O., Mohamad, K., Hadi, A.H.A., Litaudon, M. and Awang, K., 2011. Neolamarckines A and B, new indole alkaloids from *Neolamarckia cadamba*. *Chemical and Pharmaceutical Bulletin*, *59*(2), pp. 291–293.
618. Sultana, R., Rahman, M.S., Bhuiyan, M.N.I., Begum, J. and Anwar, M.N., 2008. In vitro antibacterial and antifungal activity of *Borreria articularis*. *Bangladesh Journal of Microbiology*, *25*(2), pp. 95–98.
619. Saha, K., Lajis, N.H., Israf, D.A., Hamzah, A.S., Khozirah, S., Khamis, S. and Syahida, A., 2004. Evaluation of antioxidant and nitric oxide inhibitory activities of selected Malaysian medicinal plants. *Journal of Ethnopharmacology*, *92*(2–3), pp. 263–267.
620. Wang, G.C., Li, T., Deng, F.Y., Li, Y.L. and Ye, W.C., 2013. Five new phenolic glycosides from *Hedyotis scandens*. *Bioorganic & Medicinal Chemistry Letters*, *23*(5), pp. 1379–1382.
621. Jagdishprasad, P. and Subba, R.N., 1988. Anticoagulant and anti-inflammatory and sunscreening effects of *Hymenodictyon excelsum*. *Indian Journal of Pharmacology*, *20*(2), p. 221.
622. Wickramasinghe, R., Kumara, R.R., De Silva, E.D., Ratnasooriya, W.D. and Handunnetti, S., 2014. Inhibition of phagocytic and intracellular killing activity of human neutrophils by aqueous and methanolic leaf extracts of *Ixora coccinea*. *Journal of Ethnopharmacology*, *153*(3), pp. 900–907.
623. Annapurna, J., Amarnath, P.V.S., Kumar, D.A., Ramakrishna, S.V. and Raghavan, K.V., 2003. Antimicrobial activity of *Ixora coccinea* leaves. *Fitoterapia*, *74*(3), pp. 291–293.
624. Latha, P.G. and Panikkar, K.R., 2001. Chemoprotective effect of *Ixora coccinea* L: flowers on cisplatin induced toxicity in mice. *Phytotherapy Research*, *15*(4), pp. 364–366.
625. Latha, P.G. and Panikkar, K.R., 1998. Cytotoxic and antitumour principles from *Ixora coccinea* flowers. *Cancer Letters*, *130*(1–2), pp. 197–202.

626. Xiang, W., Song, Q.S., Zhang, H.J. and Guo, S.P., 2008. Antimicrobial anthraquinones from *Morinda angustifolia*. *Fitoterapia*, *79*(7–8), pp. 501–504.
627. Akter, R., Uddin, S.J., Grice, I.D. and Tiralongo, E., 2014. Cytotoxic activity screening of Bangladeshi medicinal plant extracts. *Journal of Natural Medicines*, *68*(1), pp. 246–252.
628. Itoh, A., Tanahashi, T., Nagakura, N. and Nishi, T., 2003. Two chromone-secoiridoid glycosides and three indole alkaloid glycosides from *Neonauclea sessilifolia*. *Phytochemistry*, *62*(3), pp. 359–369.
629. Kang, W.Y., DU, Z.Z. and Hao, X.J., 2004. Triterpenoid saponins from *Neonauclea sessilifolia* Merr. *Journal of Asian Natural Products Research*, *6*(1), pp. 1–6.
630. Kumar, V., Al-Abbasi, F.A., Ahmed, D., Verma, A., Mujeeb, M. and Anwar, F., 2015. *Paederia foetida* L. inhibits adjuvant induced arthritis by suppression of PGE 2 and COX-2 expression via nuclear factor-κB. *Food & Function*, *6*(5), pp. 1652–1666.
631. De, S., Ravishankar, B. and Bhavsar, G.C., 1994. Investigation of the anti-inflammatory effects of *Paederia foetida*. *Journal of Ethnopharmacology*, *43*(1), pp. 31–38.
632. Afroz, S., Alamgir, M., Khan, M.T.H., Jabbar, S., Nahar, N. and Choudhuri, M.S.K., 2006. Antidiarrhoeal activity of the ethanol extract of *Paederia foetida* L. (Rubiaceae). *Journal of Ethnopharmacology*, *105*(1–2), pp. 125–130.
633. Baliga, M.S., Bhat, H.P., Joseph, N. and Fazal, F., 2011. Phytochemistry and medicinal uses of the bael fruit (*Aegle marmelos* Correa): a concise review. *Food Research International*, *44*(7), pp. 1768–1775.
634. Gautam, M.K., Ghatule, R.R., Singh, A., Purohit, V., Gangwar, M., Kumar, M. and Goel, R.K., 2013. Healing effects of *Aegle marmelos* (L.) Correa fruit extract on experimental colitis. *Indian Journal of Experimental Biology*, *51*(2), pp. 157–164.
635. Bafna, P. and Bodhankar, S., 2003. Gastrointestinal effects of Mebarid®, an ayurvedic formulation, in experimental animals. *Journal of Ethnopharmacology*, *86*(2–3), pp. 173–176.
636. Ansari, P., Afroz, N., Jalil, S., Azad, S.B., Mustakim, M.G., Anwar, S., Haque, S.N., Hossain, S.M., Tony, R.R. and Hannan, J.M.A., 2017. Anti-hyperglycemic activity of *Aegle marmelos* (L.) corr. is partly mediated by increased insulin secretion, α-amylase inhibition, and retardation of glucose absorption. *Journal of Pediatric Endocrinology and Metabolism*, *30*(1), pp. 37–47.
637. Chakraborty, A., Saha, C., Podder, G., Chowdhury, B.K. and Bhattacharyya, P., 1995. Carbazole alkaloid with antimicrobial activity from *Clausena heptaphylla*. *Phytochemistry*, *38*(3), pp. 787–789.
638. Sohrab, M.H., Mazid, M.A., Rahman, E., Hasan, C.M. and Rashid, M.A., 2001. Antibacterial activity of *Clausena heptaphylla*. *Fitoterapia*, *72*(5), pp. 547–549.
639. Joshi, R.K., Badakar, V.M., Kholkute, S.D. and Khatib, N., 2011. Chemical composition and antimicrobial activity of the essential oil of the leaves of *Feronia elephantum* (Rutaceae) from North West Karnataka. *Natural Product Communications*, *6*(1), pp. 141–143.
640. Kamat, C.D., Khandelwal, K.R., Bodhankar, S.L., Ambavade, S.D. and Mhetre, N.A., 2003. Hepatoprotective activity of leaves of *Feronia elephantum* Correa (Rutaceae) against carbon tetrachloride-induced liver damage in rats. *Journal of Natural Remedies*, *3*(2), pp. 148–154.
641. Rahuman, A.A., Gopalakrishnan, G., Ghouse, B.S., Arumugam, S. and Himalayan, B., 2000. Effect of *Feronia limonia* on mosquito larvae. *Fitoterapia*, *71*(5), pp. 553–555.
642. Nayak, S.S., Jain, R. and Sahoo, A.K., 2011. Hepatoprotective activity of *Glycosmis pentaphylla* against paracetamol-induced hepatotoxicity in Swiss albino mice. *Pharmaceutical Biology*, *49*(2), pp. 111–117.

643. Silambujanaki, P., Chandra, C.B.T., Kumar, K.A. and Chitra, V., 2011. Wound healing activity of *Glycosmis arborea* leaf extract in rats. *Journal of Ethnopharmacology, 134*(1), pp. 198–201.
644. Handral, H.K., Pandith, A. and Shruthi, S.D., 2012. A review on *Murraya koenigii*: multipotential medicinal plant. *Asian Journal of Pharmaceutical and Clinical Research, 5*(4), pp. 5–14.
645. Gupta, S., George, M., Singhal, M., Sharma, G.N. and Garg, V., 2010. Leaves extract of *Murraya koenigii* linn for anti-inflammatory and analgesic activity in animal models. *Journal of Advanced Pharmaceutical Technology & Research, 1*(1), p. 68.
646. Okugawa, H., Ueda, R., Matsumoto, K., Kawanishi, K. and Kato, A., 1995. Effect of α-santalol and β-santalol from sandalwood on the central nervous system in mice. *Phytomedicine, 2*(2), pp. 119–126.
647. Matsuo, Y. and Mimaki, Y., 2012. α-Santalol derivatives from *Santalum album* and their cytotoxic activities. *Phytochemistry, 77*, pp. 304–311.
648. Santha, S. and Dwivedi, C., 2015. Anticancer effects of sandalwood (*Santalum album*). *Anticancer Research, 35*(6), pp. 3137–3145.
649. Guo, H., Zhang, J., Gao, W., Qu, Z. and Liu, C., 2014. Anti-diarrhoeal activity of methanol extract of *Santalum album* L. in mice and gastrointestinal effect on the contraction of isolated jejunum in rats. *Journal of Ethnopharmacology, 154*(3), pp. 704–710.
650. Matsuo, Y. and Mimaki, Y., 2010. Lignans from *Santalum album* and their cytotoxic activities. *Chemical and Pharmaceutical Bulletin, 58*(4), pp. 587–590.
651. Waako, P.J., Gumede, B., Smith, P. and Folb, P.I., 2005. The in vitro and in vivo antimalarial activity of *Cardiospermum halicacabum* L. and *Momordica foetida* Schumch. Et Thonn. *Journal of Ethnopharmacology, 99*(1), pp. 137–143.
652. Sheeba, M.S. and Asha, V.V., 2009. *Cardiospermum halicacabum* ethanol extract inhibits LPS induced COX-2, TNF-α and iNOS expression, which is mediated by NF-κB regulation, in RAW264.7 cells. *Journal of Ethnopharmacology, 124*(1), pp. 39–44.
653. Gaziano, R., Campione, E., Iacovelli, F., Marino, D., Pica, F., Di Francesco, P., Aquaro, S., Menichini, F., Falconi, M. and Bianchi, L., 2018. Antifungal activity of *Cardiospermum halicacabum* L. (Sapindaceae) against *Trichophyton rubrum* occurs through molecular interaction with fungal Hsp90. *Drug Design, Development and Therapy, 12*, p. 2185.
654. Khan, M.J., Saraf, S. and Saraf, S., 2017. Anti-inflammatory and associated analgesic activities of HPLC standardized alcoholic extract of known ayurvedic plant *Schleichera oleosa. Journal of Ethnopharmacology, 197*, pp. 257–265.
655. Pettit, G.R., Numata, A., Cragg, G.M., Herald, D.L., Takada, T., Iwamoto, C., Riesen, R., Schmidt, J.M., Doubek, D.L. and Goswami, A., 2000. Isolation and structures of schleicherastatins 1–7 and Schleicheols 1 and 2 from the teak forest medicinal tree *Schleichera oleosa. Journal of Natural Products, 63*(1), pp. 72–78.
656. Ghosh, P., Chakraborty, P., Mandal, A., Rasul, M.G., Chakraborty, M. and Saha, A., 2011. Triterpenoids from *Schleichera oleosa* of Darjeeling foothills and their antimicrobial activity. *Indian Journal of Pharmaceutical Sciences, 73*(2), p. 231.
657. Mohod, S.M., Kandhare, A.D. and Bodhankar, S.L., 2016. Gastroprotective potential of Pentahydroxy flavone isolated from *Madhuca indica* JF Gmel: leaves against acetic acid-induced ulcer in rats: the role of oxido-inflammatory and prostaglandins markers. *Journal of Ethnopharmacology, 182*, pp. 150–159.
658. Nimbekar, T., Bais, Y., Katolkar, P., Wanjari, B. and Chaudhari, S., 2012. Antibacterial activity of the dried inner bark of *Madhuca indica* J.F. Gmel. *Bulletin of Environment Pharmacology & Life Sciences, 1*(2), pp. 26–29.
659. Kar, B., Kumar, R.S., Bala, A., Dolai, N., Mazumder, U.K. and Haldar, P.K., 2012. Evaluation of antitumor activity of *Mimusops elengi* leaves on Ehrlich's ascites carcinoma-treated mice. *Journal of Dietary Supplements, 9*(3), pp. 166–177.

660. Purnima, A., Koti, B.C., Thippeswamy, A.H.M., Jaji, M.S., Swamy, A.V., Kurhe, Y.V. and Sadiq, A.J., 2010. Antiinflammatory, analgesic and antipyretic activities of *Mimusops elengi* L. *Indian Journal of Pharmaceutical Sciences*, *72*(4), p. 480.
661. Jana, G.K., Dhanamjayarao, M. and Mamillapalli, V., 2010. Evaluation of anthelmintic potential of *Mimusops elengi* L. (Sapotaceae) leaf. *Journal of Pharmacy Research*, *3*(10), pp. 2514–2515.
662. Brahmachari, G., Mandal, N.C., Roy, R., Ghosh, R., Barman, S., Sarkar, S., Jash, S.K. and Mondal, S., 2013. A new pentacyclic triterpene with potent antibacterial activity from *Limnophila indica* Linn. (Druce). *Fitoterapia*, *90*, pp. 104–111.
663. Brahmachari, G., Mandal, N.C., Jash, S.K., Roy, R., Mandal, L.C., Mukhopadhyay, A., Behera, B., Majhi, S., Mondal, A. and Gangopadhyay, A., 2011. Evaluation of the antimicrobial potential of two flavonoids isolated from *Limnophila plants*. *Chemistry & Biodiversity*, *8*(6), pp. 1139–1151.
664. Sharma, K.R., Adhikari, A., Hafizur, R.M., Hameed, A., Raza, S.A., Kalauni, S.K., Miyazaki, J.I. and Choudhary, M.I., 2015. Potent insulin secretagogue from *Scoparia dulcis* Linn of Nepalese origin. *Phytotherapy Research*, *29*(10), pp. 1672–1675.
665. De Farias Freire, S.M., Da Silva Emim, J.A., Lapa, A.J., Souccar, C. and Torres, L.M.B., 1993. Analgesic and antiinflammatory properties of *Scoparia dulcis* L. extracts and glutinol in rodents. *Phytotherapy Research*, *7*(6), pp. 408–414.
666. Babincová, M., Schronerová, K. and Sourivong, P., 2008. Antiulcer activity of water extract of *Scoparia dulcis*. *Fitoterapia*, *79*(7–8), pp. 587–588.
667. Murali, A., Ashok, P. and Madhavan, V., 2012. Effect of *Smilax zeylanica* roots and rhizomes in paracetamol induced hepatotoxicity. *Journal of Complementary and Integrative Medicine*, *9*(1), pp. 1–17.
668. Uddin, M.N., Ahmed, T., Pathan, S., Al-Amin, M.M. and Rana, M.S., 2015. Antioxidant and cytotoxic activity of stems of *Smilax zeylanica* in vitro. *Journal of Basic and Clinical Physiology and Pharmacology*, *26*(5), pp. 453–463.
669. Murali, A., Ashok, P. and Madhavan, V., 2011. In vitro antioxidant activity and HPTLC studies on the roots and rhizomes of *Smilax zeylanica* L. (Smilacaceae). *International Journal of Pharmacy and Pharmaceutical Sciences*, *3*(1), pp. 192–195.
670. Yu, C.H., Wang, K.L., Hsu, R.L., Ho, Y.J. and Wang, P.S., 2011. Effects of diosgenin on the reproductive function of D-galactose-induced aging model of male rats. *Biology of Reproduction*, *85*, p. 550.
671. Yamada, T., Hoshino, M., Hayakawa, T., Ohhara, H., Yamada, H., Nakazawa, T., Inagaki, T., Iida, M., Ogasawara, T., Uchida, A. and Hasegawa, C., 1997. Dietary diosgenin attenuates subacute intestinal inflammation associated with indomethacin in rats. *American Journal of Physiology-Gastrointestinal and Liver Physiology*, *273*(2), pp. G355–G364.
672. Govindarajan, V.S. and Sathyanarayana, M.N., 1991. Capsicum—production, technology, chemistry, and quality. Part V. Impact on physiology, pharmacology, nutrition, and metabolism; structure, pungency, pain, and desensitization sequences. *Critical Reviews in Food Science & Nutrition*, *29*(6), pp. 435–474.
673. Kim, C.S., Kawada, T., Kim, B.S., Han, I.S., Choe, S.Y., Kurata, T. and Yu, R., 2003. Capsaicin exhibits anti-inflammatory property by inhibiting IkB-a degradation in LPS-stimulated peritoneal macrophages. *Cellular Signalling*, *15*(3), pp. 299–306.
674. Wei, Y.X., Shuai, L., Guo, D.S., Li, S., Wang, F.L., Ai, G.H., 2006. Study on antibacterial activity of capsaicin. *Food Science*, *8*, p. 12.
675. Heinrich, M., Barnes, J., Prieto-Garcia, J., Gibbons, S. and Williamson, E.M., 2017. *Fundamentals of Pharmacognosy and Phytotherapy*. Elsevier Health Sciences, London, United Kingdom.
676. Wannang, N.N., Ndukwe, H.C. and Nnabuife, C., 2009. Evaluation of the analgesic properties of the *Datura metel* seeds aqueous extract. *Journal of Medicinal Plants Research*, *3*(4), pp. 192–195.

677. Sharma, G.L., 2002. Studies on antimycotic properties of *Datura metel*. *Journal of Ethnopharmacology*, *80*(2–3), pp. 193–197.
678. Yang, B.Y., Guo, R., Li, T., Wu, J.J., Zhang, J., Liu, Y., Wang, Q.H. and Kuang, H.X., 2014. New anti-inflammatory withanolides from the leaves of *Datura metel* L. *Steroids*, *87*, pp. 26–34.
679. Wu, J., Li, X., Zhao, J., Wang, R., Xia, Z., Li, X., Liu, Y., Xu, Q., Khan, I.A. and Yang, S., 2018. Anti-inflammatory and cytotoxic withanolides from *Physalis minima*. *Phytochemistry*, *155*, pp. 164–170.
680. Ooi, K.L., Muhammad, T.S.T. and Sulaiman, S.F., 2013. Physalin F from *Physalis minima* L. triggers apoptosis-based cytotoxic mechanism in T-47D cells through the activation caspase-3-and c-myc-dependent pathways. *Journal of Ethnopharmacology*, *150*(1), pp. 382–388.
681. Choudhary, M.I., Yousaf, S., Ahmed, S. and Yasmeen, K., 2005. Antileishmanial physalins from *Physalis minima*. *Chemistry & Biodiversity*, *2*(9), pp. 1164–1173.
682. Mirjalili, M.H., Moyano, E., Bonfill, M., Cusido, R.M. and Palazón, J., 2009. Steroidal lactones from *Withania somnifera*, an ancient plant for novel medicine. *Molecules*, *14*(7), pp. 2373–2393.
683. Bhattacharya, S.K., Bhattacharya, A., Sairam, K. and Ghosal, S., 2000. Anxiolytic-antidepressant activity of *Withania somnifera* glycowithanolides: an experimental study. *Phytomedicine*, *7*(6), pp. 463–469.
684. Ambiye, V.R., Langade, D., Dongre, S., Aptikar, P., Kulkarni, M. and Dongre, A., 2013. Clinical evaluation of the spermatogenic activity of the root extract of Ashwagandha (*Withania somnifera*) in oligospermic males: a pilot study. *Evidence-Based Complementary and Alternative Medicine*, *2013*, p. 571420.
685. Jayaprakasam, B., Zhang, Y., Seeram, N.P. and Nair, M.G., 2003. Growth inhibition of human tumor cell lines by withanolides from *Withania somnifera* leaves. *Life Sciences*, *74*(1), pp. 125–132.
686. Halim, M., 2003. Lowering of blood sugar by water extract of *Azadirachta indica* and *Abroma augusta* in diabetes rats. *Indian Journal of Experimental Biology, 41*(6), pp. 636–640.
687. Khanra, R., Dewanjee, S., Dua, T.K., Sahu, R., Gangopadhyay, M., De Feo, V. and Zia-Ul-Haq, M., 2015. *Abroma augusta* leaf extract attenuates diabetes induced nephropathy and cardiomyopathy via inhibition of oxidative stress and inflammatory response. *Journal of Translational Medicine*, *13*(1), p. 6.
688. Khanra, R., Bhattacharjee, N., Dua, T.K., Nandy, A., Saha, A., Kalita, J., Manna, P. and Dewanjee, S., 2017. Taraxerol, a pentacyclic triterpenoid, from *Abroma augusta* leaf attenuates diabetic nephropathy in type 2 diabetic rats. *Biomedicine & Pharmacotherapy*, *94*, pp. 726–741.
689. Abirami, N. and Natarajan, B., 2014. Isolation and characterization of (4Z, 12Z)-cyclopentadeca-4, 12-dienone from Indian medicinal plant *Grewia hirsuta* and its hyperglycemic effect on 3T3 and L6 cell lines. *International Journal of Pharmacognosy and Phytochemical Research*, *6*(2), pp. 393–398.
690. Barbera, R., Trovato, A., Rapisarda, A. and Ragusa, S., 1992. Analgesic and antiinflammatory activity in acute and chronic conditions of *Trema guineense* (Schum. et Thonn.) Ficalho and Trema micrantha Bl. extracts in rodents. *Phytotherapy Research*, *6*(3), pp. 146–148.
691. Dimo, T., Ngueguim, F.T., Kamtchouing, P., Dongo, E. and Tan, P.V., 2006. Glucose lowering efficacy of the aqueous stem bark extract of *Trema orientalis* (Linn) Bl. in normal and streptozotocin diabetic rats. *Die Pharmazie*, *61*(3), pp. 233–236.
692. Uddin, S.N., Uddin, K.M.A. and Ahmed, F., 2008. Analgesic and antidiarrhoeal activities of *Treama orientalis* L. in mice. *Oriental Pharmacy and Experimental Medicine*, *8*(2), pp. 187–191.

693. Lin, C.C., Yen, M.H., Lo, T.S. and Lin, J.M., 1998. Evaluation of the hepatoprotective and antioxidant activity of *Boehmeria nivea* var. nivea and B. nivea var. tenacissima. *Journal of Ethnopharmacology*, *60*(1), pp. 9–17.
694. Huang, K.L., Lai, Y.K., Lin, C.C. and Chang, J.M., 2006. Inhibition of hepatitis B virus production by *Boehmeria nivea* root extract in HepG2 2.2. 15 cells. *World Journal of Gastroenterology*, *12*(35), p. 5721.
695. Liu, Y., Nielsen, M., Staerk, D. and Jäger, A.K., 2014. High-resolution bacterial growth inhibition profiling combined with HPLC–HRMS–SPE–NMR for identification of antibacterial constituents in Chinese plants used to treat snakebites. *Journal of Ethnopharmacology*, *155*(2), pp. 1276–1283.
696. Singh, A.K. and Agpawal, P.K., 1994. 16oc, 17-isopropylideno-3-oxo-phyllocladane, a diterpenoid from *Callicarpa macrophylla*. *Phytochemistry*, *37*(2), pp. 587–588.
697. Chatterjee, A., Desmukh, S.K. and Chandrasekharan, S., 1972. Diterpenoid constituents of *Callicarpa macrophylla* vahl: the sturctures and stereochemistry of calliterpenone and calliterpenone monoacetate. *Tetrahedron*, *28*(16), pp. 4319–4323.
698. Wang, Z.H., Niu, C., Zhou, D.J., Kong, J.C. and Zhang, W.K., 2017. Three new abietane-type diterpenoids from *Callicarpa macrophylla* Vahl. *Molecules*, *22*(5), p. 842.
699. Rahman, M.A.A., Azam, A.T.M. and Gafur, M.A., 2000. In vitro antibacterial principles of extracts and two flavonoids from *Clerodendrum indicum* Linn. *Pakistan Journal of Biological Sciences*, *3*(10), pp. 1769–1771.
700. Somwong, P. and Suttisri, R., 2018. Cytotoxic activity of the chemical constituents of *Clerodendrum indicum* and *Clerodendrum villosum* roots. *Journal of Integrative Medicine*, *16*(1), pp. 57–61.
701. Kaur, S., Bedi, P.M.S. and Kaur, N., 2017. Anti-inflammatory effect of methanolic extract of *Gmelina arborea* bark and its fractions against carrageenan induced paw oedema in rats. *Natural Product Research*, *32*(23), pp. 1–4.
702. Attanayake, A.P., Jayatilaka, K.A.P.W., Pathirana, C. and Mudduwa, L.K.B., 2016. *Gmelina arborea* Roxb. (Family: Verbenaceae) extract upregulates the β-cell regeneration in STZ induced diabetic rats. *Journal of Diabetes Research*, *2016*, p. 4513871.
703. Al Mahmud, Z., Emran, T.B., Qais, N., Bachar, S.C., Sarker, M. and Uddin, M.M.N., 2016. Evaluation of analgesic, anti-inflammatory, thrombolytic and hepatoprotective activities of roots of *Premna esculenta* (Roxb). *Journal of Basic and Clinical Physiology and Pharmacology*, *27*(1), pp. 63–70.
704. Dharmasiri, M.G., Jayakody, J.R.A.C., Galhena, G., Liyanage, S.S.P. and Ratnasooriya, W.D., 2003. Anti-inflammatory and analgesic activities of mature fresh leaves of *Vitex negundo*. *Journal of Ethnopharmacology*, *87*(2–3), pp. 199–206.
705. Telang, R.S., Chatterjee, S. and Varshneya, C., 1999. Study on analgesic and anti-inflammatory activities of *Vitex negundo* L. *Indian Journal of Pharmacology*, *31*(5), p. 363.
706. Ragasa, C.Y., Morales, E. and Rideout, J.A., 1999. Antimicrobial compounds from *Vitex negundo*. *Philippine Journal of Science (Philippines)*, *128*(1), pp. 21–29.
707. Díaz, F., Chávez, D., Lee, D., Mi, Q., Chai, H.B., Tan, G.T., Kardono, L.B., Riswan, S., Fairchild, C.R., Wild, R. and Farnsworth, N.R., 2003. Cytotoxic flavone analogues of vitexicarpin, a constituent of the leaves of *Vitex negundo*. *Journal of Natural Products*, *66*(6), pp. 865–867.
708. Panthong, A., Supraditaporn, W., Kanjanapothi, D., Taesotikul, T. and Reutrakul, V., 2007. Analgesic, anti-inflammatory and venotonic effects of *Cissus quadrangularis* L. *Journal of Ethnopharmacology*, *110*(2), pp. 264–270.
709. Deka, D.K., Lahon, L.C., Saikia, J. and Mukit, A., 1994. Effect of *Cissus quadrangularis* in accelerating healing process of experimentally fractured radius-ulna of dog, a preliminary study. *Indian Journal of Pharmacology*, *26*(1), pp. 44–45.

710. Aziz, A.N., Ibrahim, H., Syamsir, D.R., Mohtar, M., Vejayan, J. and Awang, K., 2013. Antimicrobial compounds from *Alpinia conchigera*. *Journal of Ethnopharmacology*, *145*(3), pp. 798–802.
711. Ibrahim, H., Aziz, A.N., Syamsir, D.R., Ali, N.A.M., Mohtar, M., Ali, R.M. and Awang, K., 2009. Essential oils of *Alpinia conchigera* Griff. and their antimicrobial activities. *Food Chemistry*, *113*(2), pp. 575–577.
712. Awang, K., Azmi, M.N., Aun, L.I., Aziz, A.N., Ibrahim, H. and Nagoor, N.H., 2010. The apoptotic effect of 1′S-1′-Acetoxychavicol acetate from *Alpinia conchigera* on human cancer cells. *Molecules*, *15*(11), pp. 8048–8059.
713. Warit, S., Rukseree, K., Prammananan, T., Hongmanee, P., Billamas, P., Jaitrong, S., Chaiprasert, A., Jaki, B.U., Pauli, G.F., Franzblau, S.G. and Palittapongarnpim, P., 2017. In vitro activities of enantiopure and racemic 1′-acetoxychavicol acetate against clinical isolates of *Mycobacterium tuberculosis*. *Scientia Pharmaceutica*, *85*(3), p. 32.
714. Baradwaj, R.G., Rao, M.V. and Kumar, T.S., 2017. Novel purification of 1′S-1′-acetoxychavicol acetate from *Alpinia galanga* and its cytotoxic plus antiproliferative activity in colorectal adenocarcinoma cell line SW480. *Biomedicine & Pharmacotherapy*, *91*, pp. 485–493.
715. Das, B.N. and Biswas, B.K., 2012. Anti-inflammatory activity of the rhizome extract of *Alpinia nigra*. *International Research Journal of Pharmacy*, *2*(3), pp. 73–76.
716. Roy, B. and Swargiary, A., 2009. Anthelmintic efficacy of ethanolic shoot extract of *Alpinia nigra* on tegumental enzymes of Fasciolopsis buski, a giant intestinal parasite. *Journal of Parasitic Diseases*, *33*(1–2), pp. 48–53.
717. Qiao, C., Wang, Z., Dong, H., Xu, L. and Hao, X., 2000. The chemical constituents of blackfruit galangal (*Alpinia nigra*). *Chinese Traditional and Herbal Drugs*, *31*(6), pp. 404–405.
718. Ghosh, S., Ozek, T., Tabanca, N., Ali, A., ur Rehman, J., Khan, I.A. and Rangan, L., 2014. Chemical composition and bioactivity studies of *Alpinia nigra* essential oils. *Industrial Crops and Products*, *53*, pp. 111–119.
719. Rajashekhara, N., Ashok, B.K., Sharma, P.P. and Ravishankar, B., 2014. The evaluation of anti-ulcerogenic effect of rhizome starch of two source plants of Tugaksheeree (*Curcuma angustifolia* Roxb. and *Maranta arundinacea* Linn.) on pyloric ligated rats. *Ayu*, *35*(2), p. 191.
720. Banerjee, A. and Nigam, S.S., 1977. Antifungal activity of the essential oil of *Curcuma angustifolia*. *Indian Journal of Pharmacy*, *39*(143), pp. 214–217.
721. Nugroho, B.W., Schwarz, B., Wray, V. and Proksch, P., 1996. Insecticidal constituents from rhizomes of *Zingiber cassumunar* and *Kaempferia rotunda*. *Phytochemistry*, *41*(1), pp. 129–132.
722. Lallo, S., Lee, S., Dibwe, D.F., Tezuka, Y. and Morita, H., 2014. A new polyoxygenated cyclohexane and other constituents from *Kaempferia rotunda* and their cytotoxic activity. *Natural Product Research*, *28*(20), pp. 1754–1759.
723. Kishore, N. and Dwivedi, R.S., 1992. Zerumbone: a potential fungitoxic agent isolated from *Zingiber cassumunar* Roxb. *Mycopathologia*, *120*(3), pp. 155–159.
724. Han, A.R., Min, H.Y., Windono, T., Jeohn, G.H., Jang, D.S., Lee, S.K. and Seo, E.K., 2004. A new cytotoxic phenylbutenoid dimer from the rhizomes of *Zingiber cassumunar*. *Planta Medica*, *70*(11), pp. 1095–1097.
725. Jeenapongsa, R., Yoovathaworn, K., Sriwatanakul, K.M., Pongprayoon, U. and Sriwatanakul, K., 2003. Anti-inflammatory activity of (E)-1-(3, 4-dimethoxyphenyl) butadiene from *Zingiber cassumunar* Roxb. *Journal of Ethnopharmacology*, *87*(2–3), pp. 143–148.

Bibliography

Ahmmed, M.R., Ahmed, S., Sunny, S.S.I., Kar, A., Mahmud, S.N., Kabir, M.H., Mahmud, S. and Shaon, S.M., 2017. A study on diversity of medicinal plant usage by folk medicinal practitioners in different villages of Dhunat Upazila, Bogra district, Bangladesh. *Journal of Pharmacognosy and Phytochemistry*, *6*(1), pp. 177–186.

Balfour, E., 1857. *The Cyclopaedia of India and of Eastern and Southern Asia*. Scottish Press: Madras.

Banik, G., Bawari, M., Choudhury, M.D., Choudhury, S. and Sharma, G.D., 2010. Some anti-diabetic plants of Southern Assam. *Assam University Journal of Science and Technology*, *5*(1), pp. 114–119.

Basu, R., 2009. Biodiversity and ethnobotany of sacred groves in Bankura district, West Bengal. *Indian Forester*, *135*(6), p. 765.

Biswas, A., Bari, M.A., Roy, M. and Bhadra, S.K., 2010. Inherited folk pharmaceutical knowledge of tribal people in the Chittagong hill tracts, Bangladesh. *Indian Journal of Traditional Knowledge*, *9*(1), pp. 77–79.

Biswas, M., Roy, D.N., Rahman, M.M. and Hossen, M., 2013. Medicinal plants for snake bite and sexual dysfunction in Jessore and Bagerhat districts of Bangladesh. *International Journal of Medicinal and Aromatic Plants*, *3*(4), pp. 486–491.

Biswas, S., Shaw, R., Bala, S. and Mazumdar, A., 2017. Inventorization of some ayurvedic plants and their ethnomedicinal use in Kakrajhore forest area of West Bengal. *Journal of Ethnopharmacology*, *197*, pp. 231–241.

Borah, P.K., Gogoi, P., Phukan, A.C. and Mahanta, J., 2006. Traditional medicine in the treatment of gastrointestinal diseases in Upper Assam. *Indian Journal of Traditional Knowledge*, *5*(4), pp. 510–512.

Bose, D., Roy, J.G., Mahapatra, S.D., Datta, T., Mahapatra, S.D. and Biswas, H., 2015. Medicinal plants used by tribals in Jalpaiguri district, West Bengal, India. *Journal of Medicinal Plants Studies*, *3*, pp. 15–21.

Chakraborty, M.K. and Bhattacharjee, A., 2006. Some common ethnomedicinal uses for various diseases in Purulia district, West Bengal. *Indian Journal of Traditional Knowledge*, *5*(4), pp. 554–558.

Chakraborty, R., Mondal, M.S. and Mukherjee, S.K., 2016. Ethnobotanical information on some aquatic plants of South 24 Parganas, West Bengal. *Plant Science Today*, *3*(2), pp. 109–114.

Chakravorty, A. and Ghosh, P.D., 2013. Exploration of the otherwise-neglected weeds of Nadia district-the treasure trove of medicines. *International Journal of Pharmaceutical Science Invention*, *2*(6), pp. 19–21.

Chowdhury, A. and Das, A.P., 2015. Ethnopharmacological survey of wetland plants used by local ethnic people in sub-Himalayan Terai and Duars of West Bengal, India. *American Journal of Ethnomedicine*, *2*(2), pp. 2348–9502.

Chowdhury, M. and Das, A.P., 2007. Folk medicines used by the Rabha tribe in Coochbehar district of West Bengal: a preliminary report. *Advances in Ethnobotany*, pp. 289–296.

Chowdhury, M. and Das, A.P., 2009. Inventory of some ethno-medicinal plants in wetlands areas in Maldah district of West Bengal. *Pleione*, *3*(1), pp. 83–88.

Chowdhury, M.S.H., Koike, M., Muhammed, N., Halim, M.A., Saha, N. and Kobayashi, H., 2009. Use of plants in healthcare: a traditional ethno-medicinal practice in rural areas of southeastern Bangladesh. *International Journal of Biodiversity Science & Management*, *5*(1), pp. 41–51.

Das, D. and Ghosh, P., 2017. Some important medicinal plants used widely in Southwest Bengal, India. *International Journal of Engineering & Science Invention*, *6*(6), pp. 28–50.

Das, D.C., Mahato, G., Das, M. and Pati, M.L., 2015. Investigation of ethno medicinal plants for the treatment of carbuncles from the Purulia district of West Bengal. *International Journal of Bioassays*, *4*(5), pp. 3896–3899.

Das, D.C., Sinha, N.K., Patsa, M.K. and Das, M., 2015. Investigation of herbals for the treatment of leucorrhoea from south west Bengal, India. *International Journal of Bioassays*, *4*(11), pp. 4555–4559.

Datta, T., Patra, A.K. and Dastidar, S.G., 2014. Medicinal plants used by tribal population of Coochbehar district, West Bengal, India–an ethnobotanical survey. *Asian Pacific Journal of Tropical Biomedicine*, *4*, pp. S478–S482.

DeFilipps, R.A. and Krupnick, G.A., 2018. The medicinal plants of Myanmar. *PhytoKeys*, (102), pp. 1–341.

Devi, N.B., Singh, P.K. and Das, A.K., 2014. Ethnomedicinal utilization of Zingiberaceae in the valley districts of Manipur. *Journal of Environmental Science, Toxicology & Food Technology*, *8*(2), pp. 21–23.

Dey, A. and De, J.N., 2012. Ethnobotanical survey of Purulia district, West Bengal, India for medicinal plants used against gastrointestinal disorders. *Journal of Ethnopharmacology*, *143*(1), pp. 68–80.

Dey, S.K., De, A., Karmakar, S., De, P.K., Chakraborty, S., Samanta, A. and Mukherjee, A., 2009. Ethnobotanical study in a remote district of West Bengal, India. *Pharmbit*, *2*, pp. 91–96.

Drury, H., 1873. *The Useful Plants of India: With Notices of Their Chief Value in Commerce, Medicine, and the Arts*. WH Allen: London.

Faruque, M.O. and Uddin, S.B., 2014. Ethnomedicinal study of the Marma community of Bandarban district of Bangladesh. *Academia Journal of Medicinal Plants*, *2*(2), pp. 14–25.

Ghosh, A., 2002. Ethnoveterinary medicines from the tribal areas of Bankura and Medinipur districts, West Bengal. *Indian Journal of Traditional Knowledge*, *1*(1), pp. 93–95.

Ghosh, A., 2003. Herbal folk remedies of Bankura and Medinipur districts, West Bengal. *Indian Journal of Traditional Knowledge*, *2*(4), pp. 393–396.

Ghosh, A., 2008. Ethnomedicinal plants used in West Rarrh region of West Bengal. *Natural Product Radiance*, *7*(5), pp. 461–465.

Golam, K.A.M.B., Rahman, S.A., Alam, M.N., Ali, M.K. and Imtiat, A., 2007. Survey, evaluation and health care of medicinal plants at rural Kishorpur village of Rajshahi. *Bangladesh Journal of Environmental Science*, *13*(2), pp. 252–259.

Hossain, M.M., 2009. Traditional therapeutic uses of some indigenous orchids of Bangladesh. *Medicinal and Aromatic Plant Science and Biotechnology*, *42*(1), pp. 101–106.

Hossan, M.S., Hanif, A., Khan, M., Bari, S., Jahan, R. and Rahmatullah, M., 2009. Ethnobotanical survey of the Tripura tribe of Bangladesh. *American Eurasian Journal of Sustainable Agriculture*, *3*(2), pp. 253–261.

Islam, M.K., Saha, S., Mahmud, I., Mohamad, K., Awang, K., Uddin, S.J., Rahman, M.M. and Shilpi, J.A., 2014. An ethnobotanical study of medicinal plants used by tribal and native people of Madhupur forest area, Bangladesh. *Journal of Ethnopharmacology*, *151*(2), pp. 921–930.

Kabir, T. and Saha, S., 2014. A study on the indigenous medicinal plants and healing practices of Murong tribe in Khagrachari district (Bangladesh). *International Journal of Pharmacognosy*, *1*(10), pp. 654–659.

Khare, C.P., 2008. *Indian Medicinal Plants: An Illustrated Dictionary*. Springer Science & Business Media: New York.

Khisha, T., Karim, R., Chowdhury, S.R. and Banoo, R., 2012. Ethnomedical studies of Chakma communities of Chittagong hill tracts, Bangladesh. *Bangladesh Pharmaceutical Journal*, *15*(1), pp. 59–67.

Kirtikar, K.R., 1918. *Indian Medicinal Plants*. Bishen Singh Mahendra Pal Singh: Dehradun.

Mitra, S. and Mukherjee, S.K., 2005. Root and rhizome drugs used by the tribals of West Dinajpur in Bengal. *Journal of Tropical Medicinal Plants*, *6*(2), p. 301.

Mitra, S. and Mukherjee, S.K., 2007. Plants used as ethnoveterinary medicine in Uttar and Dakshin Dinajpur districts of West Bengal, India. *Advances in Ethnobotany*, pp. 117–122.

Mondal, S. and Rahaman, C.H., 2012. Medicinal plants used by the tribal people of Birbhum district of West Bengal and Dumka district of Jharkhand in India. *Indian Journal of Traditional Knowledge*, *11*(4), pp. 674–679.

Monorajan, C. and Rajrupa, M., 2010. Ethno-medicinal survey of Santal tribe of Malda district of West Bengal, India. *Journal of Economic and Taxonomic Botany*, *34*(3), pp. 602–606.

Ocvirk, S., Kistler, M., Khan, S., Talukder, S.H. and Hauner, H., 2013. Traditional medicinal plants used for the treatment of diabetes in rural and urban areas of Dhaka, Bangladesh–an ethnobotanical survey. *Journal of Ethnobiology and Ethnomedicine*, *9*(1), p. 43.

O' Shaugnessy, W.B. 1852. *Bengal Dispensatory*. Thancker & Co St Andrew: Calcutta.

Pandit, P.K., 2010. Inventory of ethno veterinary medicinal plants of Jhargram division, West Bengal, India. *Indian Forester*, *136*(9), p. 1183.

Pattanayak, S., Dutta, M.K., Debnath, P.K., Bandyopadhyay, S.K., Saha, B. and Maity, D., 2012. A study on ethno-medicinal use of some commonly available plants for wound healing and related activities in three southern districts of West Bengal, India. *Exploratory Animal & Medical Research*, *2*(2), pp. 97–110.

Pattanayak, S., Mandal, T.K. and Bandyopadhyay, S.K., 2015. A study on use of plants to cure enteritis and dysentery in three southern districts of West Bengal, India. *Journal of Medicinal Plants*, *3*(5), pp. 277–283.

Rahaman, C.H. and Karmakar, S., 2014. Ethnomedicine of Santal tribe living around Susunia hill of Bankura district, West Bengal, India: the quantitative approach. *Journal of Applied Pharmaceutical Science*, *5*(2), pp. 127–136.

Rahman, A.H.M.M., 2013. Graveyards angiosperm diversity of Rajshahi city, Bangladesh with emphasis on medicinal plants. *American Journal of Life Sciences*, *1*(3), pp. 98–104.

Rahman, A.H.M.M. and Akter, M., 2013. Taxonomy and medicinal uses of Euphorbiaceae (Spurge) family of Rajshahi, Bangladesh. *Research in Plant Sciences*, *1*(3), pp. 74–80.

Rahman, A.H.M.M. and Kumar, A.K., 2015. Investigation of medicinal plants at Katakhali Pouroshova of Rajshahi District, Bangladesh and their conservation management. *Applied Ecology and Environmental Sciences*, *3*(6), pp. 184–192.

Rahman, M.A., 2010. Indigenous knowledge of herbal medicines in Bangladesh. 3. Treatement of skin diseases by tribal communities of the hill tracts districts. *Bangladesh Journal of Botany*, *39*(2), pp. 169–177.

Rahman, M.A., Uddin, S.B. and Wilcock, C.C., 2007. Medicinal plants used by Chakma tribe in Hill Tracts districts of Bangladesh. *Indian Journal of Traditional Knowledge*, *6*(3), pp. 508–517.

Rana, M.P., Sohel, M.S.I., Akhter, S. and Islam, M.J., 2010. Ethno-medicinal plants use by the Manipuri tribal community in Bangladesh. *Journal of Forestry Research*, *21*(1), pp. 85–92.

Roxburgh, W., 1874. *Flora Indica: Or, Description of Indian Plants*. Thacker, Spink and Co: Calcutta.

Saha, M.R., Sarker, D.D. and Sen, A., 2014. Ethnoveterinary practices among the tribal community of Malda district of West Bengal, India. *Indian Journal of Traditional Knowledge*, *13*(2), pp. 359–367.

Sarkar, N.R., Mondal, S. and Mandal, S., 2016. Phytodiversity of Ganpur forest, Birbhum District, West Bengal, India with reference to their medicinal properties. *International Journal of Current Microbiology & Applied Sciences*, *5*(6), pp. 973–989.

Shaw, P. and Panda, S., 2015. Spices commonly consumed in west Bengal India–an appraisal. *International Journal of Life Sciences*, *4*(2), pp. 129–133.

Shukla, G. and Chakravarty, S., 2012. Ethnobotanical plant use of Chilapatta reserved forest in West Bengal. *Indian Forester*, *138*(12), p. 1116.

Talukdar, D., 2013. Species richness and floral diversity around 'Teesta Barrage Project' in Jalpaiguri district of West Bengal, India with emphasis on invasive plants and indigenous uses. *Biology and Medicine*, *5*, p. 1.

Tumpa, S.I., Hossain, M.I. and Ishika, T., 2014. Ethnomedicinal uses of herbs by indigenous medicine practitioners of Jhenaidah district, Bangladesh. *Journal of Pharmacognosy and Phytochemistry*, *3*(2), pp. 23–33.

Uddin, S., Lee, S. and Uddin, S.B., 2016. *Ethnomedicinal Plants of Bangladesh*. Korea Research Institute of BioScience and Biotechnology Publishers: Daejeon, South Korea.

Watt, G., 1889. *A Dictionary of the Economic Products of India Vol-1*. The Superindent of Govt. Print.: Calcutta.

Wu, Z.Y., Raven, P.H. and Hong, D.Y., 2004. *Flora of China*. Missouri Botanical Garden Press: St. Louis; Missouri and Harvard University Herbaria: Cambridge, MA.

Yusuf, M., Wahab, M.A., Chowdhury, J.U. and Begum, J., 2006. Ethno-medico-botanical knowledge from Kaukhali proper and Betbunia of Rangamati District. *Bangladesh Journal of Plant Taxonomy*, *13*(1), pp. 55–61.

Index of Binomial Denominations

B

C

D

E

H

L

M

N

R

V

W

X

Z

Index of Common Names

A

B

C

V

W

Y

Index of Local Names

V

W

Y